Optimization of Aerosol Drug Delivery

OPTIMIZATION OF AEROSOL DRUG DELIVERY

Edited by

LEON GRADOŃ

University of Technology,
Warsaw, Poland

and

JAN MARIJNISSEN

Delft University of Technology,
Faculty of Appplied Sciences, Delft, The Netherlands

KLUWER ACADEMIC PUBLISHERS

DORDRECHT / BOSTON / LONDON

A C.I.P. Catalogue record for this book is available from the Library of Congress.

ISBN 1-4020-1651-4

Published by Kluwer Academic Publishers,
P.O. Box 17, 3300 AA Dordrecht, The Netherlands.

Sold and distributed in North, Central and South America
by Kluwer Academic Publishers,
101 Philip Drive, Norwell, MA 02061, U.S.A.

In all other countries, sold and distributed
by Kluwer Academic Publishers,
P.O. Box 322, 3300 AH Dordrecht, The Netherlands.

Lay out: C. Fieneke Spaans, *Delft University of Technology*
Cover design: Tomasz Sosnowski, *Warsaw University of Technology*

Printed on acid-free paper

Printed in the Netherlands.

TABLE OF CONTENTS

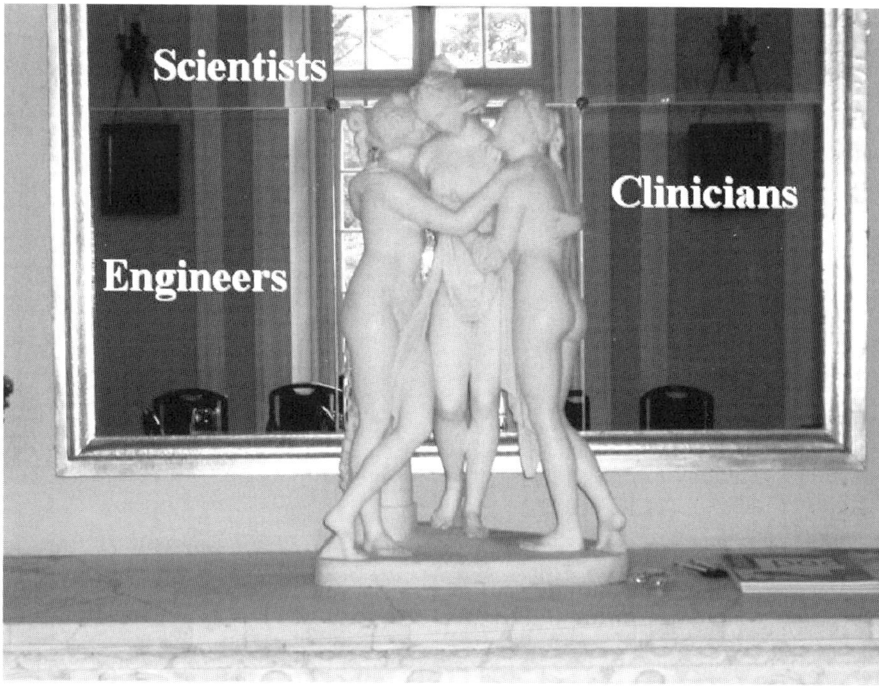

Statue, Jablonna Palace

INTRODUCTION

Aerosol therapy has significantly improved the treatment of a variety of respiratory diseases. Besides the treatment of respiratory diseases there is currently also a great interest to use the lungs as a portal to introduce drugs for systemic therapy.
The success of therapy with the application of aerosolized medicaments depends on the possibility to deliver the proper amount of drug to the appropriate sites in the respiratory system, thus limiting the side effects to a minimum.
Aerosolized delivery of drugs to the lung is optimized if, for a given chemical composition of a medicine, the target of deposition and the required mass of drug to be deposited are precisely defined.
The next step is the specification of the number of respirable particles or droplets, to be generated by appropriate devices.
Another very important factor for successful aerosol therapy is the condition of the patient coupled with his or her inhalation technique.
During June the 14th and 15th, 2002 a group of scientists from different disciplines but all in fields related to aerosol therapy, met in the beautiful Jablonna Palace near Warsaw in Poland. They dealt with some important issues and problems with the aim to find optimal solutions for aerosol drug delivery into the human respiratory system. We believe that we have dealt with important questions that can lead to new directions to develop aerosol therapy. The subjects were assembled in a few themes of problems, proceeded by a historical overview of aerosol therapy. The considered themes were: History, Molecular design, Production, Deposition, Lung interaction and Clinical aspects. The workshop was concluded with an overview and roundtable discussion.
The papers presented in this book , as well as opinions formed in the discussion contribute greatly to the understanding of aerosol therapy, as a complex and interdisciplinary process. It is very clear that the success of aerosol therapy, its extension to other then inhalation medicines and the optimisation only is possible if specialists from different fields work closely together. The workshop made also very evident that new ideas arising in one discipline have to be discussed in details with the other groups to understand the limitations and requirements of the other disciplines involved in the total area of interest.
Finally, the effect of being together socially, the fantastic Polish hospitality, the wonderful Jablonna Palace as well as the lovely musical evening realized by the Dutch pianists Mme's Jetje van Wijk and Marlies van Gent contributed highly to the pleasant atmosphere of the Workshop.
The workshop was made possible by Warsaw University of Technology, Delft University of Technology, OctoPlus, Micro Montage, Astra Zeneca, Aventis Pharma, Inamed GmbH, Intervet International B.V., and Glaxo-Wellcome Poland.

Jan C.M. Marijnissen, Delft University of Technology
Leon Gradoń, Warsaw University of Technology

OLA NERBRINK

A HISTORY OF THE DEVELOPMENT OF THERAPY BY JET NEBULISATION

1. HISTORICAL BACKGROUND TO INHALATION THERAPY

Inhalation of aerosols, fumes and vapours from plants and aromatic mixtures, often in a liquid form, have been known for a long time. One of the earliest documented use of aerosols for inhalation purpose in asthma therapy is found in the Ayurvedic literature referenced in literature in the 17th century (Gandevia, 1975). The literature refers to the smoking of a preparation from a plant of the Datura family. Gandevia noted that it has been known in Europe, from the work done by Dr. Sims in UK, since the beginning of the 19th century (Sims, 1812).

This first recordings of inhalation therapy described in Ayurvedic medicine in India is more than 4000 years old (Gandevia, 1975). It describes the preparation and use of a paste consisting of *Datura* species that was dried and fixed into a pipe. The length of the pipe controlled the strength of the inhalant. The classical Hindu physician Charaka advised to ground spices, gum resins and fragrant wood into a powder and then into a paste. This paste was then smeared over thin tubes or sticks and lighted, and the smoke was inhaled as a treatment for diseases in throat and chest (Muthu, 1922). The ancient Greeks later on also acknowledged the idea of inhaling volatile substances into the lungs. This can be deduced from the recommendation of sending consumptive patients to the pine forests of Libya (Muthu, 1922). The ancient Egyptians on the other hand relied on the inhalation of vapours when heating plants of *Hyosycamus muticus* on hot bricks (Brewis, 1973).

Even Hippocrates proposed the use of hot vapours for inhalation purposes for treatment of sore throats and lung diseases (Moeller, 1882, Miller, 1973). He was also in favour of inhaled smoke, which is described by Waldenburg (Waldenburg 1872). This inhalation therapy was not commonly practiced until Galien during the antique period started to use it for treatment of sore throat and coryza. At this time the awareness of the positive influence sea air had on chest diseases was beginning to grow. Both Aretaeus and Galien had concluded that living near the seaside had a good influence on patients suffering from "pthitis" (tuberculosis) and other illnesses of the chest (Moeller, 1882, Waldenburg, 1872). The beneficial effect comes from the sea air that contains large amounts of salt particles from the breaking of waves. These salt particles can readily be inhaled. It is not clear if this was understood at this early time period but the awareness of it is clearly

L. Gradoń and J. Marijnissen (eds.), Optimization of Aerosol Drug Delivery, 1–22.

Figure 1. Large outdoor atomiser (located at Reichenall Spa in Germany)

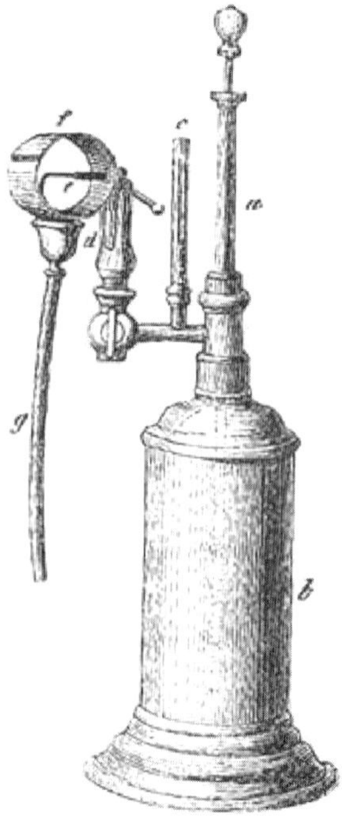

Figure 2. Sales-Girons portable pulverisator

documented in the 19th century by Moeller and Waldenburg. Antyllus used smoke as a remedy for diseases in the airways was in the 13th and 14th century (Waldenburg, 1872). During treatment the patient was instructed to spread his/her legs and bow the head downwards and start to inhale. A vessel with the contents burning was placed between the legs and the generated smoke inhaled. The material used was ether leaves of clematis, parsley, sulfur or pieces of ropes used when shipping. Mercury inhalation was used by the Italian Johann de Vigo in the end of the 15th century and beginning of the 16th century as a treatment against syphilis (Waldenburg, 1872). The therapy of using vapours for chest diseases was re-employed and continued in mid 17th century when Bennet in 1664 used vapour inhalation by steam together with infusions of aromatic herbs for the treatment of tuberculosis (Muthu, 1922). Willis further developed Bennets inhalation therapy and divided it into three degrees. He did this by applying a steam firstly consisting of balsamic mixtures, then sulphuric vapours and finally arsenic vapours (Moeller, 1882). This era involved the discovery of new gases and Priestly employed the use of newly discovered gases for therapeutic purposes in the late 18th century (Priestly, 1774). An English doctor by the name of Beddöes published remarkable work that widely contributed to popularise the therapeutic method of inhaling (Beddöes, 1803). He and his colleague Watt also used gas therapy like Priestly (Beddöes and Watt, 1796). Later on in 1810 does Laennec mention the use of vapours from aromatic plants, balsams, myhrr, sulphur etc. for the treatment of chest infections (Muthu, 1922). At this time it was becoming increasingly common to use both cold and hot mists of water for the treatment of Bronchitis, Croup and diseases in the upper airways (Waldenburg 1872). At the same time infusions of various herbs were used for inhalation purposes, although most of the active ingredients must have stayed in the liquid phase and not in the mist or steam. This must have made the therapy quite ineffective. In 1802, the Indian use of the smoke of *Datura* was introduced into the UK, which was described as mentioned earlier, by Sims (Sims, 1812). Sims had acquired an insight in the therapy from doctor Anderson working in Madras, India, where it was successfully employed against asthma (Hegewisch, 1811). It was used as when smoking and was blended with tobacco as the patient preferred. At the same time, *Datura stramonium* was substituted by *Datura ferox* in the relief of the paroxysm of asthma. Both were initially smoked in a pipe alone or in a mixture with tobacco. Later the *Datura*-tobacco mixture found its way to the cigarette, which was later on given the name asthma cigarette. The cigarette was often doped with an opium extract and it was recommended that the patient used two per day (Waldenburg 1872). Even substances like cannabis (Polak, 1860), originating from the orient, and derivatives of Picolin were smoked. Herbs like digitalis were also inhaled in a vapour or mist form. It must have worked fine, as the active compound is volatile and can reach the airways when heated and inhaled.

It is interesting to note that the use of liquids, preferably water based mixtures of various herbs or chemicals, were still used as the base for inhaling a vapour phase. This was still the predominant way in the early 19th century to administer drug compounds and other substances together with smoke, either from a cigarette or in

Figure 3. Mathieu's Néphogène (by Tirman)

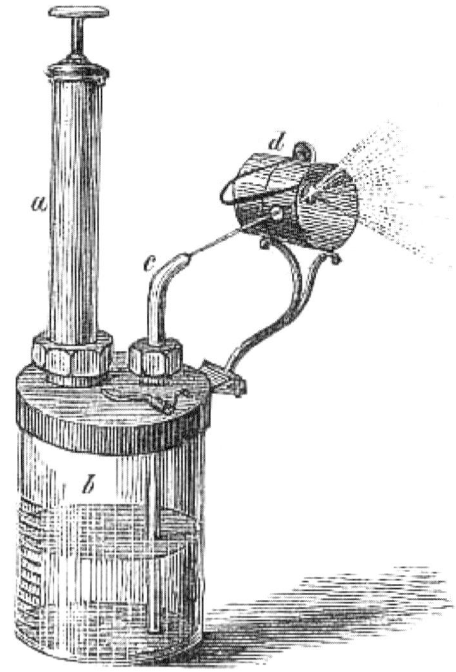

Figure 4. Lewin's device

fumes from open fire. It would not be long though, until the production and use of finely dispersed liquid droplets for inhalation would enter the scene. This new administration technology would give many new experiences and a give rise to an elaborate debate on the effect of this new therapy.

2. DROPLET GENERATION IN THE 19th CENTURY

The first published works done were on the on the concept of liquid jet disintegration and droplet formation. The initial work on water jets were performed in 1829 and 1833 by Bidone and Savart (Bidone, 1829, Savart, 1833). Bidone studied the geometric forms of jets produced by non-circular nozzles whereas Savart supplied the first quantitative data related to jet disintegration. Savart showed, for a constant jet diameter, that the jet velocity is directly proportional to the length of the smooth, continuous part of the jet before break-up into droplets. This work was basic for the understanding of the nebulising process but it was not connected to the medical area.

The driving force to develop the nebuliser for medical purposes came from the awareness of the beneficial effect sea air had on chest diseases, which probably was once again recalled from the early work by Galien. Sea air contains high levels of salt particles that are produced when waves break up and Waldenburg concluded that the same thing occurred in small scale in German brine graduation plants (Waldenburg, 1872). A large number of Spas (Rehme, Elmen, Meinberg, Kösen, Kissineg, Reichenhall etc) grew up close to these brine graduation plants where the salt containing atmosphere was used as therapy to treat airway illness. An example of one of these graduation plants is shown in Figure 1 (Heyder, 1999). It is now clear that the understanding of the beneficial effect, from a medical point of view, of this type of inhalation was beginning to be clearly recognized. In 1841 Lobethal and later on Hirzel in 1845 started to set up synthetic sea air environment in Zürich by using a fountain that sprayed seawater in a closed room (Waldenburg, 1872). Auphan built a so-called vaporatorium in Euzet les Bains in 1847 (Solis-Cohen, 1876). In his location was salt containing water splashed against the walls of a room. Two years later he built another vaporatorium in Lamothe les Bains where a number of 7 metre high waterfalls were used to break liquid into droplets when they hit the walls. The size of the majority of droplets must have been comparatively large and not suitable for inhalation so the need for a device that could decrease the size was imminent. The start to a solution came in 1829 from Schneider and Walz who constructed the first apparatus that could break up liquid into finer droplets (Solis-Cohen, 1876, Moeller, 1882). The device was intended for use in showers and baths and not at all for inhalation purposes but may well have been the development that triggered others to develop and use devices with in the medical field although no connection has yet be found in the literature.

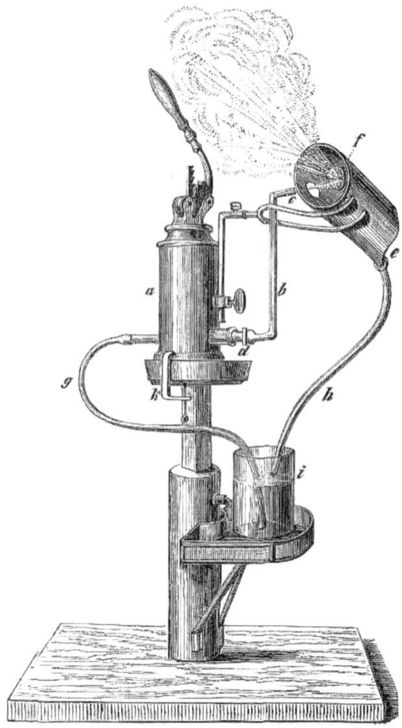

Figure 5. Mayers Uterusdouche

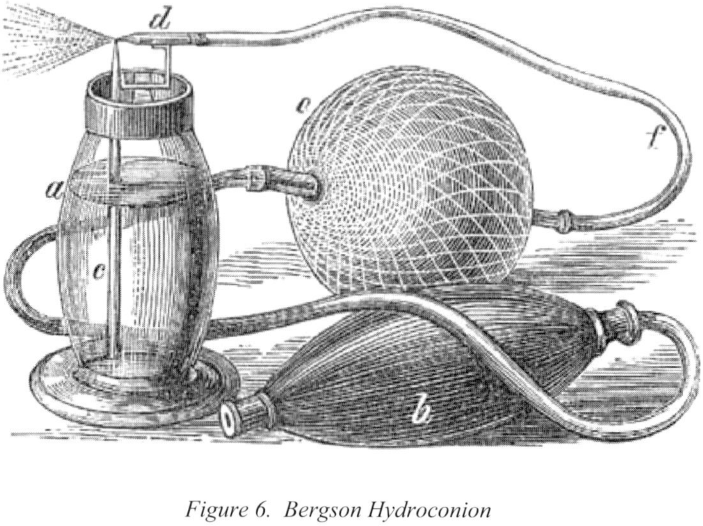

Figure 6. Bergson Hydroconion

The major breakthrough came in 1856 when Sales-Girons erected his vaporatorium together with Flubé in Pierrefonds (Sales-Girons, 1860; Solis-Cohen, 1876). The idea was similar as that to Auphan but Sales-Girons constructed a novel device that could break up the droplets into a fine inhalable mist. The vaporatorium was a seven metres long and four and a half metres wide chamber where up to 15 patients could be treated by inhalation. The chamber had windows that were open during the treatment. A suction pump outside the chamber drew mineral water to be nebulised into a tube, which passed through a heated water bath, and then entered into the chamber. It terminated in a vertical cylinder at which end the nebuliser apparatus was attached. The nebuliser consisted of six fine grooves in the cylinder, which gave exit to six compressed mineral water streams. A stopcock controlled these streams on and off function. Each stream struck a metallic plate after a short distance and a fine mist or an aerosol was produced which was carried out into the chamber. Excess liquid was drained out of the chamber and the patients were covered with an appropriate covering to prevent them from getting wet.
Some time later (1860) did Sales-Girons present a portable device, see Figure 2 (Solis-Cohen, 1876; Moeller, 1882). The device, known to be the first known to be constructed by Charrières, is a pneumatic device. It is hand driven having an air compression pump that is used to exert a high pressure (typically 2 to 4 bars overpressure) on the liquid to be nebulised. Doctor Sales-Girons could rightly claim the credit of having constructed the first nebuliser device that produced, in most part, an inhalable aerosol. Thanks to this invention and the work he published along with his documented stamina and eagerness in public discussions, he managed to start the scope of inhalation practice to move from exclusive seaside resorts to places where it could be more commonly adopted.

3. THE EARLY NEBULISERS

From this date and on a number of devices are presented in the literature and the use of liquid droplets for inhalation therapy starts to take place. The devices are often referred to as apparatus for the pulverization of liquids and it is first in 1872 the term nebuliser is defined and found in the Oxford dictionary.
In 1859 Mathieu presented a portable spray-producing device after a principle suggested by H. Tirman (Solis-Cohen, 1876). His device, which he called the Néphogène Figure 3, was exhibited to the Parisian Academy of Medicine on May 9th the same year. Again compressed air is used to drive the apparatus by exerting pressure on the liquid to be nebulised. The atomisation occurred at the exit of a tube that had a small orifice opening and consequently no plate was needed for the atomisation process to take place.
At the same time Lewin, Mayer and Baumbärtner manufactured devices with various designs using the same principle as Sales-Girons (Waldenburg, 1872). Luër constructed a nebuliser device where two opposing liquid jets under high pressure disintegrated into an aerosol at the point where they met each other (Waldenburg,

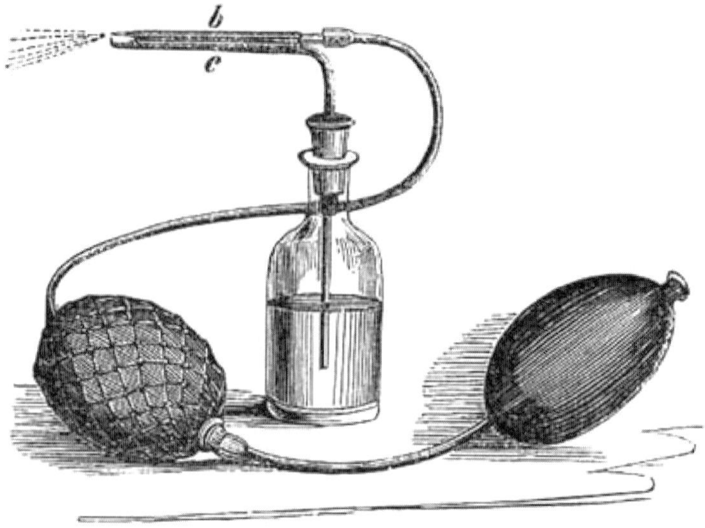

Figure 7. Winterlich device (with Bergson tubes)

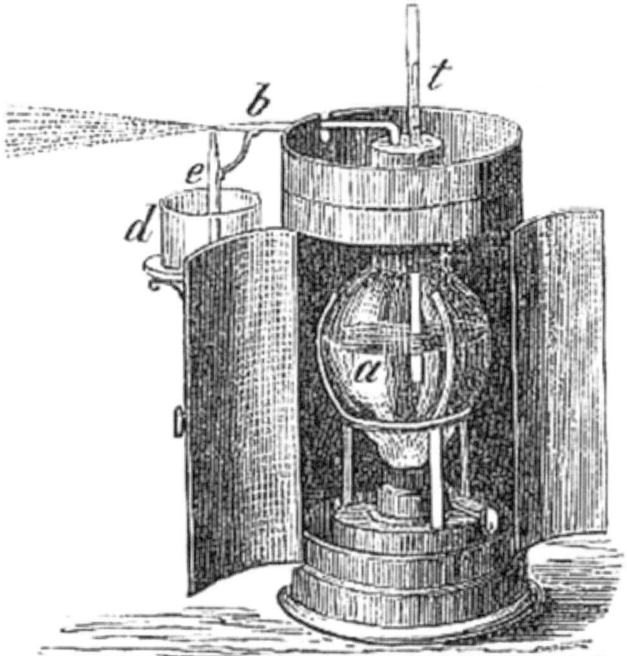

Figure 8. Siegle steam apparatus

1872). The device was later on modified by Cube to be more handy and easy to use (Waldenburg, 1872).

A rapid development now commences and three groups of device having different atomisation principles can be identified. The first group consists of devices where the liquid is ejected against a plate or another jet of liquid. This group comprise the first development of devices based on the construction suggested by Sales-Girons. The second group consists of devices where compressed air and liquid to be atomised is mixed at the atomisation point. The third group comprise devices the produce aerosols from a pressurised steam that is used to suck liquid to the atomisation point.

The device development was carried on and Lewin together with Goldschmidt suggested a modification of the Sales-Girons device. This device is displayed in Figure 4 (Waldenburg, 1872). Again compressed air is used and the vessel is in glass instead of metal the liquid is forced up through the exit tube into a drum where a metal plate is used for the disintegration of the liquid stream into droplets. Later on Mayer proposed a more complicated device that he called the Mayriche uterus douche; see Figure 5. Liquid is sucked up the atomisation point by the use of a suction-pressure pump/vessel and a stopcock is used to commence the atomisation after the required pressure is obtained. The end of the feed tube had a small orifice opening in which a thin liquid beam exited to hit a metal plate where the disintegration occurred. Excess liquid after atomisation was fed back to the liquid reservoir through a tube that was connected to the back of the drum.

Waldenburg mentions the use of a variety of different substances dissolved or mixed with water and used in inhalation therapy around 1850. Amongst these can the following be mentioned:

Silver nitrate, zinkvitriol, coppersulphate, chinin, and diluted acids like acetic-, sulphuric- and phosphorus acid. Salmiak, sodium- cloride carbonate bicarbonate and sulphate. Narcotic substances like opium, morphine, hyosycamus, stramonium, belladonna, nicotine, cannabis and Atropin. Clorine in water and iodine solutions were also used. Substances like turpentine and sage oil, carbolic acid and creosote and acetone were all preferably inhaled in the vapour phase.

Quite rapidly the development moved on to what is known as the devices classified in the second group. In these devices compressed air was used to suck up liquid from a reservoir. The credit for this modification must be given to Bergson who constructed a number of devices built on this principle, and many others were to follow (Solis-Cohen; 1876, Waldenburg, 1872; Moeller, 1882). One of his first designs can be seen in Figure 6 and is often referred to as the Hydroconion device. The device uses the rapid expansion of the air at the orifice exit of the compressed air tube to suck up the liquid from the reservoir, a feature that still is used in modern devices. Bergson and his followers adapted the principle in many ways, which can be seen in the example of the Winterlich apparatus displayed in Figure 7. Here the liquid is transported, by under-pressure, to the atomisation point in the lower tube, to end in a 90-degree bend, thereafter presenting the liquid at the air orifice expansion point.

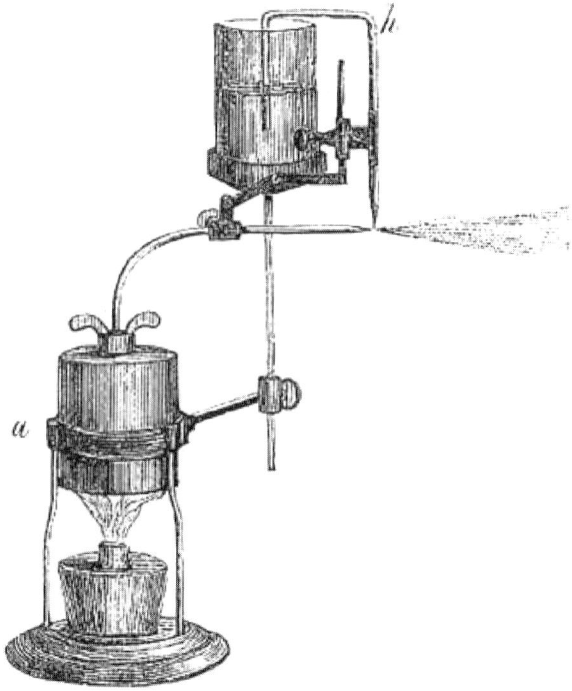

Figure 9. Pissin steam device

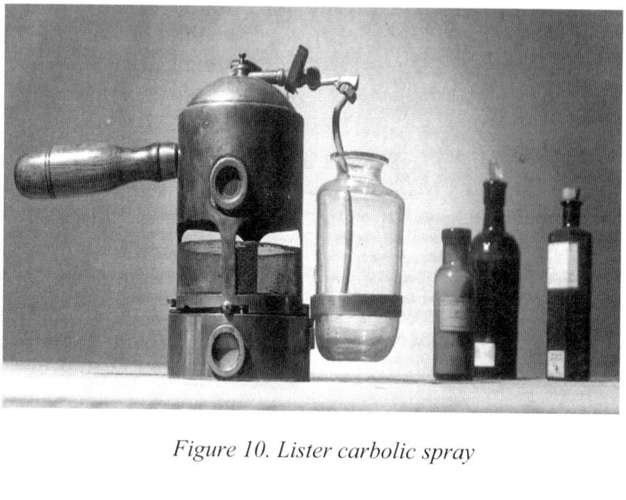

Figure 10. Lister carbolic spray

The next development of the devices was the incorporation of a heater that could produce a steam aerosol, which in turn was used to suck liquid to the steam exit point of the device. These devices are referred to as the third group. An example of a device that is steam driven is the Seeger steam apparatus (Muers, 1997). Other versions and modifications were made by Siegle (Figure 8), Bûrow and Pissin (Waldenburg, 1872). The latter used the Bergson tube arrangement (nozzle) in his device to transport the fluid to the atomisation point Figure 9.

There are many similarities between these early devices and the modern nebulisers. Many of the devices had a detached "drug" container. Liquid was sucked to the atomisation point by the negative pressure generated by steam or oxygen, under pressure, suddenly expanding at the exit nozzle. Baffle systems were not incorporated into the early nebuliser. Looking back we would now define such early devices as atomisers and modern nebulisers can be regarded as atomisers that incorporate a baffle system that select out smaller droplets or particles for inhalation. This modern type of nebuliser appeared in the early 20th century.

4. THE DEVELOPMENT OF NEBULISED THERAPIES

The indication in which these early nebuliser are used for are diversified, for example the Sales-Girons device was used for pharyngitis, laryngitis, bronchitis, pain catarrh, asthma, tuberculosis and sleeplessness (Valet, 1995). The liquids and substances used show almost the same variety as the indications: mineral water containing sulphur, iodine and chlorine, sedative and antiseptic infusions and belladonna and the treatment time varied from 10 to 20 minutes.

Simultaneously at this time there was much controversy regarding the ability of aerosols to pass into the airways together with their following deposition in the lungs. Demarquay (1861) and Siegle (1865) published experimental results indicating that aerosol particles passed into the respiratory system whereas authors like Briau (1861), Fournié (1862) and Rey (1861) argued against. Then the question if aerosols, or pulverized liquids, could reach the lungs was debated intensively around late 1850 and early 1860 according to Dautrebande (1962). It was not positively answered until 1862 when Poggiale presented his report to the Paris Medical Academy (Yernault, 1994). This triggered further investigations in the inhalation and deposition of pulverized liquids in the airways. Around 1870 Waldenburg conducted and presented several studies on the bronchial deposition of pulverized liquids. The conclusions he drew originate from both practical trials and theoretical discussions and have been given a chapter in his book from 1872. He concluded that about 25% were deposited in the mouth and larynx and about 30% passed beyond the larynx. He also recognised that bronchial deposition depended on the physical characteristics of the patient's inhalation route such as the position of the tongue in the mouth and inspiratory volume (depth) as well as the patient's pathological conditions. Consequently, he recommended gargling after the treatment as he concluded that upper airway deposition could occur. It is interesting to note that the experiments, reported by Heubner (Dautrebande, 1962) in 1923, helped to convince others that aerosolised liquids reached the lungs.

Almost parallel with the work done in the inhalation field, Lister (Lister, 1871) developed a carbolic steam spray apparatus Figure 10 for employment in the antiseptic field. The reference is interesting as it shows an apparatus used for producing aerosols thereby giving a picture of the current nebuliser development in another field at this time. Less than 10 years after its introduction it was questioned by von Bruns (von Burns, 1880). Criticism against his method grew and Lister abandoned the technique in 1887. In his paper from 1890 (Lister, 1890) he commented, "I feel ashamed that I should have ever recommended it for the purpose of destroying the microbes of the air". Still, the apparatus was used for nearly thirty years when it was abandoned due to its incapacity of showing positive results.

As said earlier, the variety of substances used in inhalation therapy in the late 19th century was quite impressive. Some of these have been used in vapour inhalers whereas some have been used in nebuliser set-ups. At this time period it was still common to use vapourizers as well as nebulisers. For instance, in 1882 Yeo (Muthu, 1922) prescribed the use of a mixture of creasote, carbolic acid, eucalyptus or turpentine with equal parts of spirits and chloroform in his oro-nasal inhaler. Earlier, in 1878 Lee (Muthu, 1922) performed antiseptic experiments using a jet of steam containing a mixture of phenol and water. Iodoform, iodine, thymol and terebene was also used (Muthu, 1922).

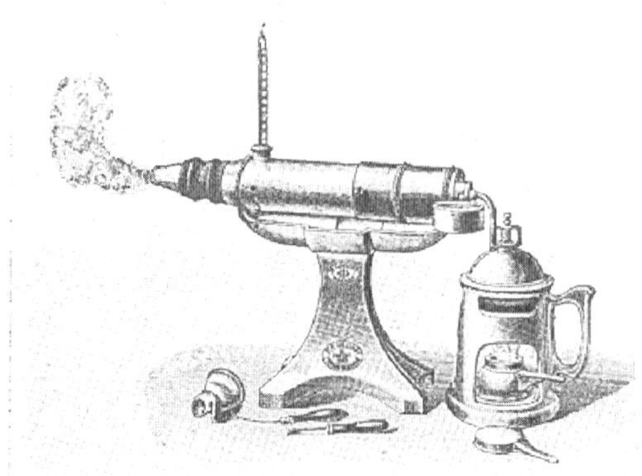

Abb. 7: Thermovariator nach Bulling, entnommen aus Seyderhelm, Neue Deutsche Klinik V p. 360 (1930)

In the late 1890's Bulling constructed a device that he called the thermovariator (Schiessle and Steiner, 1961), see Figure 11. The device was driven by steam and the atomisation was managed by Bergsons tube arrangements. The steam was confined in a tube and was inhaled at a high temperature, often up to 85 degrees centigrade. One of the first references containing data published on nebuliser characteristics is given in this paper that states the output of the device to be 100 grams per 12.5 minutes. In early 1900 Heryng (Heryng1906) commented on the device and showed a further development of the thermovariator concept that had a higher output (100 grams per 3 minutes) and a lower temperature of the steam (Schiessle and Steiner, 1961). A reference is also given on the performance and some critical dimensions such as exit orifice (0.7 to 0.9 mm diameter) stated. Heryng concluded that the droplet size was dependent on the steam exit orifice size predominantly and not the driving pressure.

The development and use of adrenaline in the treatment of asthma in the early 1900 was a major advance. In 1900, Solis-Cohen (Solis-Cohen, 1900) injected crude adrenal extract into patients with asthma and hayfever. Shortly afterwards Bullowa and Kaplan (Bullowa, 1903) reported the successful use of adrenaline injection that then became established as a standard therapy for relief of severe attacks of asthma.

In 1911 Zeulzer (Zeulzer 1911) refers to the treatment of acute catarrh in the upper airways by inhalation of nebulised substances using a Draeger nebuliser constructed by Professor Spiess. Unfortunately the Draeger nebuliser mentioned in the paper is not described but a later reference (Heubner, 1925) has indicated that it is derived from the Bergson atomisation principle. The paper describes the use of glycerinan in the nebuliser in 25 cases of chronic Bronchitis, chronic Laryngitis, croup related pneumonia, interstitual pneumonia and TBC. Glycerinan is a mixture of Epirenan, an adrenaline analog, in water and glycerine. The treatment time was about 5 minute and this was repeated 2-3 times a day depending on diagnose. The output of the nebuliser was stated to be 1.3 g/min at an airflow rate of 5 to 12 l/min.

Much of the work around 1920 to 1930 was focussed on characterisation of the devices used at that time. Parameters such as "fog thickness", " fog dosis", "fog amount" and atomisation degree was investigated (Schiessle and Steiner, 1961). This resulted in the conclusion that the dimensions and drilling of the air and liquid feed tubes were important for the characteristics. Tancré, Regener and Spiess-Draeger proposed a number of different designs of the nozzles. Again the forms were derived from the Bergson principle. Electricity was now more commonly available thereby enabling heating of liquids to steam and also the introduction of the air compressor that can be seen in Figure 12. In the same Figure (12 top) a portable medical nebuliser is displayed. The device consists of a nebuliser fitted with a Bergson type nozzle and a 5 horsepower, electically driven compressor.

Another interesting use of nebulisation technology is the multiple large nebulisers with substantial output capabilities. These were used together as a collective

Schiessle, Steiner, Geschichte der Tröpfcheninhalation 273

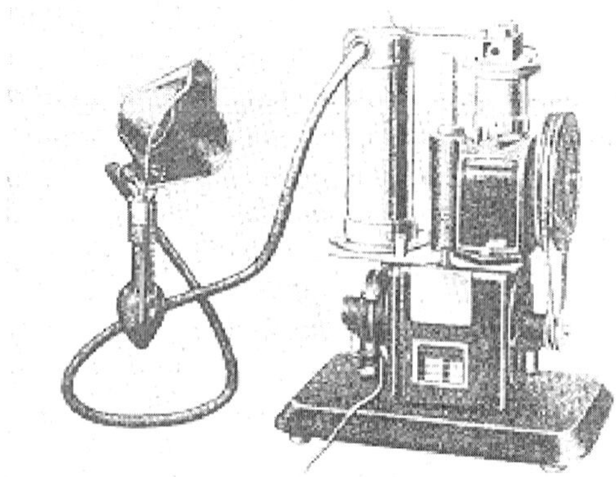

Abb. 13: Transportabler Medikamentenvernebler, entnommen aus Sydenheim, Neue Deutsche Klinik V, p. 390 (1930)

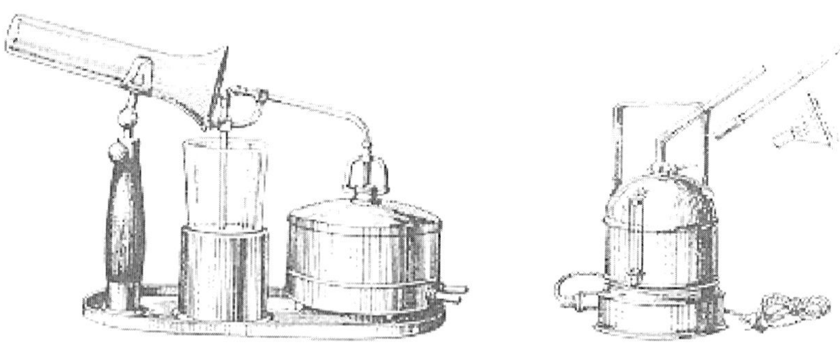

Elektrisch beheizter Inhalierapparat.

Bronchitiskessel mit elektrischem Anschluß.

Abb. 14: Feuchtzerstäuber, entnommen aus Sydenheim, Neue Deutsche Klinik V, p. 369 (1930)

Figure 12. Device development at 1930

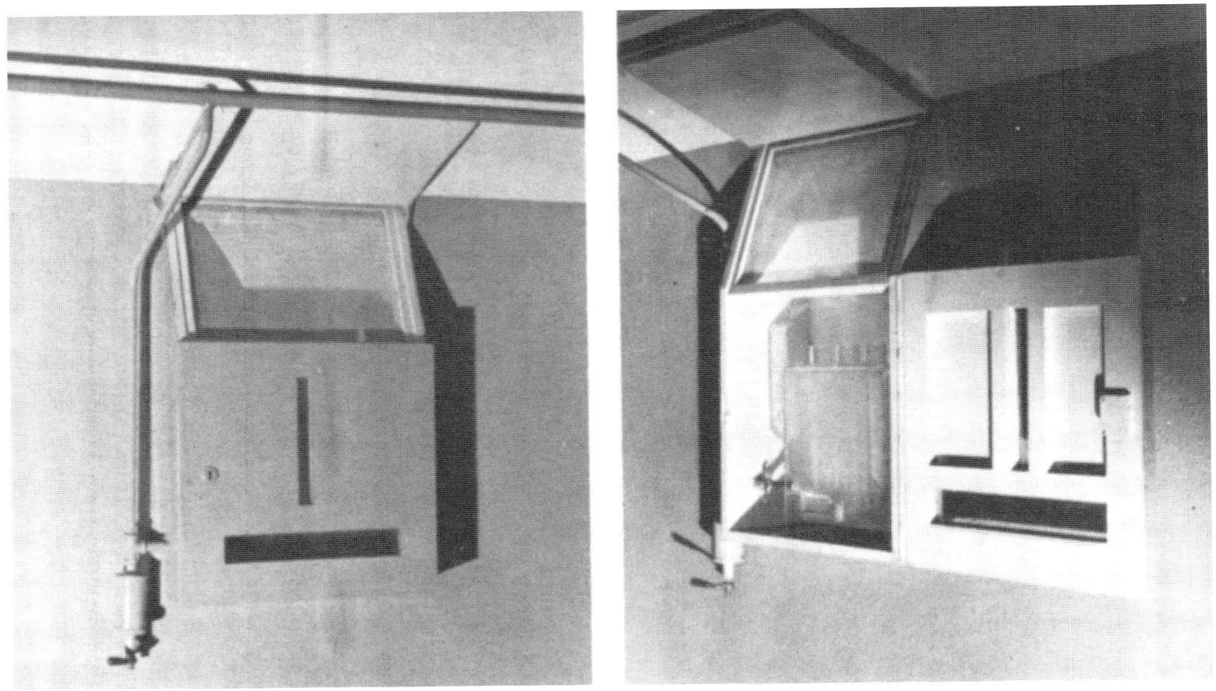

Figure 13. Multi Nozzle device used in coalmines in Germany

inhalation chamber in the coal mines in South Africa and Germany in the 1930s (Dautrebande, 1962). The nebuliser used was an 8 or 16 nozzle device placed in a cabinet in a corridor Figure 13. The corridor could be several hundred meters long and could contain a number of nebuliser cabinets Figure 14. The nebuliser was filled with what was considered to be a mucolytic and bronchodilating solution. At the end of the miners' shift, nebulisation was commenced, and the miners walked slowly through the mist-containing corridor inhaling the aerosol.

Figure 14. Collective inhalation device, the exposure corridor

The Draeger or Hirth apparatus (Camps, 1929) and also a nebuliser called Apneu (Graeser, 1935) became popular nebulisers. Bulb nebulisers, where the patient operated the device by pressing a rubber bulb connected to the nebuliser to

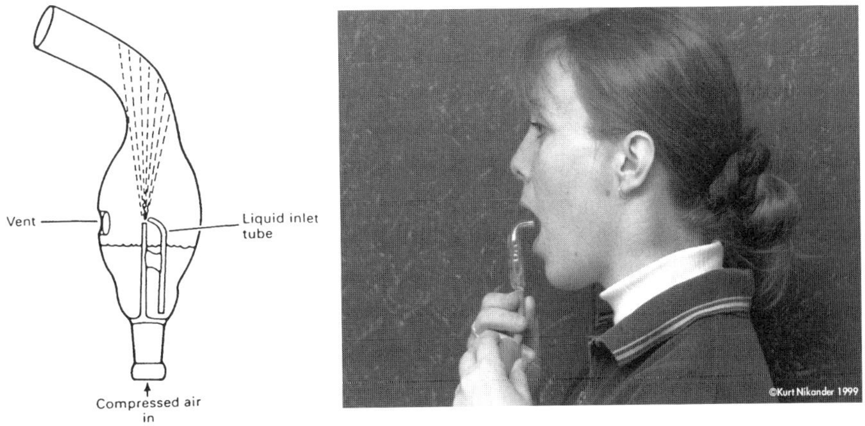

Figure 15 De Vilbiss 40 nebuliser

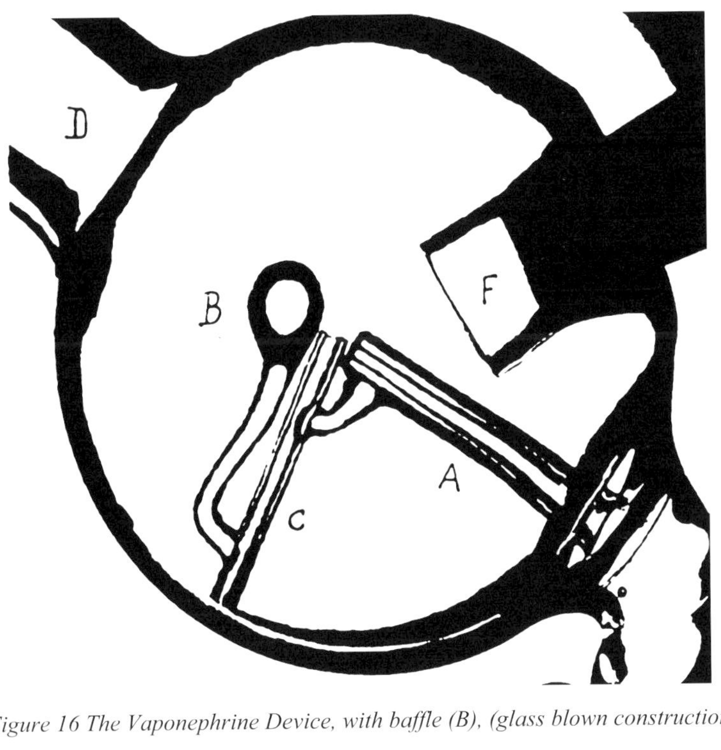

Figure 16 The Vaponephrine Device, with baffle (B), (glass blown construction)

generate the gas flow, became popular (24). The nebulisers generated an aerosol with a wide droplet size distribution and a large fraction of non-respirable droplets. In the 1930s, the glass bulb nebulisers such as the DeVilbiss No 40 and equivalent Figure 15 and the Vaponephrine nebulisers Figure 16 were popular. In 1944, Bryson (Bryson, 1944) published particle size distribution data on the DeVillbis no 40 nebuliser and the Vaponephrine nebuliser. The glass-spray type of nebuliser had two serious drawbacks. Firstly, the sprays varied in their efficacy because they were 'hand' made and thereby variable in design. Secondly, the feed pipe of the spray was small and liable to become blocked after a period of use and attempts to clear it were often associated with breakage. A curiosity about the DeVillbis no 40 is that it was pressurised by a hand operated rubber bulb, giving great variability in the operating pressure. The Vaponephrine nebuliser was one of the first nebulisers with a baffle close to the generation point. Not surprisingly, subsequent investigations showed that the DeVilbiss device produced larger droplets than the Vaponephrine device (Palmer, 1952).

In 1932 Collison introduced a nebuliser bearing his name at the British Medical Association meeting. The use of the device was described in Collison's book "Inhalation Therapy Technique" (Collison, 1935). Unfortunately no technical data were given in the book but May published data in 1973 on the device (May, 1973). Initially the construction had one atomisation point that was later on modified to three points and the nebuliser unit was the size of a small glass jar. The Collison had one serious defect: the liquid was fed into the spray through three blind holes in an ebonite spray unit. Since most liquids used in nebulisers tended to form a deposit, these holes invariably became blocked and were difficult to clean. In the mid 1940's electrical pumps providing a continuous flow of air were getting more commonly used and the Collison nebuliser became the most popular nebuliser in the UK.

The next major advance in inhalational therapy for asthma was the introduction in 1951 of isoprenaline. In 1940 Konzett of Boehringer Ingelheim found that an analogue of adrenaline, its N-isopropyl derivative (Scheuing, 1942), helped to relieve the bronchospasm of asthma when inhaled. Knowledge of its development became available when the US State Department investigated work carried out by German chemical manufacturers during the war. The drug, named isoprenaline, gave the bronchodilating effect of adrenaline, but was relatively free from troublesome pressor activity (Siegemund, 1947). The compound was introduced in 1951 into clinical use with the name of isoprenaline. Examples of broncodilating drug formulations used at this time period are Epinephrine; 1:100 in water (ex. Suprenaline by Aurmour), racemic Epinephrine hydrocloride (ex. Vaponephrine supplied by Vaponephrine Co) and the mixture of Isoprenaline in water; 1:200 (ex. Isoprel from Winthrop) (Harris, 1957).

5. DEVELOPMENTS POST WWII

In 1945, the introduction of penicillin was followed swiftly by attempts to nebulise it directly to the lungs. A device know as the "Deedon" inhaler (Moore Medicinal Products Ltd, Aberdeen, Scotland) was a neat hand held inhaler constructed entirely from plastic intended for the administration of penicillin or anti-spasmodics. It was suggested that the fine mist of penicillin was likely to penetrate as far as the small bronchi and bronchioles. Trials showed that penicillin, in 30 per cent glycerine, placed in the reservoir was administered in 6 – 7 minutes by squeezing the rubber bulb at each inspiration. The makers also advertised an electric pump at a cost of £20, which was considered expensive by the Lancet (Anon, 1946). Other compounds, such as Streptomycin and Terramycin, were also nebulised.

In 1958 Wright (Wright, 1958), working in the pneumoconiosis research unit of the Medical Research Council, presented a new nebuliser bearing his name that was robust and easy to clean. The nebuliser was entirely made of perspex. It was practically unbreakable and easy to clean, also one of the criterions used as an indicator of good performance was the ability of the nebuliser to produce droplets ranging in size between 1 and 6 microns in diameter at a given gas flow. His nebuliser concept was the beginning of the modern era of nebuliser design and size.

6. THE PRESENT

Today there are a number of nebulisers available for patients and hospitals. They are manufactured in large numbers in plastic material. The jet nebuliser is by far the most common and they still often have a compressor or high-pressure gas flask to obtain the air supply. They are used by patient groups ranging from infants to elderly but are predominantly used by patient that have coordination problems or do not have sufficient inhalation effort capabilities. Nebuliser therapy is today commonly used in many clinical areas and inhalation treatment is used for delivery of various agents including bronchodilators, corticosteroids, anticholinergics, antibiotics, mucolytics and anaethstetics (Heslop, 2000). The physicochemical properties such as viscosity and surface tension may vary extensively for formulations made when using these agents. Concurrently, the output characteristics of a given nebuliser equipment and the therapeutic response will be affected (Muers, 1997b) and even more if different nebuliser equipments are used together with a single drug formulation. This has been identified and has been addressed by the European standard EN 13544-1 and in the ERS guidelines for use of nebulisers.

A modern example of collective inhalation originating from the thoughts of Galien and from the graduation plans described by Waldenburg can still be found in Germany (Heyder, 1999). The large so-called aerosol generator is seen in Figure 1. It is located along a hallway about 100 m long with piles of tree branches at one side. Salt solutions are allowed to drip onto the branches and each time a falling drop strikes a branch, smaller droplets are generated. Patients walk in the hallway inhaling the particles. This is still considered a simple and efficient way to produce large volumes of therapeutic aerosols for inhalation and is nowadays a part of some modern Spa facilities located in Europe.

In the late 1950 and early 1960 the development of the ultrasonic nebuliser took place. These were initially just as large and bulky as the jet nebuliser equipments used but their improvement has proceeded and today they are being developed into smaller units that can be handheld and more portable than the conventional jet nebuliser. Examples of these are the NE-U14 and the NE-U22 from Omron Healthcare Company in Japan. They still use ml quantities of liquid and have similar characteristics as the conventional jet nebuliser systems. During the 1990s there has been a development of ultrasonic and pneumatic technologies that are targeted to deliver a defined amount of liquid to the airways during a single breath. These devices have many similarities to the Metered Dose Inhaler (MDI) but lack some of the disadvantages that the MDI has. None of these novel liquid nebuliser devices have been marketed yet but they may well reach the market within a couple of years. Examples of these are the Boeringer Ingelheim Respimat® soft mist inhaler, the Aradigm AERx® and the Aerogen AeroDose®. If they are successful in penetrating the market they will surely change the focus of liquid drug delivery to the upper airways and possibly question the more than 150 year old conventional jet nebuliser therapy.

7. REFERENCES

Anon. (1946) Inhaler for penicillin: Notes and News. Lancet, 1, p 986.

Beddöes T. (1803) Ueber die neusten methoden, die schwindsucht zu heilen. Ueberzets von Kühn.

Beddöes T., Watt J. (1796) Betrachtungen über den medicinschen gebrauch küstlicher luftarten and die methode,sie in grossen quantitäten zu bereiten; Ueberzetzt von Zollikofer.

Bidone G. (1829) Expériences sur la forme at sur la direction des vienes et des courants d'eau lances par diverse ouvertours. *Imprimerie Royale*, Turin, pp 1-136.

Boe J., Dennis J. H., O'Driscoll B. R. (2001) European Respiratory Society Guidelines on the use of nebulizers. Eur Respir J. 18, p 228-242.

Brewis R. A. L. (1990) *Classic papers in asthma.* Science Press, London.

Briau (1861) Des effects de la respiration de léau minérale pulverisée. *Gaz. hebdomadaire méd. chir.* 8, Nos 44 and 45.

Bryson V., Sansome E. and Laskin S. (1944) Aerosolization of penicillin solutions. *Science,* 100, p 33.

Bullowa G.G.M., Kaplan D. (1903) Med News, 83, p 789.

Camps P. W. L. (1929) A note on the inhalation treatment of asthma. Guy´s Hospital Reports, 79, pp496-498.

Collison W. E. (1935) *Inhalation Therapy Technique.* Heineman, London.

Dautrebande L. (1962) *Microaerosols.* Academic Press Inc, New York.

Demarquay (1861). Mémoire sur la pénétration des liquides pulvérisés dans les voies respiratoires et de leur application au traitment des yeux, du pharynx et du larynx. *Asselin*, Paris.

European standard EN 13544-1 (2001) Respiratory therapy equipment-Part1: Nebulizing systems and their components

Fournié E. (1862) De la pénétration des corps pulvérulents gazeux, solides et liquides dans les voies respiratoires au point de vue de l´hygiéne et de la thérapeutique. *Adrien Delahaye*, Paris.

Gandevia B. (1975) Historical review of the use of parasympathicolytic agents in the treatment of respiratory disorders. *Postgrad Med J.* 51, pp 13-20.

Graeser JB and Rowe AH. (1935) Inhalation of epinephrine for the relief of asthmatic symptoms. J. Allergy, 6, pp 415 420.

Harris c. M., Shure N., (1957) *Practical Allergy*, Philadelphia, F.A. Davis Company.

Hassal A.H. (1885) *The inhalation treatment of diseases of the organs of respirations including consuption.* London, Longmans, Green, and Co.

Hegewisch. Communications relative to the Datura Stramonium or thornaplle as a cure or relif of Asthma. Edinbourg Medical J. April and July 1811 No.XXVI and XXVII: Hufeland's J. XXXVI; 2, p 82, 1813.

Heryng T. (1906) Ueber neue inhalationsmethodenund neue inhalationsapparate. *Berliner Kliniche Wochenshr.* 11, pp 323-326.

Heryng T. (1906) Ueber neue inhalationsmethodenund neue inhalationsapparate. *Berliner Kliniche Wochenshr.* 12, pp 358-363.

Heryng T. (1906) Ueber neue inhalationsmethodenund neue inhalationsapparate. *Berliner Kliniche Wochenshr.* 11, pp 323-326.

Heryng T. (1906) Ueber neue inhalationsmethodenund neue inhalationsapparate. *Berliner Kliniche Wochenshr.* 30, pp 1028-1029.

Heslop K., Harkawat R. (2000)Nebulizer therapy from a practical perspective. Eur Respir Rev, 10:(72), pp 13-215.

Heubner W. (1925) *Durchlässligheit der Lunge für fremte Stoffe. In "Handbuch der normalen und pathologischen Physiologie*". (edited by A. Bethe et al.), Vol II, 473, Springer, Berlin.

Heubner W. (1925) Über Inhalation, insbesonderes Insulin. *Münch. Med. Wochensr.*, No. 244/245.

Heyder J. (1999) Personal communication.

Lister J. (1871) On a case illustrating the present aspect of the antiseptic treatment in surgery. *British Med J. i*, pp 30-32.

Lister J. (1890) An address on the present position of antiseptic surgery. Brit Med J, pp 377-379.

May K. R. (1973) The Collison nebuliser: Description, Performance and Application. *Aerosol Sci*, 4, pp235-243.

Mercer T. T., Goddard R. F. and Flores F. L.(1965) Output characteristics of several commercial nebulizers. *Ann. of Allergy*, 23, pp 314-326.

Miller W. F. (1973) Aerosol therapy in acute and chronic respiratory disease. *Arch Intern Med,* 131, pp 148-155.
Moeller. (1882) *Thérapeutique locales des maladies de l'appareil respiratoire par les inhalations médicamenteuses et les pratiques aérothérapiques.* J-B Baillière et fils, Paris.
Muers M. F. (1997) Overview of nebuliser treatment. *Thorax,* 52, pp 25-30.
Muers M. F. (1997b) The rationale use of nebulizers in clinical practice. Eur Respir Rev, 7:44, pp 189-197.
Muthu D. C. (1922) *Pulmonary tuberculosis its etiology and treatment, record of twent-two years observation and work in open-air sanatoria..* Bailliére, Tindall and Cox, 8 Henrietta Street, Covent Garden, London.
Palmer F, Kingsbury S. S. (1952) Particle size in nebulized aerosols. Am J Pharm, 124, pp 112-124.
Polak. (1860) Bereich der Deutchen naturforcherversamlung zu Köningsberg.
Priestly (1774) Experiments and observations of different kinds of air. London.
Rey A. (1861) La pulvérisation. *Union méd. No* 139, p 346.
Sales-Girons. (1860) *Traitement de la Phthisie Pulmonaire par L'inhalation des Liquides Pulvérisés et par les Fumigations de Goudron..* F. Savy, Libraire Éditeur, Paris.
Savart F. (1833). *Ann. Chim. Phys.* Vol 53, pp 337-386.
Scheuing, (1942) Thoma, Ger. Pat., pp 723-728.
Schiessle, W and Steiner H. (1961) *Zentralblatt für biologische aerosol forschung, 9, 3, geschichte der tröpfcheninhalation.* pp 365-382 and 468-487.
Siegemund O. H. Grainger H. R. and Lands A. M.(1947) J Pharmacol Exp Ther., 90, p 254
Siegle E. ed (1865) *Die Behandlung und Heilung der Hals- und Lunglieden durch Einathmungen.* 2nd edition, Stuttgart.
Sims J. (1812) Datura stramonium or thorn apple as a cure or relief of asthma. *Edinburgh Med Surg J*, 8, pp 364-367.
Solis-Cohen S. 1900 J Am Assoc, 34, p 1164.
Solis-Cohen J. (1876) *Inhalation in the treatment of disease: Its therapeutics and practice.* Lindsay and Blakiston, Philadelphia.
Valet J. S. (1995) To breathe or not to breathe. Ancora SA, Barcelona.
von Bruns V. (1880) Fort mit dem spray! Berliner Klinische Wochenschrift, 17, pp 609-611.
Waldenburg L. (1872) *Die locale behandlung der krankenheit der athmungsorgane. Lehrbuch der respiratorischen therapie.* G. Reimer, Berlin.
Wright B. M. (1958) New inventions: A new nebuliser. *Lancet* ii, pp 24-25.
Yernault J. C. (1994) Inhalation therapy: a historical perspective. *Eur Respir Rev,* 4, pp 65-67.
Zeulser G. (1911) Die Behandlung akuter Katarrhe der oberen Luftwege durch Inhalation von Nebennierenpräparaten. *Berliner Klin. Wochenschr.*

9. AFFILIATIONS

Ola Nerbrink, NovoNordisk A/S, 3400 Hillerød, Denmark

BOGDAN LESYNG AND WITOLD RUDNICKI

MOLECULAR MODELING IN DRUG DESIGN

1. INTRODUCTION

Molecular modelling provides indispensable methods and simulation tools for the rational design of novel, biologically active compounds. These methods can be applied to quite a broad class of problems, where the desired effect is related to blocking or enhancing the activity of a target macromolecule (receptor) by a small molecule (ligand), which specifically binds to the receptor. Typically, a ligand is an inhibitor molecule that binds either to an enzyme active site, a regulatory site, or a given sequence of DNA. In the last case it can play the role of a gene repressor. A ligand can also be an activator molecule binding to a regulatory site. In all cases the goal is to find a small molecule, that binds in *a specific way* and *with high affinity* to the given receptor. Specificity is required to avoid side effects, and high affinity is required to ensure the desired treatment effects. High affinity is related to a high release of the free energy upon binding. Therefore, looking at the drug design process from the point of view of chemical physics, we are looking for a fast and efficient way of predicting the free energy of binding of various molecules to the given receptor(s).

Molecular modelling can be applied at several stages of the design process. First, it is very useful to know an approximate 3D structure of the target macromolecule, if possible also in complexes with ligands. Molecular modelling methods are then applied in the process of refinement and analysis of the available experimental data leading to the precise 3D model of the macromolecule. This refers to the location of ionisable hydrogen atoms (protons) at given pH and ionic strength, that in turn is required for the precise determination of free energy differences. This is called *virtual titration*. For details see (Antosiewicz et al., 2000). In situations were the structure of the target macromolecule is unknown, molecular modelling methods can be used to predict its structure. Current methods in bioinformatics - in particular *homology modelling* or *ab initio* protein folding techniques - are capable of predicting approximate 3D macromolecular structures. Once the structure is known, one looks for best ligands, exploring the various procedures that allow the free energy of binding to be estimated. For an overview of molecular modelling as well as computational chemistry methods and applications see, for example, Lesyng & McCammon 1993; Jensen 1998; Leach 2001 and Young 2001. For an example of the application of molecular modelling

L. Gradoń and J. Marijnissen (eds.), Optimization of Aerosol Drug Delivery, 23–48.

methods in the study of HIV-1 protease and its inhibitors see for example Geller et al, 1997.

2. SOURCES OF STRUCTURAL DATA

There are two main experimental techniques that provide structural data for biologically active macromolecules: X-ray and/or neutron diffraction methods and Nuclear Magnetic Resonance spectroscopy (NMR).

Diffraction is carried out on crystals formed from macromolecules and/or their complexes. However, crystallization is a bottleneck for many biologically important targets, because getting good-quality crystals requires time, and in many cases proteins simply do not crystallize. This refers, for example, to membrane-bound proteins. Nevertheless, for smaller globular proteins crystallization is relatively easy, and this technique provides an excellent source of high quality structural data, with standard resolution around 2 Å (0.2 nm).

NMR, it can be applied for macromolecules in solution. However, its application is limited to smaller systems with a molecular weight of below 30 kD. On the other hand solid state NMR experiments can reveal structures of membrane proteins, that are inaccessible to X-ray or neutron diffraction techniques.

Most of the structural data that has been published on biological macromolecules are stored in the Protein Data Bank (PDB) <http://www.rcsb.org/pdb/>, or the Nucleic Acid Database Project (NDB) <http://ndbserver.rutgers.edu/>. The Protein Data Bank was established in 1972 at the National Brookhaven Laboratory in the United States of America. It is now maintained by a Research Collaboratory for Structural Biology (RCSB), a joint initiative of several US institutions. These databases are publicly available.

For a short summary of the experimental techniques see the RCSB site: <http://www.rcsb.org/pdb/experimental_methods.html>. For more detailed information see for example, Fersht, 1999, Glasel & Deutscher, 1995.

3. PHYSICAL FOUNDATIONS OF MOLECULAR MODELLING METHODS

3.1 Born-Oppenheimer Approximation

3D representation of macromolecules consisting of atoms suggests the existence of a mechanical model describing macromolecular structures. For example, we can ask what happens if we deform the structure by shifting selected atoms? Classical physics says that the potential energy changes, and if the reference structure is a stable one, the deformation leads to an increase of the potential energy. Such an argument is correct, even though the basic theory describing molecular systems is not classical mechanics but quantum mechanics. The fact that we can treat atoms like point-like or ball-type objects connected by chemical bonds ("mechanical sticks") results from the Born-Oppenheimer approximation, see for example, Lesyng & McCammon 1993; Leach 2001. Roughly speaking, this approximation assumes that positions of nuclei, q_{nuc}, can be treated as classical degrees of freedom, whereas electronic degrees of freedom, r_{el}, are responsible for quantum effects. In particular the deformation of the molecular structure is represented by motions of the atomic nuclei whereas the electron charge distribution is the source of atomic forces.

The potential for the nuclei motion, $U(q_{nuc})$ is the eigenvalue of the electronic time-independent Hamiltonian, H_{el}.

$$H_{el}(r_{el};q_{nuc})\Psi(r_{el};q_{nuc})=U(q_{nuc})\Psi(r_{el};q_{nuc})$$

For simplicity we neglect dependence of the wave function on spin. Formally, coordinates of the nuclei are parameters of the electronic Hamiltonian. Note that there exists an infinite number of eigenvalues representing the electronic ground and excited states. In biomolecular applications we are most interested in is the ground state. Therefore we assume that $U(q_{nuc})$ represents the hypersurface of the ground state.

3.2 Mechanical Model of Macromolecular Structures

The potential energy function, $U(q_{nuc})$, allows us to formulate the new, classical Hamiltonian of the molecular system consisting of atoms. Since the masses of atoms and their nuclei are practically the same, the index "nuc" can be skipped. Thus we will simply write $q_{nuc} = q$, where q represent coordinates of atoms and p their momenta

$$H(p,q) = T(p)+U(q),$$

T(p) is the classical kinetic energy of atoms treated as point-like or ball-like objects. Usually the kinetic energy of the system is given in the form:

$$T = \sum_i \frac{p_i^2}{2m_i},$$

where p_i is the momentum and m_i is the mass of the *i-th* atom.

The mechanical, molecular model assumes that the potential energy function $U=U(q)$ can be approximated by analytical functions with parameters being parameterised either on the basis of quantum mechanical calculations or available from, mostly spectroscopic, experimental data. Thus, the potential energy is usually represented as a sum of several easily interpreted and experimentally verified terms: bond stretching, bond angle deformation, torsional rotation around bonds as well as the long range interactions – electrostatic and Van der Waals interactions. Typically potential energy is written as:

$$U = \sum_{bonds} k_b \left(r_b - r_b^o\right)^2 + \sum_{angles} k_a \left(\varphi_a - \varphi_a^o\right)^2 + \sum_{torsions} \sum_{n=1}^{3} k_t^n \cos\left(n\theta_t + \delta_t^n\right) + \sum_{i,j<i} \left(\frac{q_i q_j}{\varepsilon r_{ij}} + \varepsilon_{ij}^{L-J} \left(\frac{\sigma_{ij}^{12}}{r_{ij}^{12}} - 2\frac{\sigma_{ij}^{6}}{r_{ij}^{6}} \right) \right),$$

where,

k_b, r_b, and r^0_b are, respectively, the force constant, actual length and optimal length for a *b-th* bond,

k_a, φ_a, and φ^0_a are, respectively, the force constant, actual value and optimal value for an *a-th* angle between bonds,

k^n_t, θ_t, and δ^n_t are, respectively, the force constant, torsion angle value and offset for a *t-th* torsion angle,

q_i, and q_j are, respectively, partial charges on the i-th and j-th atom,

r_{ij} is the distance between the i-th and j-th atom,

σ_{ij} is the distance between the *i-th* and *j-th* atom, at which the Van der Waals energy is at minimum and,

ε^{L-J}_{ij} is the minimal Van der Waals interaction energy.

The potential energy function shown above is neither unique nor universal. However, it can be regarded as a generic one. In many cases additional terms are also included in the parameterisation of the potential energy function like, for example, cross terms between bonds and bond angles. Also the formulae presented above are not unique, in many cases they are supplemented by more precise, alternative formulae which, however, require more parameters. For example, the bond stretching potential can be supplemented by a potential consisting of terms proportional to the second and fourth power of bond deformation. Usually parameterisation of the potential energy function is derived for a set of small molecules and used for all molecular systems built up from small molecular fragments. Such parametrization is called *force field*. The assumption underlying this approach is simple and intuitive – atoms embedded in similar molecular fragments will have similar properties. According to this assumption it is possible to find a set of parameters for a particular atom, bond, bond angle or torsion angle, that will reproduce the basic properties of a small set of molecules and that are transferable to a much larger class of molecules built from similar building blocks. This approach is quite successful, since macromolecules of biological importance consist of a limited number of small molecular fragments (20 amino acids or 5 nucleotides), and ligands are usually small organic molecules built of carbon, oxygen, nitrogen and hydrogen. An example of a study where the validity of this approach is tested for nucleic acids can be found in Rudnicki & Lesyng, 1995.

Biological macromolecules *in vivo* are immersed in a complex solution with many species - ions, small molecules, and macromolecules floating around. In computer simulations only a small fragment of solvent can be included. There are two possible approaches that can be used for solvent treatment in simulations. Either it can be included explicitly in the simulations or solvation effects can be simulated with one of the many approximate models.

Approximate models will be discussed later. However, an explicit solvent model also requires some approximations. The introduction of periodic boundary conditions is necessary to simulate properties of molecules in solution, as opposed to water droplets. In reality biological systems are not periodic. One should note, however, that superficial ordering effects are not significant for most problems of biological significance, in particular for receptor - ligand interactions.

The second issue that needs to be addressed is the problem of calculating long range forces. The Van der Waals forces can be simply cut-off at a certain distance and can be adjusted to match desired accuracy because their attractive component decays as r^{-6}. For example at the distance 5σ the attractive component of the Van der Waals potential will be only 0.0064% of the value at 1σ. On the other hand, electrostatic interactions decay much more slowly and their range is much longer than the size of the simulation box. Any practical approach must cut the interactions at a smaller distance because computations of the long-range interaction are the most time consuming part of simulations and number of interactions that need to be computed increase with the sixth power of the cut-off distance. A simple, atom-based cut-off introduces significant distortions into the behaviour of the system and should never be used. The group based cut off performs much better, but still ignores presence of solvent outside the cut-off radius, and this approach should also be regarded as a history.

Currently there are three methods for treating the long-range part of the electrostatic interactions in such a way that they model interactions of the system with the infinite solvent. Ewald summation is based on smart summation of an infinite series of interactions in a periodic system. A cell multipole algorithm is based on the treatment of long-range interactions as the interactions of a central charge with cell-based multipoles, with the cell size increasing with the distance from the centre. The reaction field model is based on an approximation of the long-range part of interactions of the charge in a spherical cavity in the dielectric medium with its reaction field. All three methods give comparable results and work better than the group-based cut-off. Therefore any of them can be used in realistic simulations of bimolecular systems. Readers interested in this particular topics can refer to one of textbooks available on molecular modelling, for example, Leach 2001, or Haile 1997 .

Three main simulation methods are used to derive the properties of the molecular system in the classical mechanistic model – molecular mechanics (energy minimization), molecular dynamics and Monte Carlo. Each method is based on well - defined physics and each has its own area of applications. We will not be describing commonly used, standard Monte Carlo procedures (see Leach, 2001; Landau & Binder, 2000 for representative references).

4. CONVENTIONAL MOLECULAR MODELLING METHODS

4.1 Molecular Mechanics

Molecular mechanics (MM) is based on the minimization of the potential energy of the system $U = U(\mathrm{q})$. On the basis of statistical physics, one may assume that the properties of the system will be dominated by the properties of the dominant state of the system, which should be located near the global minimum of $U(\mathrm{q})$. A number of minimization techniques are used to find the minimum. Unfortunately, in a general case, an algorithm for finding the global minimum is unknown. Most often one applies local minimizers, in particular steepest descent, conjugate gradient, and Newton or Newton-Raphson algorithms. Minimization techniques are useful for interpreting the properties of a molecular system as a function of several conformational degrees of freedom. Good examples of this approach are Ramachandran maps for peptides which allow us to understand how the properties of the polypeptide chain depend on the underlying geometric properties of aminoacids.

The following pseudocode describes the steepest descent (SD) algorithm:

```
read initial structure
for the current structure
 compute the energy gradient
 if gradient is larger then preset value
   find minimum of the energy in the direction of
   the gradient by displacing atoms in this
   direction.
   replace the initial structure with the current
   one
 else
   minimum is achieved;
```

The conjugate gradient algorithm is very similar to the SD one with the exception that the direction of search is a combination of the current gradient and the preceding direction of search. This modification avoids the main problem of the steepest descent algorithm, namely that in the long valleys the steepest descent algorithm tends to oscillate across the valley instead of taking a direction along the valley. Both steepest descent and conjugate gradient algorithms are first order algorithms and use information exclusively from the first derivative to look for the minimum. They are both robust, but rather slowly convergent, especially close to the minimum. One can also use higher order derivatives to increase the speed of search. An example of this approach is the Newton algorithm, which uses information of the second derivatives to find the minimum. The algorithm is based on the formula, which is exact for the second order polynomial:

$$\vec{x} - \vec{x}_{\min} = \Delta\vec{x} = \mathbf{U}''^{-1}(\vec{x}) \bullet \mathbf{U}'(\vec{x})$$

Since the potential energy in molecular modelling applications is not a second order polynomial, the solution will not be exact, and it should be applied iteratively. The Newton-Raphson algorithm is quadratically convergent in the vicinity of the minimum. However, it requires NxN computations of the energy derivatives that need to be stored in a matrix. Therefore, even for small proteins it requires large amounts of the system memory and computational time. Computing each matrix element requires quite a few floating-point operations, and computation of the inverse matrices scales as N^3. Even applications of the approximate methods, that do not compute the second derivatives at each step, are still limited to at most a few hundred atoms. Moreover, for systems far from the minimum the Newton-Raphson algorithm becomes unstable, especially where there is a steep potential resulting from Van der Waals repulsion.. A standard approach initiates minimization with steepest descent, then switches to the conjugate gradient method. For smaller systems one finally switches to the Newton-Raphson algorithm for quick convergence near the minimum.

4.2 Molecular dynamics

Molecular dynamics (MD) is the standard method for exploring the dynamic and thermodynamic properties of molecules at finite temperatures. This approach has significant advantages over molecular mechanics. First, molecular mechanics determines properties of the system at T = 0, whereas real biological systems exist at temperatures around 300 K. Atoms in molecules are in constant motion, with kinetic energy equal to ½ k_BT per degree of freedom (3/2 k_BT per atom). Due to equipartition of the energy a similar amount is stored in the potential form, however, with the opposite sign.

Molecular dynamics is conceptually very simple. One generates an initial structure of the molecule, which is usually either an experimental structure or a theoretical one. Then random velocities are assigned to atoms and Newtonian dynamics is applied to the system. The resulting trajectory, if sufficiently long, forms a representative sample of the phase space of the system. The initial velocities are derived from the Maxwellian distribution, and the system in the course of dynamics equilibrates to the Boltzmann distribution.

Newtonian, second order equations of motion are:

$$\vec{F} = m\ddot{\vec{x}}$$

or in the explicit form

$$-\frac{\partial U(x)}{\partial x_{ji}} = m_j \frac{\partial^2 x_{ji}}{\partial t^2}$$

With $j=1,2,3,....N_{atom}$ and $i=1,2,3$.

This equation is replaced by a set of first order equations

$$\left[\begin{array}{c} -\dfrac{\partial U(\vec{x})}{\partial x_{ji}} = m_j \dfrac{\partial v_{ji}}{\partial t} \\ \dfrac{\partial x_{ji}}{\partial t} = v_{ji} \end{array}\right]$$

that are solved numerically using a finite difference approach

$$\left[\begin{array}{c} -\dfrac{\partial U(x)}{\partial x_{ji}} = m_j \dfrac{\Delta v_{ji}}{\Delta t} \\ v_{ji} = \dfrac{\Delta x_{ji}}{\Delta t} \end{array}\right] \Rightarrow \left[\begin{array}{l} \Delta v_{ji} = -\dfrac{1}{m_j}\dfrac{\partial U(x)}{\partial x_{ji}}\Delta t \\ \Delta x_{ji} = v_{ji}\Delta t \end{array}\right] \Rightarrow$$

$$\left[\begin{array}{l} v_{ji}(t+\Delta t) = v_{ji}(t) - \dfrac{1}{m_j}\dfrac{\partial U(x)}{\partial x_{ji}}\Delta t \\ x_{ji}(t+\Delta t) = x_{ji}(t) + v_{ji}\Delta t \end{array}\right]$$

The algorithm derived above is not used in actual simulations because of numerical instabilities, However, it does illustrates the applied methodology.

A much better algorithm can be obtained from a simple Taylor series expansion of the position at time t (for one degree of freedom).

$$x(t+\delta t) = x(t) + x'(t)\delta t + \frac{1}{2}x''(t)\delta t^2 + \frac{1}{6}x'''(t)\delta t^3 + \frac{1}{24}x^{IV}(t)\delta t^4 + O(\delta t^5)$$

$$x(t-\delta t) = x(t) - x'(t)\delta t + \frac{1}{2}x''(t)\delta t^2 - \frac{1}{6}x'''(t)\delta t^3 + \frac{1}{24}x^{IV}(t)\delta t^4 + O(\delta t^5)$$

After summation of the two expansions one gets

$$x(t+\delta t) + x(t-\delta t) = 2x(t) + x''(t)\delta t^2 + \frac{1}{12}x^{IV}(t)\delta t^4 + O(\delta t^5)$$

which results in

$$x(t+\delta t) = 2x(t) - x(t-\delta t) + \frac{F(x(t))}{m}\delta t^2 + O(\delta t^4).$$

This equation defines the Verlet algorithm, which is routinely used in most molecular dynamics codes either in this, or in a few equivalent formulations.

The time-step used for integrating the MD equations of motion depends on the properties of the system. It never should exceed 10% of the shortest oscillation period in the system. For a system that includes explicit hydrogen atoms this means that the simulation time-step should not exceed 0.5-1.0 fs.

MD simulations for realistic systems, including several thousands of atoms, require significant computing resources, therefore, it is very important that the algorithm is well selected. An ideal algorithm should be stable (it should keep the total energy of the system constant), accurate (it should follow exact trajectories) and effective (it should use the least computer time for the given simulated time). In the case of MD simulations with long-range interactions, it means that only a single computation of the forces per time-step is allowed, since the force calculation step is by far the most intensive procedure in computational terms. It turns out that the Verlet algorithm derived above fulfils these requirements best, and this is the reason for its widespread adoption.

Although the MD method is quite easy, the user requires a few hints about its practical application. First, the initial model might contain strained molecular fragments, which during simulations would turn into "hot spots" with high kinetic energy that might disrupt the molecular structure. Therefore, it is necessary to minimize the structure before performing MD. However, even minimization may not be able to remove such strains. Therefore, the molecular system should be thermalized starting from the lowest temperatures, with many restarts at each temperature using random sets of velocities. This procedure should remove all strains. There are many more practical problems that need to be taken into account in each simulation. This accounts for specialized [NOT CLEAR] algorithms to compute long-range forces and algorithms used for restraints and constraints. In every case one should decide in which statistical ensemble simulation is performed – most popular are NVT (constant number of particles, constant volume, and constant temperature) or NVE (constant total energy). There are also various algorithms that can be applied for simulations with varying number of atoms or varying pressure.

4.3 Microscopic and Mesoscopic Description of Intermolecular Interactions

As mentioned above molecular mechanics and molecular dynamics are based on the microscopic, molecular mechanical model. The basic objects of microscopic models are atoms, small molecular groups and subatomic particles - electrons and nuclei. In practical applications microscopic models have obvious limitations. The number of atom-atom type interactions is large and simulation takes a lot of computer time. In addition, MD trajectories became numerically uncertain at longer time-scales. The longest MD simulations for small proteins do not exceed 10^{-6} s. Therefore, a number of mean-field, mesoscopic models and applications are being currently developed.

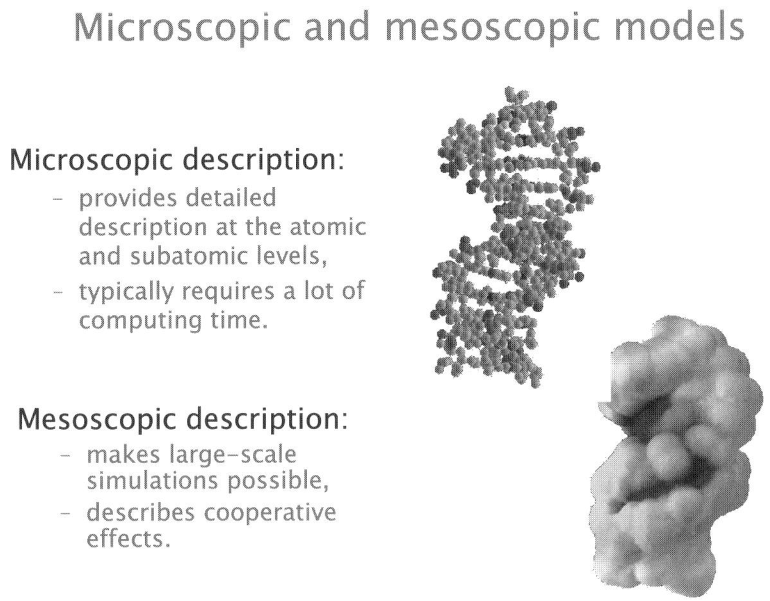

Figure 1. Microscopic and mezoscopic representations of a DNA molecule.

The difference between microscopic and mesoscopic approach has been represented visually in Figure 1 which shows the same molecular object in microscopic and mesoscopic representations.

The microscopic approach can be applied to compute the intermolecular interaction energy between a ligand and its target. In the simplest approximation the interaction energy can be computed using the Lennard-Jones potential. In turn, based on a mesoscopic approach one can estimate the free energy of binding. The free energy of binding can be represented as a sum of the following contributions,

$$\Delta G_{INT} \cong \Delta G_{EL}^{PB} + \Delta G_{HYDROPHOBIC} + \Delta E_{V.REP.}$$

where the first term describes the mean-field, Poisson-Boltzmann electrostatic interaction energy (see, for example, Briggs & McCammon, 1992, for more details), the second is the hydrophobic interaction potential and the third the valence repulsion term. The PB energy accounts for Coulombic interactions as well as electronic and solvent polarization effects, and also includes the presence of counterions. In turn, the hydrophobic potentials account mostly for entropic effects, and are responsible for forcing the clustering of non- interacting or weakly interacting atoms or molecular groups in water solution. It should be noted that ideal hydrophobic objects are empty holes in water. Water tends to bring them into close contact that minimizes the surface of the cluster. This, in turn, minimizes the rearrangement of the water molecules in the vicinity of the hydrated cluster.

The valence repulsion term results from quantum-mechanical principles that prohibit the collapse of interacting objects. Such decomposition of the binding free energy makes it possible to rationalize experimentally observed binding modes of ligands and their receptors (see Leach, 2001; Schlick 2002; Frenkel 2001; Haile 1997 or Rapaport 1997 for further reading).

5. THERMODYNAMICS AND STATISTICAL PHYSICS IN DRUG DESIGN

The concentration of the complex is formed by ligand and its receptor, and is related to the binding constant k_{RL}:

$$[RL] = k_{RL}[R][L],$$

where [RL], [R] and [L] are the concentrations of complex and the concentration of receptor and concentration of ligand respectively. Binding constant k_{RL} is related to the free energy of binding

$$k_{RL} = e^{\frac{-\Delta G_{RL}}{k_B T}} \text{ or } \Delta G_{RL} = -k_B T \ln(k_{RL}).$$

A negative value of ΔG_{RL} favours binding while a positive value favours dissociation. The free energy of binding is the difference between the free energies of the dissociated subunits and the complex:

$$\Delta G_{RL} = G_{RL} - (G_R + G_L)$$

Molecular modelling is used for predicting the free energy of binding. It is given by the equation

$$G = U - TS$$

where U is the internal energy of the system, S is the entropy of the system and T is temperature.

Estimating the free energy of molecular systems or the differences between free energies of systems is the main objective of many techniques. Formal statistical physics helps us to link thermodynamic concepts with computer simulations.

Statistical physics describes the properties of ensembles of macroscopically indistinguishable, nevertheless microscopically distinct states of the system. The state of the system at any moment is a point in the 6N dimensional phase space, and the evolution of the system is given by the trajectory in the phase space.

The probability of finding the system in the microscopic state (q,p), where q and p are coordinates and momenta is given by the Boltzmann distribution

$$\rho(q,p) = \frac{1}{Z} e^{\frac{-H(q,p)}{k_B T}}$$

where

$$Z = \int_Q \int_P dq dp e^{\frac{-H(q,p)}{k_B T}}$$

is integral over the phase space, with the normalization condition

$$\iint_{Q,P} dq dp \rho(q,p) = \iint_{Q,P} dq dp \frac{1}{Z} e^{\frac{-H(q,p)}{k_B T}} = 1$$

Integral Z, called the partition function, is related to the free energy of the system by relation

$$G = -k_B T \ln(Z) = -k_B T \ln\left(\int_Q \int_P dq dp e^{\frac{-H(q,p)}{k_B T}} \right)$$

For any function A(q,p) we can find the average value (macroscopic observable)

$$\langle A \rangle = \iint_{Q,P} dq dp A(q,p) \rho(q,p) = \iint_{Q,P} dq dp \frac{A(q,p)}{Z} e^{\frac{-H(q,p)}{k_B T}}$$

Such integrals can be evaluated using a Monte-Carlo sampling procedure or, on the basis of the ergodic hypothesis, can be evaluated as a time average over MD trajectories.

$$\langle A \rangle = \frac{1}{T} \int_0^T A(q(t), p(t)) dt$$

In practical applications one assumes that the simulated trajectory is a representative sample of the ensemble. Using the tools and methods of statistical physics one can derive system properties from the simulations. This approach is not the most optimal one for a direct computation of the free energy of a system because it requires a precise sampling of the phase space. There are, however, methods that do allow the differences of free energies to be computed. These methods are not elementary, and they will be discussed later (see also for example, Leach 2001; Landau & Binder 2000, Chandler 1987; Hill 1986)

6. PREDICTION OF MACROMOLECULAR 3D STRUCTURES

On the one hand, the sequencing of proteins and nucleic acids has become a routine task performed on an industrial scale by high-throughput robotic machines. On the other hand, however, there are many cases where resolving the structure of a protein molecule is a difficult task requiring a great deal of work and effort from highly qualified researchers. Neither the crystallization of proteins nor determining their structures is a routine operation. Despite the growing number of known protein structures, the gap between the number of known sequences and the number of known structures is increasing. Currently there are about 18200 structures deposited in the Protein Data Bank as compared with 623672 entries in the protein sequence database

TrEMBL maintained by European Molecular Biology Laboratory. Therefore, methods that could predict three dimensional protein structures, taking their sequences as inputs, are being currently developed in many laboratories around the world. There are two classes of methods – *ab initio* protein folding and homology modelling. *Ab initio* methods exclusively use sequence information, while homology based methods use a similarity measure of the given protein sequence to other sequences for which protein structures are known.

6.1 Ab initio methods

There are two basic approaches. The first starts with a random conformation of the protein chain and, by applying a specially tailored, global minimization algorithm leads to an ordered, stable structure of the global minimum. The main features of algorithms belonging to this class are the following: a simplified representation of the protein (often single pseudoatom per residue); the empirical potential energy function based on the known structural motifs and various global minimization algorithms, such as simulated annealing, diffusive operators and genetic algorithms. [CHECK PUNCTUATION]

Another class of *ab initio* algorithms is based on the reverse folding principle. It is based on the assumption, that there is a limited number of possible folds (general topological organization of the protein chain) and that most of these folds are already known. This assumption is probably correct, since the number of new folds is growing at a much slower rate than the number of new protein structures. In this approach all known folds are tried for the protein chain, and one that gives the lowest energy (or a highest score) is selected as the probable structure of the protein. Accuracy of the *ab initio* methods is still too low for being routinely applied.

6.2 Homology modelling

In this class of methods one uses all available information resulting from the protein sequence, conservation of particular regions in similar proteins, structure of similar proteins and so on.

The model is constructed in several stages. In the first stage all sequences are identified that are similar to the protein of interest (finding the family of homologous proteins). At this stage an initial alignment of the sequences is carried out. In the next stage the structurally conserved regions (SCR) are identified. Subsequently, the sequences are aligned with the structural motifs of SCRs. This is a very important stage, because shifting alignment by just one residue can sometimes completely distort interactions within the protein exposing, for example, hydrophobic side-chains to solvent or burying hydrophilic side-chains into the core of the protein. Next the conformation of the peptide chain in the SCRs is fixed. Then the conformation of loops connecting SCRs is generated either by using similar loops found in the structural database, or generating them *de novo*. Finally, the side chains are added to the main chain, and the structure is relaxed with the help of restrained molecular mechanics and/or molecular dynamic methods.

Homology modelling methods work best if structures of proteins with high sequence similarity are known, but it can also give satisfactory results if similarity is as low as 25% (see, for example, Friesner et al. 2002, Sternberg 1997, Tsigelny 2002, Webster 2000).

7. DOCKING

Let us assume that the structure of a given receptor is known including its active site region. We would like to know how a particular ligand would bind to the active site, and we would like to be able to estimate the binding constant. A generic algorithm for solving the problem is the following:

One defines a penalty function for the receptor-ligand interactions. One looks for the minimal penalty as a function of the relative position and orientation of the receptor and ligand. In some algorithms one can also use the conformation of the ligand and, although in very few instances, the conformation of the receptor. The penalty function is in fact a crude representation of the free energy of binding, depending parametrically on many factors including surface exposed to the solvent, number of hydrogen bonds, number of good/bad atomic contacts and steric clashes. The penalty function can also be computed as the interaction energy obtained from the standard force field, augmented with additional heuristic terms, like for example, additional hydrogen bonding terms and solvent exposed area. When one uses a force-field-like interaction energy function to compute the energy of binding, the number of non-bonded atomic interactions is the speed-limiting factor. For rigid fragments, one can avoid computations of the internal interactions both for receptor and ligand. For example in the system consisting of N_A=N_R+ N_L = 4000+50 atoms in the receptor and ligand respectively, computing all non-bonded interactions requires computing 4050x4050 = 16 402 500 computations of atom-atom interactions. If both ligand and receptor are rigid, one needs 4000x50 = 200 000 computations of atom-atom interactions (almost hundredfold reduction in the number of operations). Also, the number of degrees of freedom is reduced to
six - three rotational and three translational degrees of freedom. If one allows flexibility of the ligand, the number of required elementary computations does not increased significantly since only internal interactions of the relatively small ligand are added.

However, in general the number of degrees of freedom grows from 6 to $3xN_L$, in our example 150 degrees of freedom.

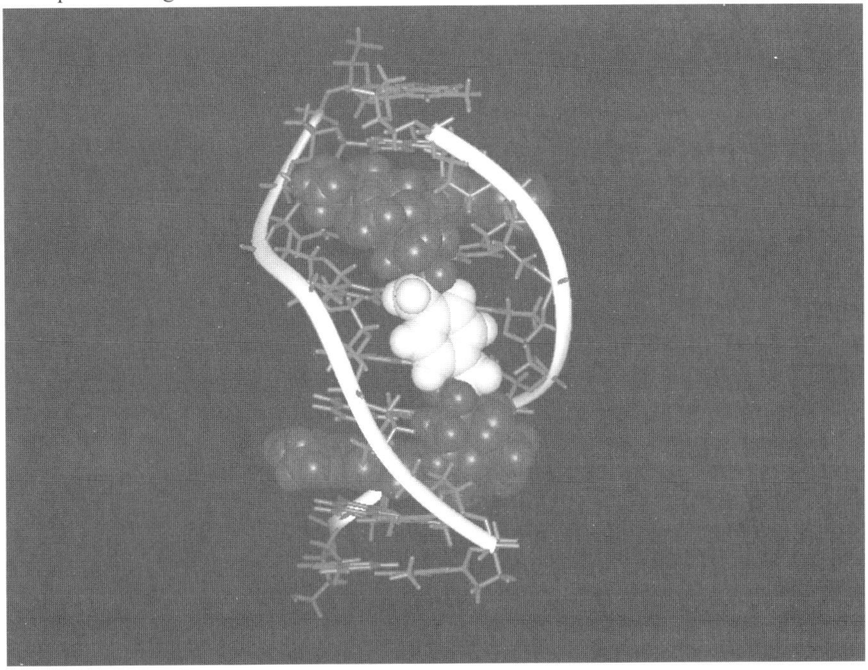

***Figure 2**. Anthracyclic antibiotic intercalated into a double-stranded DNA molecule.*

To reduce the conformational space, usually both bonds and bond angles are fixed at their optimal positions and only torsional angles around single bonds are allowed to vary. In more general cases one can also allow for a limited flexibility of atoms in the active site of receptor, which enables conformational changes to accommodate ligand. The number of interactions then grows substantially, resulting in an increased complexity of the conformational space.

To speed up computations for a rigid receptor model it is possible to precompute the electrostatic potential and the van der Waals interaction energy, with all possible ligand atom types, on a cubic grid surrounding the active site. The interaction energy of an atom is computed as a linear combination of the energy at eight corners of the cube in which the atom is located. This trick makes it possible to reduce the number of interactions from N_RxN_L to $8xN_L$ for the price of computing earlier the "potential" on the N_G grid points ($k x N_G x N_R$ computations required). The gain may not be dramatic for a single ligand (it may be 10-20 times). However, very often one needs to dock several (even thousands) potential ligands to the same receptor, with the same precomputed "potential". In such cases the speedup will be large. In practical applications, the approach most frequently used is either with flexible or rigid ligands.

Figure 2 shows an antracycline antibiotic molecule interacting in a specific way with a DNA fragment. Antracycline consists of two intercalating agents connected with a saccharide linker. Since intercalation prefers the GC nucleic acid bases, and the linker

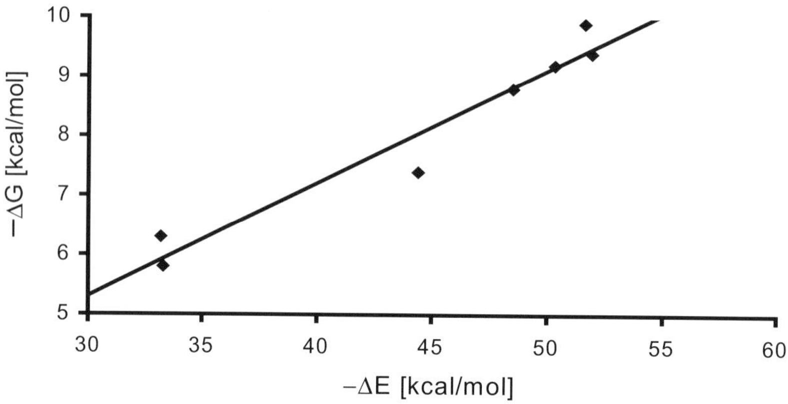

Figure 3. *Correlation between the free energy of binding and force field interaction energy.*

exhibits specific interactions with hydrogen bond donors and acceptors in the minor groove of DNA, the whole interaction process is much more specific than the interactions of the individual elements.

Experimentally determined binding constants and the corresponding free energies of binding for a number of antracycline derivatives were available. These data could be compared with the computed interaction energies using molecular mechanics methods. Such calculations and analysis were carried out (Rudnicki et al. 2000), and the comparison is presented in the Figure 3.

One can see good linear correlation between the binding free energies, ΔG, and the interaction energies, ΔE. The magnitudes of these energies are, however, different. The interaction energies are approximately five times larger than the corresponding free energies of binding. Justifications for such correlation were given, and we have not repeated them here. The conclusion which can be drawn from these studies is as follows. In cases where the binding modes of a set of ligands/inhibitors/repressors is the same, when environmental conditions are similar and the structures of the complexes and the interacting objects are relatively rigid, fast energy minimization within the MM formalism can provide a fast tool for estimating the free energies of binding and can be easily applied for an extended search of best ligands of potential pharmacological effects. For recent reviews of the ligand docking methods see, for example, Taylor et al. 2002; Shoichet et al. 2002 and Carlson 2002.

8. LIGAND DESIGN

Ligand design is a central procedure in the process of drug design. It is essential to note that the ligand design process should lead to discovery of all possible ligands, even at the price of creating many surplus molecules. There are two possible starting points in the ligand design process: either the 3D structure of the receptor molecule is known or it is not yet known. In the latter case a model of the binding site has to be built before the ligand itself can be modelled. An abstract concept of a *pharmacophore* is used as a representation of the binding site of the receptor. Once the model of the binding site is known, we can apply either traditional *structure-based ligand design*, or *virtual screening* methods. Possible ligands are selected with the help of a *scoring function*. This is a crude approximation of the free energy of binding, based on a number of hydrogen bond interactions, hydrophobic contacts, aromatic contacts – similar to the scoring functions used in docking.

8.1 Pharmacophore model

A pharmacophore is a set of features common to a series of active molecules. Typically it consists of functional groups common to all molecules and the geometric relations between them. The typical groups used in the pharmacophore definition are hydrogen-bond donors and acceptors, positively and negatively charged groups as well as the hydrophobic regions. A pharmacophore is constructed in the following way. First, for all known ligands the corresponding possible receptor interaction sites are generated. In the next step all ligands with the modelled receptor interaction sites are overlaid in a way that maximizes a common subset of interaction sites. This subset is stored as a model of the receptor interaction site. Further procedure is analogous to the one described for the structure based design.

8.2 Structure-based approach

When all possible interaction sites are known (generated either from the pharmacophore model, or from a 3D structure of the receptor site) two *de novo* ligand design schemes can be applied.

In the *outside-in* approach, molecular fragments from available databases are fitted into the interaction site. In the next step an algorithm attempts to connect interacting groups with linkers. Fragments that fit well to the binding site and are connected without inducing significant strain are stored for further processing.

In the *inside-out* approach molecules are "grown" in the binding site on a suitable molecular scaffold (for example, the polypeptide main chain) with conformations selected by a conformation search algorithm. Functional groups are attached to the scaffold at the predefined sites and contacts with the interaction sites are checked. Molecules with good scores are stored for further processing.

8.3 Virtual screening

Virtual screening utilises a combinatorial chemistry approach for ligand design. First, one builds a large set of molecules containing desired functional groups and in the next step an attempt is made to dock them to the binding site. The combinatorial approach to building the test set of molecules is based on a simple principle. Let us assume we have a molecule with N sites that can be substituted with K various groups. By exploring all possible combinations of substituents we can generate K^N different chemical compounds. For example, if N=4, and K=40, we can generate 40^4=2 560 000 compounds. Real synthesis of all possible combinations would cost time and money. Molecular modelling techniques, in combination with statistical models (QSAR – see the next section), can be used to optimise usage of real world resources. First, careful analysis of variability of compound properties can be used to reduce the number of compounds. In the example presented above, if we could reduce the number of substituents for each site to 10, 15, 20 and 15, respectively, the total number of compounds would be reduced to 10x15x20x15 = 45 000 – less than 2% of the original number. The following step is then to apply the docking procedure to the whole combinatorial library and suitable compounds can be identified selected for real world synthesis and testing. For further reading we recommend Gubernator & Bohm 1998 and Klebe 2000.

9. QUANTITATIVE STRUCTURE ACTIVITY RELATIONSHIP (QSAR)

This method does not strictly belong to the molecular modelling methods. However, in many ways it uses techniques adopted from molecular modelling and is often used as a part of the design process [NOT CLEAR]. Therefore, it is useful to describe its underlying assumptions, methodology and applications here.

QSAR is a statistical method that relates molecule biological activity with chemical structure and properties without looking at the possible mechanisms responsible for such relationships. After a suitable statistically validated equation is found, it is the role of structural biologists, chemists or biophysicists to propose mechanisms that explain the empirical relations and to explore further possibilities.

QSAR methodology is applicable when the biological activity of numerous chemical compounds is known even though these are not necessarily structurally related or similar to each other. Biological activity can be related to many factors – substituent types, substituents electronegativity, solubility in water and lipids, number bonds, types of bonds, aromaticity, polarizability, molecular weight, molecular surface, electrostatic potential in selected molecular points, and others. There are two main problems that need to be solved in finding QSAR relationships. First, a quantification procedure of various chemical properties should be applied, and identification of most important properties should be done. Second, analytical forms of the relationships should be proposed.

The chemical properties of the molecules are translated into various sets of descriptors, related to various classes of chemical properties. Most popular classes of descriptors are listed below:

- **Molecular fragments constants** – properties induced by adding/removing a particular fragment. Examples: sigma electrons effects, polar and/or resonance parts of the electronic effects, hydrophobicity, donor/acceptor properties of hydrogen bonds, steric properties.
- **Conformational descriptors:** energy of conformational states, minimal energy.
- **Electronic descriptors:** total charge, sum of the atomic polarizabilities, dipole moment. HOMO/LUMO energies.
- **Receptor descriptors** (if receptor available): binding energy, electrostatic contribution to the binding energy, Van der Waals contribution to the binding energy.
- **QM descriptors** (computed with the quantum chemistry codes, usually semiempirical): HOMO/LUMO energy, dipol moment, heat of formation.
- **Graph theory descriptors:** Wiener index (number of bonds in molecule), Zagreb index (sum of atomic valences), Hosoya index, Molecular connectivity index, atomic composition index.

- **Molecular shape descriptors** (defined in relation to a reference molecule): common volume, volume difference, root mean square difference of atomic positions.
- **Spacial descriptors:** effective radius, shade index, surface area, density, volume, moment of inertia.
- **Jurs descriptors** (describing the spatial charge distribution): negatively (positively) charged surface area, total charge weighted positive (negative) surface area, atomic charge weighted area (positive/negative), differences between positive and negative descriptors, total polar surface area $|q|>0.2$, total hydrophobic surface area $|q|<0.2$, total solvent accessible surface area, relative polar/hydrophobic surface area.
- **Structural descriptors:** atomic mass, rotatable bonds number, number of H-bond acceptors/donors.
- **Thermodynamical descriptors:** logarithm of the partition coefficient water/octanol, molecular refractivity, desolvation free energy (in water and or octanol).
- **Molecular field descriptors:** electrostatic potential, Van der Waals "potential".

Linear QSAR relations can be obtained with a standard least square fitting procedure. Often one would like to include non-linear terms. In many cases they can be linearized using suitable transformation and a standard procedure can be applied to the transformed variables. Alternative procedures include global optimisation techniques, like Monte Carlo simulated annealing or genetic algorithms for deriving the non-linear QSAR relationship.

QSAR models are useful for the virtual screening of large quantities of compounds generated with virtual combinatorial chemistry.

As can be seen the number of descriptors used is very large, resulting in a multidimensional parameter space sparsely populated with points representing descriptors.

In order to compare compounds and study similarity, there is often the need to reduce the dimensionality of the parameter space. Here the technique most often applied is the Principal Component Analysis. This is a multidimensional equivalent for the diagonalisation of the inertia tensor. Assuming we have N compounds described with M descriptors, we can code all information in the matrix form. This can then be transformed to a new parameter space with the following procedure:

$$X = \begin{bmatrix} x_{11} & x_{12} & \cdots & x_{1M} \\ x_{21} & x_{22} & \cdots & x_{2M} \\ \vdots & \vdots & \ddots & \vdots \\ x_{N1} & x_{N2} & \cdots & x_{NM} \end{bmatrix}$$

with the average value of each descriptor given by

$$\overline{x}_j = \frac{\sum_i x_{ij}}{N}$$

Matrix X can be transformed to the centered version

$$\hat{X} = \begin{bmatrix} x_{11} - \overline{x}_1 & x_{12} - \overline{x}_2 & \cdots & x_{1M} - \overline{x}_M \\ x_{21} - \overline{x}_1 & x_{22} - \overline{x}_2 & \cdots & x_{2M} - \overline{x}_M \\ \vdots & \vdots & \ddots & \vdots \\ x_{N1} - \overline{x}_1 & x_{N2} - \overline{x}_2 & \cdots & x_{NM} - \overline{x}_M \end{bmatrix}$$

The covariance matrix

$$\hat{S} = \hat{X}^T \hat{X}$$

with eigenvectors

$$\hat{S}\vec{V}_k = \lambda_k \vec{V}_k ,$$

leads to the new "composite" descriptors for each compound:

$$x'_{ij} = \sum_m v_{mj} x_{im} ,$$

with variance for the *k-th* descriptor equal to λ_k.

One can also group compounds using grouping algorithms – for example based on the number of common neighbours in the parameter space, or the distance between compounds in the parameter space. Then one tries to reduce the problem size by removing redundant substituents. This can be done using a trial and error method, namely by removing all compounds generated with a selected substituent, and then analyzing (in the descriptor space) changes of variability and the population of clusters If no significant reduction in variability is observed, the substituent is redundant and can be removed. This procedure is applied iteratively until no further reduction of substituents is possible without significantly reducing variability.

For recent reviews in this field see, for example, Besalu et al. 2002; Buchwald & Bodor 2002; Hansch et al. 2002, Kubinyi et al., 1998.

10. CONCLUDING REMARKS

Molecular modeling techniques are indispensable tools in the process of developing modern drugs. Due to limitations of space only the most important techniques have been reviewed here. In most cases we have only been able to provide a sketch of the topic. We have included references in the text to guide the reader to fuller sources of information.

In our opinion there are two main areas of interest for molecular modelers that should have a significant impact on the process of drug design in the near future. These are developments of new computational models for intermolecular interactions, and the integration of molecular modeling methods with the related disciplines of bioinformatics and cheminformatics.

In the post-genomic era, when the sequence of the human genome is already known, there has been a shift in interest from sequencing genes to finding their function and the structure of proteins coded by respective genes. This shift will increase the challenges for molecular modeling techniques, especially in the area of predicting 3D structures from sequences, as well as ligand design technologies – either based on conventional structure-based design or based on virtual screening. In both cases the development of computationally efficient and physically correct representations of the interaction energy is required, and this is presently the subject of intensive studies in many laboratories.

Two bottlenecks face molecular modeling: finding computationally feasible ways of estimating the free energy of the system, and finding a reliable method of representing the effects of a solvent without the explicit presence of solvent molecules. These specific issues have not been discussed here due to the limited size of this review. To learn more about them, readers is refered to the relevant sections of, for example, Leach, 2001; Gilson et al. 1997; Reddy & Erion, 2001.

Another field where intensive development is taking place and one that will have a significant impact on the drug development process is the effective integration of various information technology tools, such as cheminformatics, bioinformatics and molecular modeling. In fact, all these fields are strongly related. For example, the prediction of 3D protein structures belongs both to bioinformatics and molecular modelling. One expects that links between these fields will become more visible and more effective.

Acknowledgement. This work is supported by the Warsaw University, ICM BST 2003 funds.

11. REFERENCES

Antosiewicz, J., Błachut-Okrasińska, E., Grycuk, T., and Lesyng, B., "*A Correlation Between Protonation Equilibria in Biomolecular Systems and their Shapes: Studies using a Poisson-Boltzmann Model*", in *Free Boundary Problems. Theory and Applications*, Ed. N. Kenmochi, Gakuto International Series, Mathematical Sciences and Applications, 14, 11-17, 2000.

Besalu, E., Girones, X., Amat, L, Carbo-Dorca, R., "Molecular quantum similarity and the fundamentals of QSAR" *Acc. Chem. Res.* 35 (5): 289-295, 2002

Briggs, J.M., McCammon, J.A., "Computation unravels mysteries of molecular biophysics". *Comp. Phys.*, 6(3): 238-43, 1992.

Buchwald, P. & Bodor. N., "Computer-aided drug design: the role of quantitative structure-property, structure-activity and structure-metabolism relationships (QSPR, QSAR, QSMR)" *Drug Future* 27 (6): 577-588, 2002

Carlson, H.A., "Protein flexibility and drug design: how to hit a moving target" *Curr. Op. in Chem. Biol.* 6 (4): 447-452, 2002

Chandler D., *Introduction to Modern Statistical Mechanics* Oxford University Press; ISBN: 0195042778; 1987

Fersht, A., *Structure and Mechanism in Protein Science: A Guide to Enzyme Catalysis and Protein Folding*, W H Freeman & Co.; ISBN: 0716732688; 1999

Frenkel D. & Smit B.,*Understanding Molecular Simulation,* Academic Press; ISBN: 0122673514; 2001

Friesner, R., Prigogine, I. & Rice, S.A., Eds, *Computational Methods for Protein Folding (Advances in Chemical Physics, Vol* 120*)*, Wiley-Interscience; ISBN: 0471209554; 2002

Geller, M., Trylska, J., Antosiewicz, J., „HIV-1 Protease and its inhibitors" in *Theoretical and Computational Methods in Genome Research*, Suhai, S. Ed., Plenum Press; ISBN 030645503X; 1997

Gilson, M.,K., Given, J.A., Bush, B.L., McCammon, J.A., "The statistical-thermodynamic basis for computation of binding affinities: A critical review", *Biophys. J.* 72 (3): 1047-1069, 1997

Glasel, J.A & Deutscher M.P., *Introduction to Biophysical Methods for Protein and Nucleic Acid Research,* Academic Press; ISBN: 0122862309; 1995

Gubernator K., Bohm H.-J., Eds., *Structure-Based Ligand Design (Methods and Principles in Medicinal Chemistry, Vol 6)* VCH Publishing; ISBN: 3527293434; 1998.

Haile J.M., *Molecular Dynamics Simulation: Elementary Methods,* Wiley-Interscience; ISBN: 047118439X; 1997

Hansch, C., Hoekman, D., Leo, A., Weininger , D.,;Selassie, C.D., "Chem-bioinformatics: Comparative QSAR at the interface between chemistry and biology", *Chem. Rev.* Vol 102(3): 783-812, 2002,

Hill, T.L., *An Introduction to Statistical Thermodynamics* Dover Publications; ISBN: 0486652424; 1986

Jensen. F. *An Introduction to Computational Chemistry*, John Wiley & Son Ltd; ISBN: 0471984256; 1998

Klebe, G. Ed. , *Virtual Screening: An Alternative or Complement to High Throughput Screening?,* Kluwer Academic Publishers; ISBN: 0792366336; 2000

Kubinyi, H., Martin, Y.C., Folkers, G. "*3D Qsar in Drug Desig: Ligand-Protein Interactions and Molecular Similarity,(**Vol 2**) /Recent Advances, (**Vol 3**)* Kluwer Academic Publishers; ISBN: 0792347900, 1998

Landau, D. & Binder, K., *A Guide to Monte Carlo Simulations in Statistical Physic*, Cambridge Univ Press; ISBN: 0521653665; 2000

Leach, A.R., *Molecular Modelling: Principles and Applications (2nd Edition)*, Prentice Hall; ISBN: 0582382106, 2001

Lesyng, B. & McCammon J.A., "Molecular modeling methods. Basic techniques and challenging problems", *Pharmac. Therap.* 60(2):149-67, 1993.

Rapaport, D.C.,*The Art of Molecular Dynamics Simulation,* Cambridge University Press, ISBN: 0521599423;1997

Reddy M.R. & Erion M.D. *Eds. Free Energy Calculations in Rational Drug Design*, Kluwer Academic Publishers; ISBN: 0306466767, 2001.
Rudnicki, W.R. & Lesyng, B., "Applicability of commonly used atom-atom type potential energy functions in structural analysis of nucleic acids. The role of electrostatic interactions" *Computers & Chemistry* 19(3): 253-258, 1995.
Rudnicki, W.R., Kurzepa, M., Szczepanik, T., Priebe, W., Lesyng, B., "A simple model for predicting the free energy of binding between anthracycline antibiotics and DNA", *Acta Biochim. Polon.* 47 (1): 1-9, 2000
Schlick, T., *Molecular Modeling and Simulation,* Springer Verlag; ISBN: 038795404X; 2002
Shoichet, B.K., McGovern, S.L., Wei, B.Q., Irwin, J.J., "Lead discovery using molecular docking" *Curr. Op. in Chem. Biol.* 6 (4): 439-446, 2002
Sternberg, M.J.E., *Protein Structure Prediction: A Practical Approach* Irl Press; ISBN: 0199634971; 1997
Taylor, R.D., Jewsbury, P.J., Essex, J.W., "A review of protein-small molecule docking methods" *J. Comp.-Aid. Mol. Des.* 16 (3) 151-166, 2002
Tsigelny, I.F., *Protein Structure Prediction: Bioinfomatic Approach,* International University Line; ISBN: 096368177X; 2002
Webster D.M., Ed. *Protein Structure Prediction: Methods and Protocols,* Humana Press; ISBN: 0896036375; 2000
Young, D., *Computational Chemistry: A Practical Guide for Applying Techniques to Real World Problems,* Wiley-Interscience; ISBN: 0471333689; 2001

12. AFFILIATIONS

Prof.dr. Bogdan Lesyng and Dr. Marek Henczka work at the Department of Chemical and Process Engineering, Warsaw University of Technology, Waryńskiego 1, 00-645 Warsaw, Poland

J. BAŁDYGA and M. HENCZKA

PARTICLE ELABORATION USING SUPERCRITICAL MEDIA FOR DRUG DELIVERY INTO THE HUMAN LUNGS

1. INTRODUCTION

Application of aerosol powders enables fast, controlled delivery of therapeutic agents into the human lungs, and then into the blood. Hence, the method is used to deliver inhaleable drugs to the deep lung for the treatment of systemic and respiratory diseases. However, it requires applying of drug particles with controlled particle size, particle size distribution (PSD) and particle morphology. To introduce drugs into the tracheobronchial and pulmonary regions one needs to use micronized pharmaceuticals with strictly defined particle size ranges between 0.1 μm and 5 μm. There are many traditional industrial techniques to produce powdered solids (grinding, jet milling), however, they do not guarantee the proper particle size. Particle design using supercritical fluids represents an interesting alternative to milling techniques. Design of particles is considered presently as a major development of supercritical fluids applications, which results in large number of publications and patents – see (Jung and Perrut, 2001) for recent literature and patent survey. These methods are based on crystallization of a solid phase from a solution with the use of supercritical fluids with the goal to obtain high purity materials and narrow particle size distribution.

The aim of this chapter is to present methods that are currently applied to produce particles using supercritical fluids and to show their relation to the traditional crystallization and precipitation processes. The material is organized as follows: we start from fundamentals of crystallization and precipitation, and present kinetic effects on particle size and morphology. Then we present some specific features of supercritical fluids and resulting methods of particle design using supercritical or, better to say, pressurized fluids in relation to the traditional methods. Finally, we present in detail the Solution Enhanced Dispersion by Supercritical Fluids Process (SEDS-Process) and give some conclusions.

2. FUNDAMENTALS OF CRYSTALLIZATION IN RELATION TO PARTICLE SIZE DISTRIBUTION AND PARTICLE MORPHOLOGY

Consider production of powders by particle precipitation or crystallization from solution. The process of crystallization involves the nucleation and growth of crystals (particles) from supersaturated solution, and is often followed by

L. Gradoń and J. Marijnissen (eds.), Optimization of Aerosol Drug Delivery, 49–73.

agglomeration of crystals. When the crystallizing compound is transferred from the one phase to another there is a change in its chemical potential (the molar Gibbs free energy) from μ_1 in supersaturated solution to μ_2 in the crystal state

$$\Delta\mu = \mu_1 - \mu_2 \tag{1}$$

When $\Delta\mu < 0$ there is spontaneous crystallization, $\Delta\mu = 0$ defines an equilibrium and $\Delta\mu > 0$ means that an opposite transformation (dissolution) becomes spontaneous. The sign of $\Delta\mu$ shows direction of possible phase transformation, whereas modulus of $\Delta\mu$ expresses distance from the equilibrium. $\Delta\mu$ represents the thermodynamic driving force for crystallization, which means that the rate of nucleation and the rate of crystal growth depend on $\Delta\mu$. Since the chemical potential of the dissolved compound is directly related to its activity, the driving force can be expressed using activity coefficient

$$S_a = \frac{a}{a_{eq}} \approx S = \frac{c}{c_{eq}} \tag{2}$$

where a represents the activity of dissolved component of concentration c, c_{eq} is the equilibrium solubility and $S_a \approx S$ is the supersaturation ratio that is also known as relative supersaturation or saturation ratio. Supersaturation is also sometimes expressed as a difference in concentration

$$\Delta c = c - c_{eq} \tag{3}$$

or dimensionless difference in concentration

$$\sigma = S_a - 1 \approx S - 1 \tag{4}$$

The solution is supersaturated when $S_a > 1$, $S > 1$, $\Delta c > 0$ or $\sigma > 0$. The particles are created in supersaturated solution by nucleation events. There are various mechanisms of nucleation that are usually grouped into two categories known as primary and secondary nucleation. In primary nucleation creation of crystals is not affected by the presence of the crystallizing material itself. Primary nucleation may be homogeneous when it occurs in the absence of any solid phase, and heterogeneous when nucleation is facilitated by the presence of foreign solid particles. In secondary nucleation formation of solid phase is affected by the presence of solid phase of the crystallizing material including secondary nuclei formed by collisions between crystals, attrition and needle fraction, as well as formation of nuclei in the adsorption layer on the surface of growing crystals, and washing out of nuclei into the solution. In precipitation of powders being sparingly soluble substances, secondary nucleation either does not take place or its effect is negligible (Söhnel and Garside, 1992), so kinetics of secondary nucleation will not be considered in what follows. The rate of the primary homogeneous nucleation can be approximated by

$$J_{homo} = J_{max} \exp\left\{- A[\ln(S)]^{-2}\right\} \tag{5}$$

where A is a constant proportional to the solid surface free energy per unit area and inversely proportional to T^3, whereas J_{max} is the maximum possible nucleation rate (as $S \to \infty$). The rate of heterogeneous nucleation can also be expressed by Eq. (5). However, as heterogeneous nucleation takes place on a foreign surface, it may have much lower surface energy than that of solute crystals, which means that the A value may be several orders of magnitude smaller in the case of heterogeneous nucleation. Another difference between both nucleation kinetics is that the heteronuclei are used up, limiting this way the maximum possible rate of heterogeneous nucleation, J_{max}. The above mentioned features of nucleation kinetics may be discussed in relation to the phenomenon of metastable state that is observed for supersaturated solutions. Namely, the metastable regions of concentration are observed in supersaturated solutions, where the solution contains more solute than in equilibrium at the given temperature, but nucleation of the solid phase, however, does not occur until a critical value of concentration (and supersaturation) is reached. Using Eq. (5) one can find regions where no metastable nucleation occurs and regions where nucleation is very fast. These two regions are separated by the critical supersaturation S_C. Because there is an obvious relation between the constants A in Eq. (5) for heterogeneous and homogeneous nucleation, namely $A_{homo} >> A_{hetero}$, it results from eq. (5) that heterogeneous nucleation takes place at lower critical supersaturation than homogeneous nucleation (i.e. $S_{C,homo} > S_{C,hetero}$) as shown in Figure 1 – see curves II and III and regions above these curves.

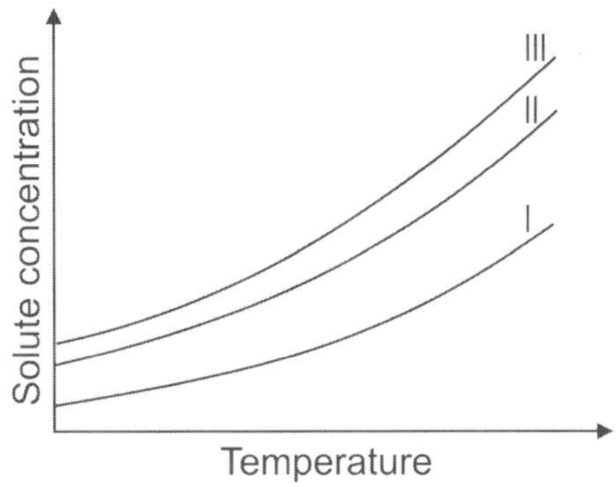

Figure 1. Nucleation in metastable zone: I – solubility, II – primary heterogeneous nucleation boundary of metastable zone, III - primary homogeneous nucleation boundary.

More about mechanisms and kinetics of nucleation the reader can find in books (Nielsen, 1964; Randolph and Larson, 1971; Söhnel and Garside, 1992; Mullin,

1993; Mersmann, 1995) and review articles (Dirksen and Ring, 1991; Garside, 1985).

Let us consider now the growth of crystals in a supersaturated solution. This is a very complex process that occurs in many stages that may all be taking place simultaneously (Mullin, 1993; Elwell and Scheel, 1975):

1. Transport of solute from the bulk solution through the diffusion boundary layer.
2. Transport of solute through the adsorption layer and adsorption on the crystal surface.
3. Surface diffusion.
4. Attachment to a step on the crystal surface and diffusion along a step.
5. Integration into the crystal lattice.
6. Diffusion of coordination shell of solvent molecules through the adsorption and diffusion layer.
7. Liberation and transport of heat of crystallization.

The simplest description of this complex process is based on the diffusion-reaction theories that date back to the works by Berthoud (1912) and Valeton (1924). According to these theories the solute is transported to the crystal surface with driving force $c - c_i$ and then integrated into the surface with the driving force $c_i - c_{eq}$ as shown in Figure 2.

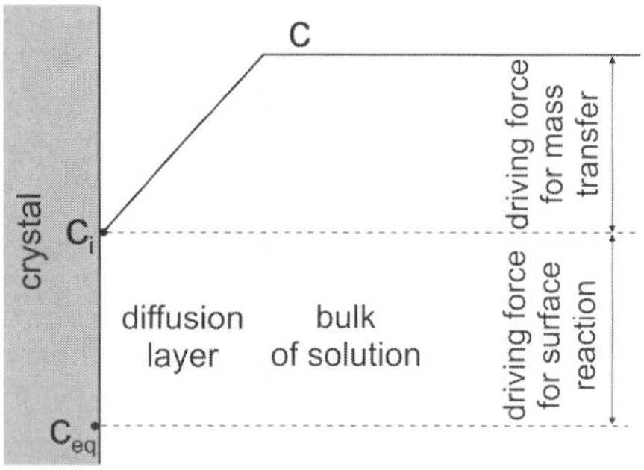

Figure 2. Concentration profile for crystal growth.

These two stages can be represented by

$$N = k_D(c - c_i) = k_r(c_i - c_{eq})^g \tag{6}$$

where N is the molar flux i.e. number of moles diffusing and integrating per unit time and unit crystal area, k_D is a mass transfer coefficient, k_r represents a rate constant for the surface integration step and g is an order of the surface reaction (often $g = 2$).

For very intensive mass transfer (high k_D) and small bulk concentration eq. (6) gives $c_i \approx c$ and the crystal growth is controlled by the surface reaction step. On the other hand, when the solute concentration is very high and the mass transfer coefficient is small, $c_i \approx c_{eq}$ and the growth is controlled by transport processes. The rate of growth is represented in Eq. (6) by the mass flux. In practice the rate of crystal growth can be expressed as the rate of displacement of a crystal surface or the rate of increase of particle mass. Most frequently the growth rate is defined as

$$G = \frac{dL}{dt} \tag{7}$$

where L is a characteristic crystal length.

Precipitation results in formation of a solid-liquid dispersion. Small particles forming this dispersion have tendency to aggregate, i.e. to form larger entities called agglomerates. The rate of particle agglomeration depends on particle size, shape and number concentration, composition of the solution and structure of the flow. Agglomeration is affected by the attractive and repulsive colloidal forces, hydrodynamic forces and inertia of the particles. Smoluchowski (1917) reported two limiting types of agglomeration for colloidal particles:

1. Perikinetic, where agglomeration takes place as a result of Brownian motion. Collision rate of particles can be expressed by Eq. (8) proposed by Smoluchowski (1917)

$$N_{ij,P} = 4\pi D_{ij}^{\infty}(a_i + a_j)c_i c_j \tag{8}$$

where $N_{ij,P}$ is the number of collisions of particles of radii a_i and a_j respectively per unit time and unit volume, c_i represents the number concentration of particles of radius a_i and D_{ij}^{∞} is the mutual Brownian diffusion coefficient

$$D_{ij} = D_i + D_j = \frac{kT}{6\pi\mu}\left(\frac{1}{a_i} + \frac{1}{a_j}\right) \tag{9}$$

2. Orthokinetic where agglomeration is dominated by convection, that is the particles are transported by movement of the surrounding fluid (Smoluchowski, 1917). Analysis by Smoluchowski was limited to the laminar shear flow and was extended to a turbulent flow by Saffman and Turner (1956)

$$N_{ij,O} = 1.3\left(\frac{\varepsilon}{\nu}\right)^{1/2}(a_i + a_j)^3 c_i c_j \tag{10}$$

where ε is the rate of energy dissipation per unit mass and ν represents the kinematic viscosity.

Both equations, (9) and (10), neglect effects of interparticle forces. In the case of static fluid effects of interparticle forces can be included by dividing the

Smoluchowski relation (8) by the Fuchs' stability ratio W as given by Spielman (1970). The stability ratio, W, depends then on properties of the solution including its composition as well as properties of particle surface. When the fluid is in motion the stability ratio, W, becomes dependent on the flow as well (Bałdyga and Orciuch, 2001). In the limit of a very high rate of energy dissipation and for large enough particles (i.e. for large enough values of the Peclet number, $Pe >> 1$)

$$Pe = \frac{\left(\frac{\varepsilon}{\nu}\right)^{1/2} (a_i + a_j)^2}{D_{ij}} \tag{11}$$

one gets $N_{ij,O} \approx N_{ij,P}/W$. This means that application of the extended concept of stability ratio, W, enables us to model simultaneously the perikinetic, orthokinetic and intermediate regime of collisions. When collisions of particles in liquid solutions are considered, effects of particle inertia are usually negligible. However, when particle density is large compared to fluid density effects of inertia have to be included. This can be done by using theories by Saffman and Turner (1956), Abrahamson (1975) and Kruis and Kusters (1997).

Up till now we have discussed collisions of particles. However, not every collision leads to agglomeration. When there is enough time during the collision and enough material (i.e. the solution is supersaturated) the crystalline bridge can be built between the particles forming this way agglomerates bounded by strong bonds. Otherwise either the collision is unsuccessful or the weak aggregate bounded by weak bonds (van der Waals forces) is formed. To express the rate of agglomeration, β_{ij}, in relation to the rate of collision, N_{ij}, the concept of probability or efficiency of agglomeration, Pa, has to be used as well

$$\beta_{ij} = N_{ij} \cdot Pa \tag{12}$$

The efficiency of agglomeration, Pa, depends on the supersaturation, collision geometry and details of the flow (David et al., 1991; Hounslow et al., 2001; Bałdyga et. al., 2001), namely Pa increases with increasing the supersaturation ratio S and decreases with increasing the rate of energy dissipation, ε.

Analysis of the above presented kinetics of nucleation and growth of crystals, as well as crystal agglomeration, unables us to interpret effects of supersaturation on the particle size (see Figure 3). At low supersaturation heterogeneous nucleation dominates. The rate of nucleation is then less sensitive to supersaturation than the rate of crystal growth (A_{hetero} is small), so increasing of initial supersaturation affects more the rate of crystal growth than the rate of nucleation, and larger crystals are produced. At higher initial supersaturation homogeneous nucleation dominates and due to large A_{homo} value the rate of nucleation is very sensitive to the supersaturation. Larger number of smaller particles is thus produced if the initial supersaturation is continually icnreased. At still higher initial supersaturation a large number of small particles [high c_i and c_j in Eqs (8) and (9)] results in large number of collisions. Due to high supersaturation also agglomeration efficiency [*Pa* in Eq. (11)] is high and an increase of agglomerate size with increasing initial supersaturation is thus observed.

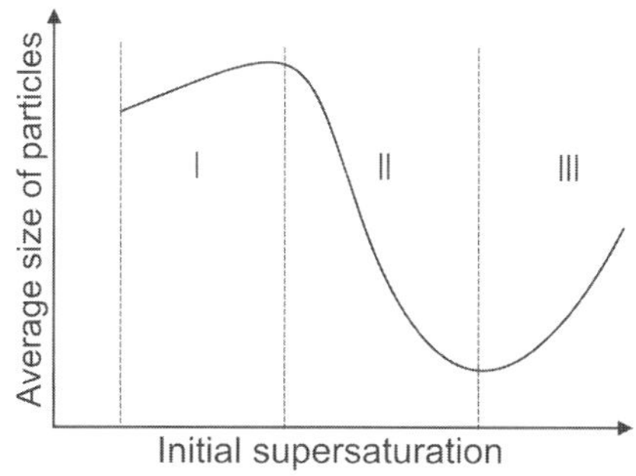

Figure 3. Effect of initial supersaturation on the average particle size: I – subrange of heterogeneous nucleation, II – subrange of homogeneous nucleation dominates, III – agglomeration dominates.

Agglomeration of large particles in the orthokinetic regime is affected by hydrodynamics; at high agitation power, P, the rate of energy dissipation, ε, is high

$$\varepsilon = \frac{P}{V\rho} \tag{13}$$

where V is the volume of the fluid in the system and ρ represents the fluid density. Increasing the rate of energy dissipation increases then the rate of collision, but decreases agglomeration efficiency, *Pa*, limiting this way the maximum attainable size of agglomerates. The initial level of supersaturation affects also particle morphology. At low supersaturation (the left hand side part of region I in Figure 3) the surface integration controlled growth takes place and crystals of compact shapes are formed. The shape is then often similar to the "equilibrium" shape that is determined by the minimum overall energy of the crystal surface. The kinetic shape of crystals is observed when the rate of slowest growing crystal face determines its shape. A sensitivity of the individual growth rates to the supersaturation is often different, therefore the relative growth rates of different faces may depend significantly on supersaturation. The rate of growth is thus anisotropic, which means that depending on the supersaturation value different shapes such as needles and plates may be produced. At higher supersaturation, but still in the region of heterogeneous nucleation, the rate of mass transfer to the crystal face and the rate of transfer of the released heat of crystallization into surrounding solution strongly affect crystal growth. As shown in Figure 4a the heat and mass transfer are the fastest at the crystal edges as they reach into the regions of high concentration and low temperature. The edges grow faster than the middle of face, which can lead to dendritic growth as shown in Figure 4b. This explains why elongated crystals and dendrites are observed at high supersaturation. If the initial supersaturation is still higher (region II of the heterogeneous nucleation) crystallization is so fast that there is not enough time to develop well defined crystal faces and the small isometric crystals are formed.

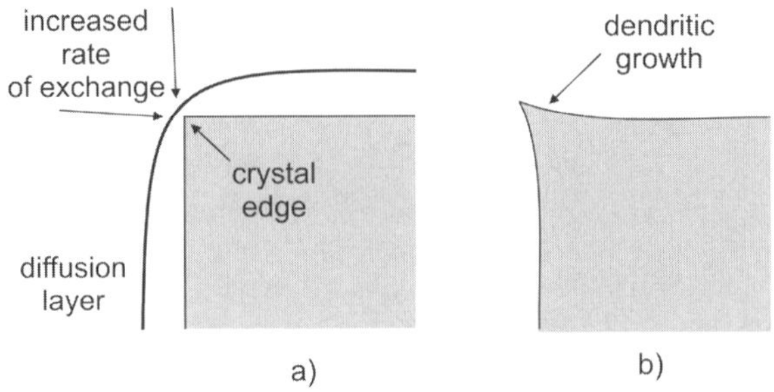

Figure 4. Diffusion controlled growth to formation of dendrids: a) effect of diffusion layer, b) growing crystal.

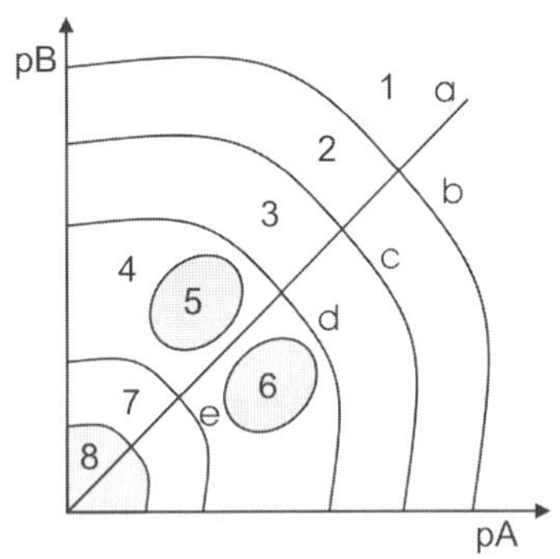

Figure 5. Schematic of precipitation diagram for $A+B \rightarrow \downarrow C$ precipitation with possible complex formation. Curves: a – stoichiometric composition (equivalence), b – solubility boundary, c - boundary of metastable composition, d – boundary of crystal shape region, e – homogeneous nucleation boundary. Regions: 1 – unsaturated solution, 2 – metastable supersaturated solution, 3 – compact crystals, 4 – dendrids, 5 – positively charged sol, 6 - negatively charged sol, 7 – small isometric crystals, 8 – formation of metastable phase.

For certain composition of the supersaturated solution the colloids (positively or negatively charged sols) may be formed as well. At still higher initial supersaturation an amorphous precipitate (the metastable solid phase) is usually formed and it transforms afterwards slowly into the crystalline solid phase (Gibbs, 1928; Dirksen and Ring, 1991; Söhnel and Garside, 1992).

The isometric and amorphous particles often agglomerate (region III in Figure 3) forming particles of a very complex structure. The aggregate shapes are often characterized by agglomerate fractal dimensions. A mass fractal dimension of an aggregate, D_F, is related to its mass, m, by

$$m \sim L^{D_F} \tag{14}$$

where L is the aggregate size. The lower is the fractal dimension, the more open is the agglomerate structure. In practice the fractal dimensions from the range $1.8 \div 3$ are observed with D_F strongly related to agglomeration mechanism (Elimelech et al., 1995).

The characteristic features of precipitates can be conveniently represented by the precipitation diagrams (Težak, 1968; Söhnel and Garside, 1992). The diagrams are constructed by plotting contours of equal values of chosen experimental parameter (solubility, supersaturation ratio, morphology, crystal growth rate) as a function of initial concentrations of components, pH or temperature. Often the boundaries of concentration regions within which precipitates with similar characteristic properties prevail are plotted as well. A typical precipitation diagram illustrating precipitation of a solid product C from A and B component solutions is shown in Figure 5. The molar concentrations of reacting components A and B are represented by negative logarithms $pA = -log(c_A)$ and $pB = -log(c_B)$. All the discussed above features of the precipitation process are presented in Figure 5 in a condensed form.

To carry out the crystallization or precipitation process one needs to create supersaturation. Conventional methods of the creation of supersaturation from solutions are as follows:

a) cooling crystallization,
b) evaporative crystallization,
c) drowning-out crystallization or precipitation,
d) reaction crystallization or precipitation.

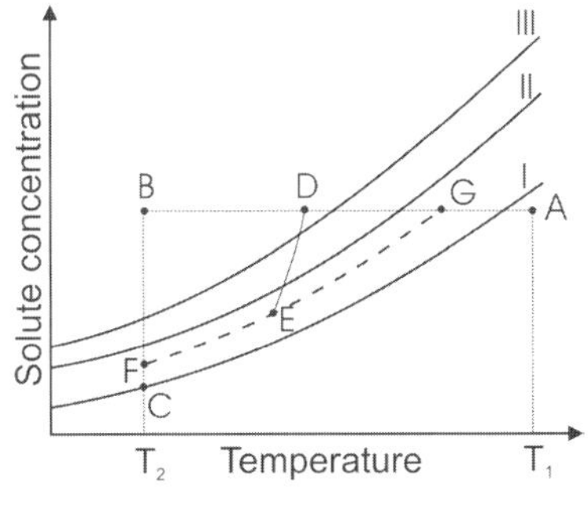

Figure 6. Schematic of cooling crystallization.

Detailed description of these methods one can find for example in (Mersmann, 1995). Here a brief presentation of crystallization methods are given with an intention to compare them in following section with methods based on using of supercritical fluids. The cooling crystallization is applied when the substrate solubility significantly increases with temperature (see Figure 1). Figure 6 illustrates the batch cooling crystallization process. The unsaturated solution of temperature T_1 (point A) is cooled in the crystallizer to temperature T_2. For very fast cooling one can reach point B characterized by a very high supersaturation, which may result in a fine crystal size at point C. Note that B-C line represents here the process of crystallization. For slower cooling crystallization starts earlier (point D). Crystallization competes then with cooling and the resulting trajectory D-E-F-C yields a coarse crystal product. Another option is to inoculate or seed a supersaturated solution in the metastable region (point G) with small particles of material to be crystallized. One can control this way the product size using the trajectory A-G-E-F-C as shown in Figure 6.

Evaporative crystallization can be applied when the solubility (curve I in Figure 1) is hardly increases with temperature, or even decreases with temperature. The unsaturated solution is heated to boiling point and then the rate of crystallization is controlled by the rate of evaporation. By proper setting of the rate of evaporation one can keep supersaturation in the metastable region and produce a coarse crystal product.

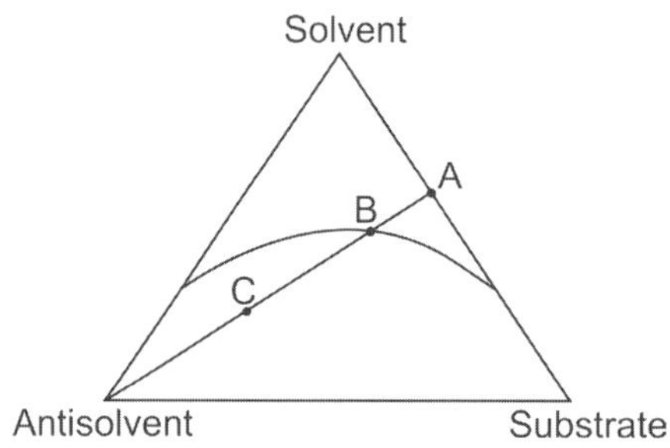

Figure 7. Schematic presentation of Anti-Solvent or Drowning-out crystallization.

Drowning-out crystallization is presented schematically in Figure 7. The unsaturated solution of the substrate (point A) is mixed with the antisolvent as shown in triangular diagram. The solution composition varies then along a straight line A-B-C-Antisolvent. The solution becomes saturated at point B and further addition of the antisolvent increases supersaturation. The rate of supersaturation creation depends then on the amount of added antisolvent, feeding rate and intensity of mixing. Drowning-out precipitation produces usually crystals having a smaller size than those produced by cooling or evaporation at the same supersaturation (Söhnel and Garside, 1992). The process is called salting-out when supersaturation is created by adding a solution of another salt.

In the process of reactive crystallization or precipitation supersaturation is created by chemical reaction. In homogeneous reactive precipitation the process involves contacting of reactant solutions by mixing, instantaneous or fast chemical reaction and subsequent crystallization of the product. The single feed precipitation process $A + B \to\downarrow C$ is presented in Figure 8. Figure 8a shows precipitation trajectory on precipitation diagram presenting the solution composition along the reaction zone and related values of supersaturation. The trajectory passes by the concentration regions with characterized by specific particle morphology; resulting particles are presented in Figure 8b. The double feed semibatch precipitation process is interpreted in a similar way in Figure 9a. One can see that in this case the precipitation trajectories pass by the concentration regions that are different than in the case of the single feed precipitation. As shown in Figures 9b and 9c this results in production of particles of different shape than those in Figure 8b. Note, that the distance between feeding points affects particle shape (see Figures 9b and 9c); this is because of separation of feeding points decreases created supersaturation due to dilution by the bulk solution. This explains why effects of mixing on the molecular scale are often considered in relation to the reactive precipitation process (Bałdyga et al., 1995; Bałdyga and Bourne, 1999). In the case of heterogeneous reactive precipitation one of reactants may be added in gas, solid or vapor form. Heterogeneous precipitation process is then affected by the mass transfer rate between the phases containing reactants.

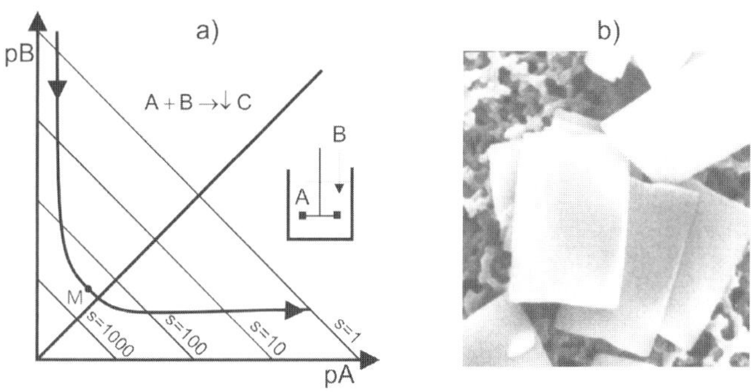

Figure 8. a) Variation of composition of supersaturated solution during mixing-precipitation process when the small value of concentrated A solution is added to the large volume of less concentrated B-solution. b) $BaSO_4$ crystals obtained in the single-feed semibatch reactor (Podgórska, 1993).

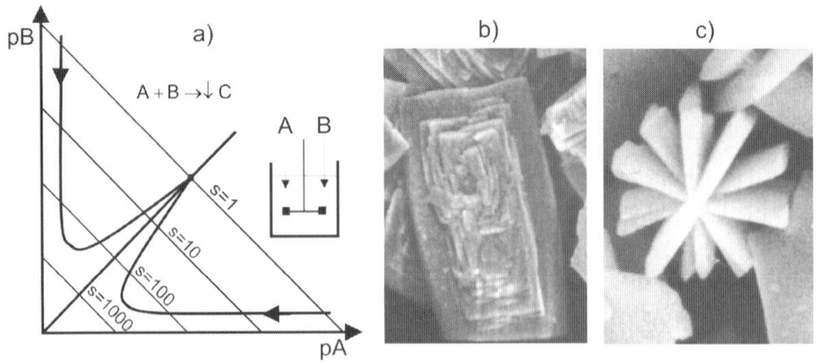

Figure 9. a) Variation of composition of supersaturated solution during double feed precipitation process. b) $BaSO_4$ crystals obtained in the case of large distance between feeding points (Podgórska, 1993). c) $BaSO_4$ crystals obtained in the case of small distance between feeding points (Podgórska, 1993).

3. APPLICATION OF SUPERCRITICAL FLUIDS

3.1. Properties of supercritical fluids

According to classical thermodynamics the supercritical fluid (SCF) state of a pure component is defined as a region of temperatures and pressures that are greater than the critical pressure, p_C, and critical temperature, T_C. The value of the pressure, p, and temperature, T, divided by its corresponding critical value is called the "reduced" pressure, p_r, and temperature, T_r.

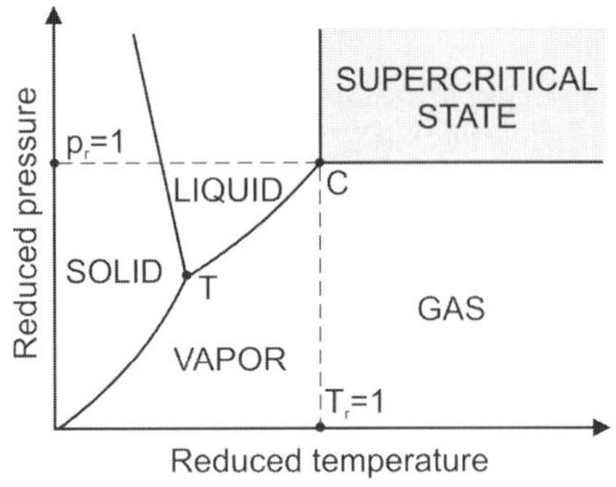

Figure 10. Temperature-pressure diagram for a single-component system; C – critical point, T – triple point.

As shown schematically in Figure 10 under supercritical conditions, i.e. when $T_r > 1$ and $p_r > 1$ there is only one phase and the two-phase region in the (T_r, p_r) plane is then not possible. This single phase is fluid that is neither a gas nor a liquid; instead it is a fluid in a supercritical state (SCF). This fluid is highly compressible, especially in the region close to the critical point, which means that a small increase of pressure or small decrease of temperature result in this region in significant increase of fluid density. The supercritical fluids exhibit properties of both liquids and gases; they may combine for example the liquid-like density with almost gas-like viscosity, conductivity and diffusivity. When the SCF is used as a solvent its solvent power is directly related to its density, which means that a solubility in the SCF solvent can be varied continuously between a low solubility at low densities and a high solubility at high, liquid-like densities. This means that one can modify the solvent power by changing temperature and pressure. Thus in many cases SCFs have the solvent power of liquids, however, SCF can also be used as antisolvents. For example solubilities of many organic solutes in supercritical carbon dioxide are orders of magnitude smaller than in many organic liquid solvents and dilution of such concentrated solution in the supercritical carbon dioxide results in dramatic decrease of solubility. An important feature of SCF is related to their transport properties, namely SCFs viscosities, diffusivities and conductivities are intermediate to those properties for liquids and gases.

The gas viscosities are almost independent of pressure at low pressure, whereas above $P_r = 1$ the viscosity increases with pressure being however smaller than the viscosity for similar density liquid. In the region $T_r > 1$ and $p_r > 1$ the SCF viscosity decrease with an increase of temperature (liquid-like behavior).

Diffusion coefficients in a gas at low pressure are inversely proportional to the pressure. Above the critical pressure the diffusion coefficient still decreases with pressure, but pressure effect is much smaller. As a result the diffusivity coefficient is larger than for liquid solution but lower than for low-pressure gas mixtures.

Similarly as in the case of viscosity, the gas conductivity is independent of pressure at low pressures and significantly increases with pressure above $P_r = 1$. Conductivity similarly like viscosity decreases with temperature at moderate temperatures in the region $T_r > 1$ and $p_r > 1$. Relations between transport properties in SCF, gases and liquids are shown in Figure 11.

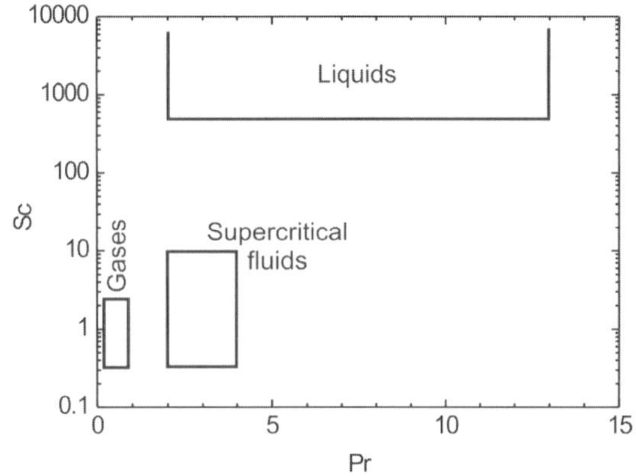

Figure 11. Relation between the Schmidt and Prandtl number for gases, liquids and supercritical fluids.

The Schmidt number, $Sc \equiv \nu/D_m$, represents relative importance of momentum diffusion to the mass diffusion, where D_m and ν denote the molecular diffusivity and kinematic viscosity respectively. For $Sc > 1$ the fluid deformation is faster than molecular diffusion, which means that spots, threads and slabs of unmixed component are produced during intensive mixing. This phenomenon is observed usually in liquids ($Sc >> 1$). Figure 11 shows that in many cases it can happen also during mixing in SCF. The Prandtl number, $Pr \equiv \nu/D_T$, is a measure of the importance of momentum diffusion to the heat diffusion. For supercritical fluids $Pr > 1$, which means that heat diffusion is slightly slower than momentum diffusion and some hot spots may be possibly observed during mixing. The relative importance of heat diffusion to the mass diffusion is expressed by the Lewis number, $Le \equiv D_T / D_m$. In gases $Le \approx 1$ whereas in liquids Le is between 10 and 100 indicating faster heat diffusion. Harstad and Bellan (1999) have studied the Lewis number problem under supercritical conditions. They have shown that the effective Lewis number can be $2 \div 40$ times larger than the traditionally calculated Lewis number, which results from combined effects of small mass diffusion factor and the difference between the specific enthalpies of mixed compounds. As shown recently by Miller (2000) and Lou and Miller (2001) "nonequilibrium" thermodynamic effects can be significant during mixing of SCF. It means that diffusion of mass may be generated by temperature and pressure gradients and diffusion of energy may be due to concentration gradients (Soret effect, Dufour effect).

Consider now the phase equilibrium behaviour of the binary mixture at elevated pressures. Figure 12 shows that in this case a single-phase fluid is observed above the mixture critical point C_m.

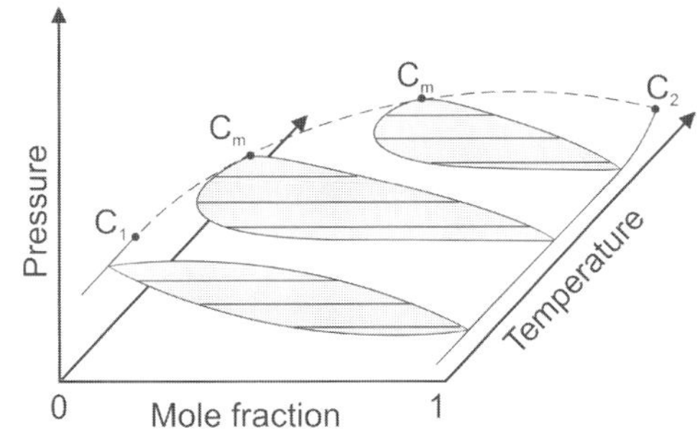

Figure 12. The pressure-temperature-mole fraction diagram. C_1 and C_2 are critical points for pure components. The dashed line represents, C_m, vapor-liquid critical point.

In the temperature range $T_{C1} < T < T_{C2}$ SCF may be mixed with the liquid in the double phase ($p < p_{Cm}$) or a single phase ($p > p_{Cm}$) system. In the first case there is an interfacial tension between the phases and mixing mechanism is based on atomization to small drops whilst energy and species are transferred between the drops and their environment. In the second case there is no interfacial tension, and mass transfer is determined by the flow - molecular diffusion interactions during mixing. Such processes are usually carried out in the turbulent flow systems (Shekunov et al., 2001; Bellan, 2000).

3.2. Methods of application

Supercritical fluids have a potential for wide field of applications. Current applications include extraction processes, reaction chemistry and polymerisation, food fractionation, waste recycle, soil remediation, cleaning of electronic and optical parts of equipment, impregnation, dry powder coating, areogels, nanotechnology, crystallization and particle formation of pharmaceuticals and many others (Perrut, 2000). In what follows we give a brief review of methods for particle design using supercritical fluids. More complete material the reader can find in recently published review articles (Jung and Perrut, 2001; Marr and Gamse, 2000; Subra and Jestin, 1999). A particle design is presently treated as a major challenge and a major development of supercritical fluids applications.

The Rapid Expansion of Supercritical Solutions (RESS-Process) is presented schematically in Figure 13. The process consists in solving the substrate to be powdered in a supercritical fluid and rapid expansion of this solution through the heated nozzle.

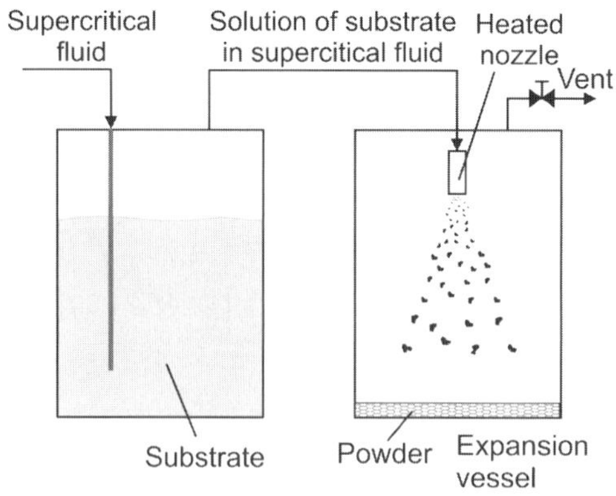

Figure 13. Rapid Expansion of Supercritical Solution Process (RESS-Process).

When the saturated solution is quickly depressurized into a low pressure chamber, solvent power decreases immediately, and a very high supersaturation is created. At high initial supersaturation very fine particles that can form small agglomerates (crystallize or amorphous) are created. As the result a fine particles of the range 0.5 – 20 μm with a narrow size distribution are produced in the case of pure substrate components. In the case of composite particles (microspheres, microcapsules) a mixture of substrates is processed in the RESS-process. Morphology of particles depends on nozzle design and geometry of the expansion vessel, RESS parameters (temperature, pressure and pressure drop), physical and chemical properties of substrate. The method can be compared with the classical method of evaporative crystallization, especially spray crystallization. The main difference is that in classical method a mass and energy transfer between liquid and vapor control creation of supersaturation, whilst depressurising (see Figure 10) does not create any energy and mass transfer barriers. The RESS method can be used for substances that are well soluble in SCF, which often requires the high pressure ranges.

The antisolvent methods are based on the same principle as the classical drowning-out method presented in Section 2. **The Gas Anti-Solvent Process** or **Supercritical Anti-Solvent Process (GAS/SAS-Process)** is presented in Figure 14. The principle of the antisolvent processes is to decrease the solvent strength by adding SCF as an antisolvent in which the substrate is insoluble or hardly soluble. This creates supersaturation and enables precipitation of the substrate. The substrate (solute) is dissolved in some suitable liquid solvent. The saturated substrate solution is then contacted with a supercritical antisolvent (supercritical fluid). The antisolvent should have a low affinity for solutes and a very good mutual solubility with the solvent. In the SAS method the vessel is partially filled with the substrate solution. The antisolvent is introduced into the vessel from the bottom to intensify mass transfer.

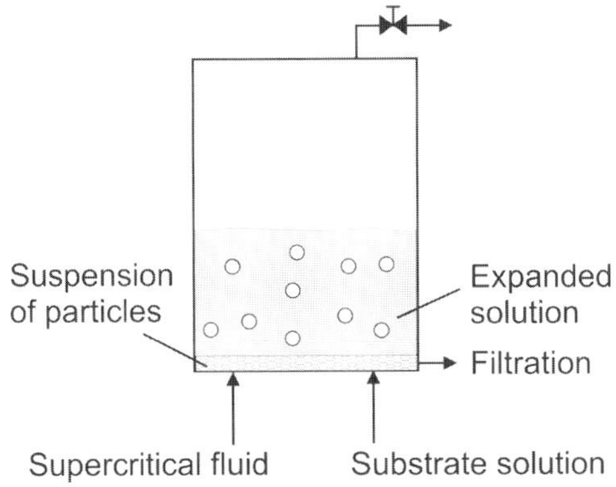

Figure 14. Gas Anti-Solvent Process or Supercritical Anti-Solvent Process (GAS/SAS-Process).

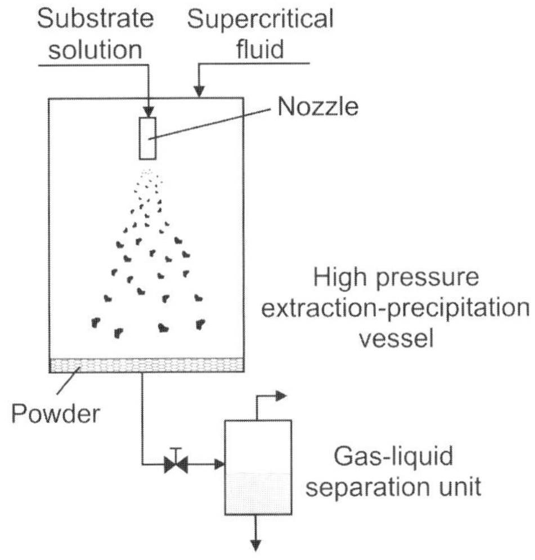

Figure 15. Aerosol Solvent Extraction System Process (ASES-Process).

Due to the dissolution of the compressed gas there are expansion and evaporation of the solvent, whereas the expanded solvent has a lower solvent power than the pure solvent. Supersaturation is thus created and the substrate precipitates forming microparticles. Particle size and morphology depend on materials, pressure and temperature and can be manipulated by controlling the rate of supersaturation creation. This can also be done by varying the antisolvent feeding rate and agitation rate. For gradual addition of the antisolvent to the solution the SAS process enables production of larger crystals for slow addition and smaller ones when addition is fast.

Faster creation of supersaturation is possible in **the Aerosol Solvent Extraction System Process (ASES-Process)** shown in Figure 15. In this process the substrate solution is sprayed into the supercritical fluid. There is then an intensive mass transfer between droplets and SCF and fast reduction in the liquid solvent power resulting in sharp rise of supersaturation within droplets. As the result small particles of uniform size are formed. Figure 16 shows combination of the SAS and ASES methods where particles are initially formed in the sprayed droplets and then grow in the expanded solvent.

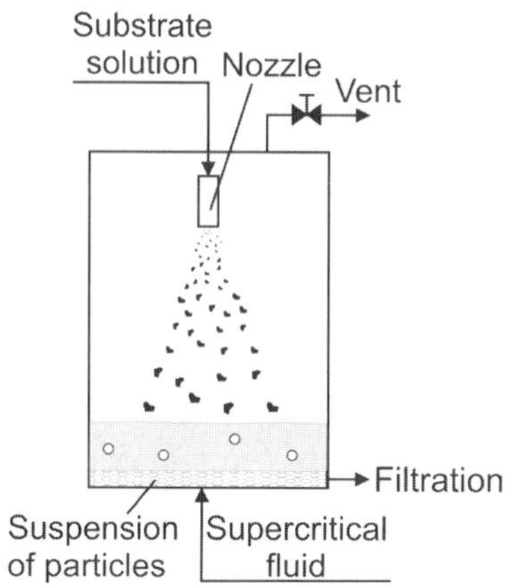

Figure 16. Combination of SAS and ASES Processes.

Next application of the supercritical antisolvent is called **Solution Enhanced Dispersion by Supercritical Fluids Process (SEDS-Process)** and was developed by the Bradford University Group (Hanna et al., 1994). The method is presented in Figure 17 and enables to achieve an intensive mixing of the fresh substrate solution with the supercritical antisolvent in the coaxial nozzle with two coaxial passages for the solution and the solvent respectively. The most intensive premixing is observed within the nozzle mixing chamber (Figure 24) and the highest supersaturation is produced there. Afterwards, beyond a distance of several nozzle diameters the supersaturated jet fluid starts to mix with the fluid present in the vessel. Figure 18 shows the phase diagram for ethanol - carbon dioxide system calculated using the Peng-Robinson equation of state. The area within the envelope corresponds to the vapor-liquid system at a given temperature. The broken line in Figure 18 shows that at the same pressure precipitation may be carried out in the two-phase system of two partly miscible fluids (T = 363 K), and in the single phase system of the completely miscible compounds (T = 313 K). In the first case the coaxial nozzle acts as "spray enhancer". The phenomena in the particle formation vessel are thus very similar to observed in the ASES process with the only difference that the droplets are finer and mass transfer faster.

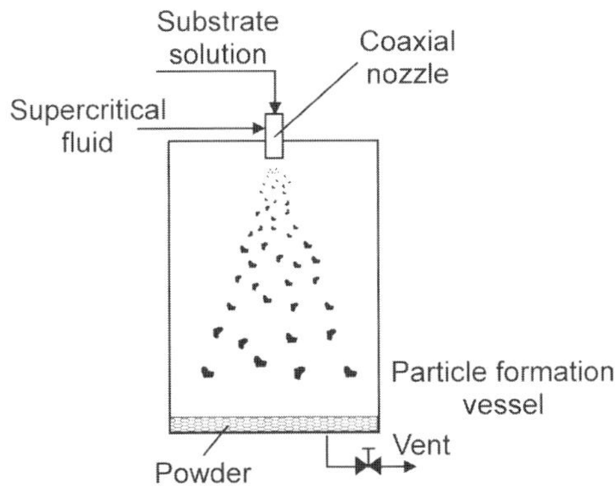

Figure 17. Solution Enhanced Dispersion by Supercritical Fluid Process (SEDS-Process).

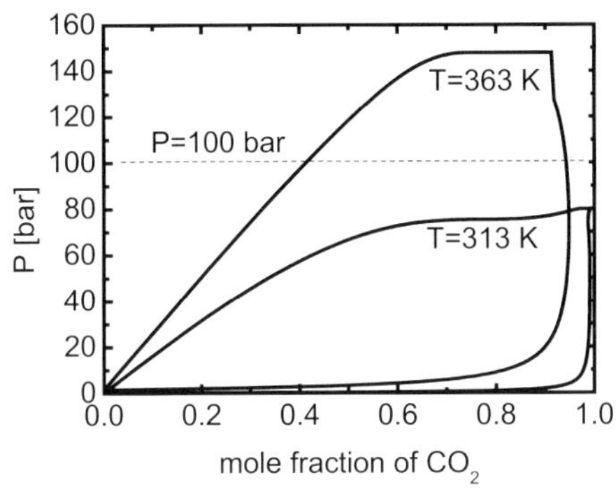

Figure 18. Effect of pressure and temperature on carbon dioxide- ethanol equilibrium.

In the second case the coaxial nozzle acts as "fast mixer". In this case decompression is rather small and the process is similar to the homogeneous reactive precipitation (single phase mixing creates supersaturation). This process will be discussed as a case study in more detail in Section 3.3.

Another improvement of the ASES process is based again on enhancement of spraying. In the **Supercritical Anti-Solvent Enhanced Mass Transfer Process (SASEM-Process)** the jet of solution is deflected by a surface vibrating in an ultrasonic frequency and atomizing the jet into very small droplets (Chattopadhyay and Gupta, 2002). The process is shown schematically in Figure 19. The ultrasound field generated by the vibrating surface enhances also mass transfer and prevents agglomeration. As a result submicron size particles of the size significantly dependent on the ultrasound power are produced.

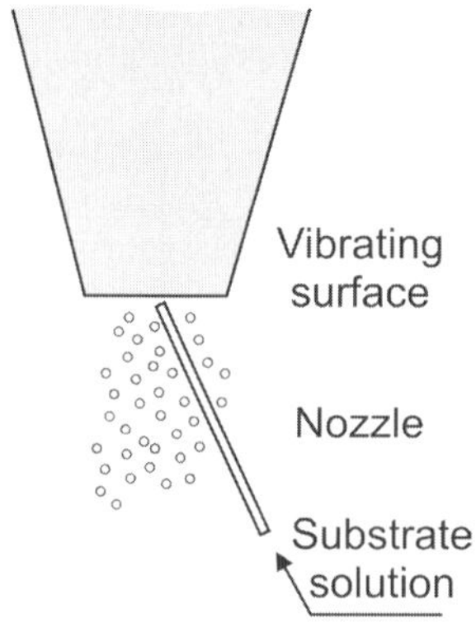

Figure. 19. Supercritical Anti-Solvent with Enhanced Mass Transfer Process (SASEM Process).

Fine **Particles** may be formed from **Gas Saturated Solutions (PGSS-Process)** as proposed by Graser and Wickenhauser (1982) and Weidner (1999). The process shown in Figure 20 consists in dissolving of a pressurised gas in melted (or liquid suspended) substrates. The gas saturated solution is then expanded through the nozzle, which results in formation of solid particles or droplets. Supersaturation is created by cooling (Joule-Thomson effect) and volume increase of the released gas.

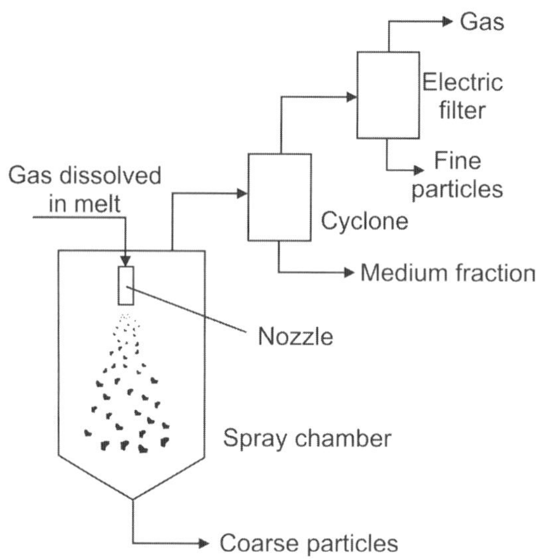

Figure 20. Production of fine Particles from Gas Saturated Solutions (PGSS-Process).

The methods presented above were discussed in relation to production of the single component powders. Figure 21 shows that depending on the level of supersaturation and the methods that are used to create supersaturation various shapes can be produced by using supercritical fluids.

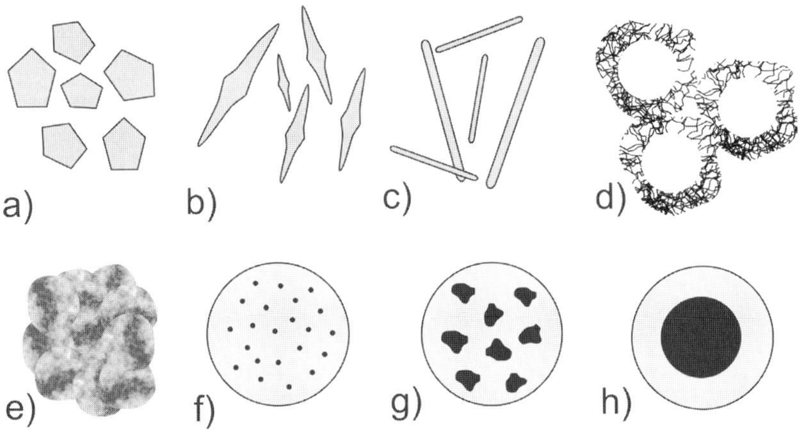

Figure 21. Structure of microparticles: a) compact crystals, b) dendrids, c) needle-like or rod-like particles, d) balloon-like structures, e) single substrate agglomerates, f) microspheres with molecularly dispersed active substance, g) microspheres with large particles of active substance, h) microcapsules.

Morphologies presented in Figure 21 abce were already discussed in Section 3.1. The baloon-like particles are produced by fast expansion of droplets in the ASES process (Reverchon et al., 2001). The composite microparticles (microspheres and microcapsules) presented in Figure 21 fgh are also produced by applying the previously described processes. Microspheres contain an active component embedded into the matrix. Additives forming the matrix are often mixed directly into the product solution or melted product before processing. In the SAS methods an active component and a carrier are both dissolved in liquid solvent and mixed with the supercritical fluid. They are then either injected in the same stream or added separately. Microcapsules are produced in the process of particle coating. The previously described methods can be used in this process. Usually the core particles are created initially and the process of deposition of the coating agent on the core material is carried out afterwards. Supersaturation of the coating agent is achieved by using one of the methods previously presented.

3.3. Case study - Particle formation by mixing with supercritical antisolvent in the system of the completely miscible compounds (Shekunov et al., 2001).

The SEDS experimental apparatus presented schematically in Figure 17 has been employed to crystallize paracetamol (interpreted as a model compound) from ethanol solution using supercritical carbon dioxide. The main aim of this study was to define effect of the flow rate (expressed through the nozzle Reynolds number,

$Re \equiv u_0 d_0/\nu$, where u_0 is the nozzle fluid velocity, d_0 is the nozzle diameter and ν represents kinematic viscosity) on particle size in a coaxial mixing jet. The important case is studied where the solvent and antisolvent are completely miscible, i.e. remain above the mixture critical point at all solvent-antisolvent proportions (see Figures 12 and 18). Details of the experimental method are presented by (Shekunov et al., 2001).

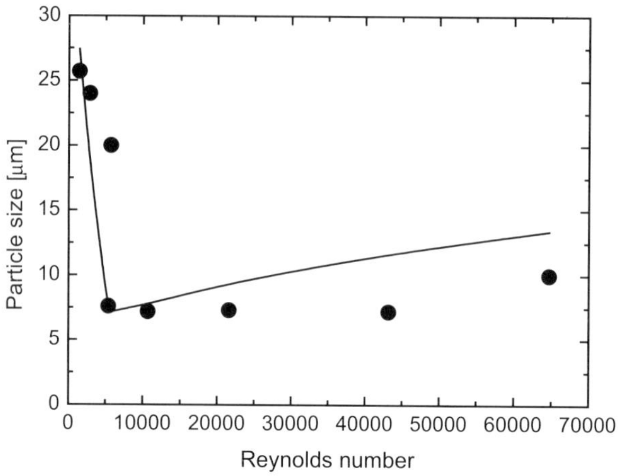

Figure 22. Comparison of experimental data (•) with model predictions (continuous curve).

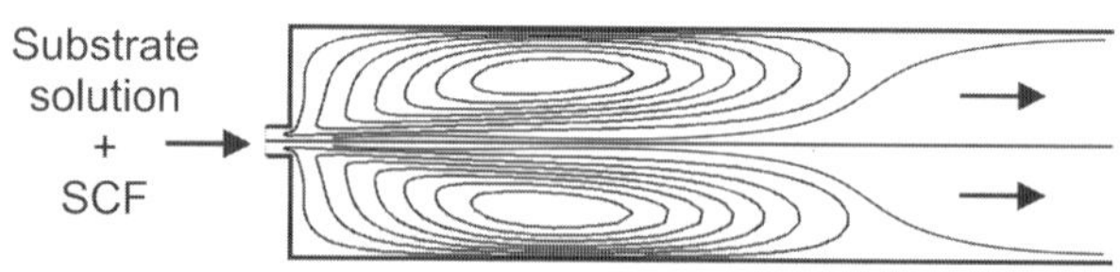

Figure 23. Structure of the flow in SEDS particle formation vessel presented using the stream lines

Figure 22 shows results of experiments obtained at a fixed flow composition (i.e. for the same proportion of mass fluxes of the carbon dioxide and paracetamol solution in ethanol) but at different flow velocities expressed through the Reynolds number, *Re*. The particle size exhibits a rapid decrease leading to a minimum particle size, whereas at higher *Re* the particle size increases slightly with *Re*. To analyze the SEDS process and explain experimental observations we have performed the Computational Fluid Dynamics (CFD) simulations (Bałdyga et al., 2002ab). The k-ε model available in the Fluent 6.0 code together with made in house mixing models were applied in computations. Examples of computations are presented in Figures 23 and 24. Figure 23 shows that in the particle formation vessel there is an intensive circulation in the region of the jet, which means that in the system there is backmixing of the fresh fluid from the nozzle with the reservoir fluid. Figure 24 shows temperature distribution in the coaxial nozzle and in the nozzle neighborhood.

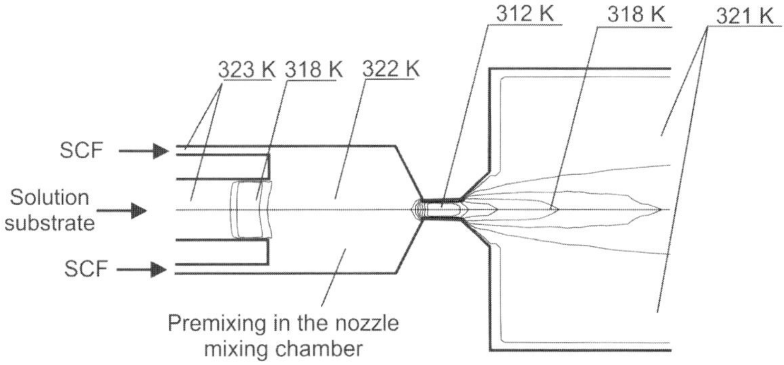

Figure 24. Temperature distribution in the coaxial nozzle of the jet mixer (SEDS-Process).

As the single phase system is considered, observed decrease of temperature is not related to the drop evaporation (which is often observed in the process of spraying) but results from heat of mixing and Joule-Thomson effect. Note, that the supercritical carbon dioxide penetrates into the core of the paracetamol flow. The fluids are then at least in part premixed in the nozzle mixing chamber and cooled in the nozzle due to the Joule-Thomson effect. Premixing of the fresh solution and fresh antisolvent occurs predominantly within the nozzle mixing chamber, which produces supersaturation. Supersaturation is also affected by cooling. A rapid decrease of particle size at $Re < 10\ 000$ is caused by establishment of supersaturation profile (up to $Re \approx 10\ 000$ maximum supersaturation increases with increasing *Re*). Afterwards the supersaturated jet fluid that enters the vessel becomes diluted with the reservoir fluid (CO_2 and solvent mixture saturated with paracetamol), which reduces supersaturation. Some increase of particle size observed at high *Re* is related to this phenomenon. After reaching $Re \approx 10\ 000$ further increase of *Re* does not enhance mixing, so the supersaturation can not be increased due to mixing anymore, and just small increase of supersaturation may result from cooling effect. On the other hand increase of *Re* decreases the residence time of fluid elements in the region of high supersaturation, so decreases the time available for nucleation process in the zone of high supersaturation. As the result smaller number of larger particles is produced. Model simulations (Shekunov et al., 2001) well predict these effects, which is also shown in Figure 22. Increase of *Re* also affects the shape of particles, at *Re* much smaller than 10 000 the plate-like crystal are produced, whereas at $Re >$ 10 000 more compact, isometric crystals are observed.

Presented examples show that modelling, including application of the Computational Fluid Dynamics can be useful for understanding of the process and helpful in process optimization and scale-up.

4. CONCLUSIONS

Material presented in Sections 2 and 3 of this Chapter shows that the process of particle formation is based on the same fundamentals as the traditional methods of crystallization and precipitation. Hence, the same method of process analysis based on studying creation and reduction of supersaturation and effects of supersaturation level and solution composition on particle size and morphology can be used in the case of powder production with supercritical media. What really differs the methods based on using supercritical fluids from the traditional ones is the possibility to create very large values of supersaturation in a very short time, which gives a useful method for controlling particle size and morphology. Moreover, one is able to reduce supersaturation in a short time as well. This may be in some cases limitation of the method, but generally it gives another opportunity to affect powder quality. Development of comprehensive models of particle creation with the use of supercritical media should be helpful in design, optimisation and scale-up of the industrial powder production processes.

5. REFERENCES

Abrahamson, J., "Collision Rates of Small Particles in a Vigorously Turbulent Fluid". *Chem. Eng. Sci.*, 30, 1371-1379 (1975).

Bałdyga, J. and Bourne, J.R., *Turbulent Mixing and Chemical Reactions,* John Wiley & Sons: Chichester, 1999.

Bałdyga, J., Jasińska, M., Krasiński, A., Kubicki, D., Franke, D., Goesele, W., Welker, S. and Zauner, R., *Mixing Effects in Precipitation-Aggregation Problem.* Proceedings of 4th International Symposium on Mixing in Industrial Processes, Toulouse, France, paper 11, 75 (2001).

Bałdyga, J., Henczka, M. Czarnocki, R. and Shekunov, B.Yu., „Scale-up of the supercritical antisolvent precipitation process". *Inż. i Ap. Chem.*, 41, 20-21 (2002).

Bałdyga, J., Henczka, M. Kubicki, D. and Shekunov, B.Yu., „Modeling of turbulent mixing with application of supercritical antisolvent". *Inż. i Ap. Chem.*, 41, 22-23 (2002).

Bałdtga, J. and Orciuch, W., "Some hydrodynamic apects of precipitation", *Powder Technol.*, 121, 9-19 (2001)

Bałdyga, J., Podgórska, W. and Pohorecki, R., „Mixing-precipitation model with application to double feed semibatch precipitation". *Chem. Eng. Sci.,* 50, 1281-1300 (1995).

Bellan, J., „Supercritical (and subcritical) fluid behavior and modeling: drops, streams, shear and mixing layers, jets and sprays". *Prog. Ener. Comb. Sci.*, 26, 329-366 (2000).

Berthoud, A., "Theorie de la formation des faces d'un crystal". *Z. Kristallogr.,* 10, 624 (1912).

Chattopadhyay, P. and Gupta, R.B., "Protein Nanoparticles Formation by Supercritical Antisolvent with Enhanced Mass Transfer". *AIChEJ,* 48, 235-244 (2002).

David, R., Marchal, P., Klein, J.P. and Villermaux, J., "Crystallization and precipitation engineering – iii. a discrete formulation of the agglomeration rate of crystals in a crystallization process". *Chem. Eng. Sci.,* 46(1), 205-213 (1991).

Dirksen, J.A. and Ring, T.A., "Fundamentals of Crystallization: Kinetic Effects on Particle Size Distributions and Morphology". *Chem. Eng. Sci.,* 46, 2389-2427 (1991).

Elimelech, M., Gregory, J., Jia, X. and Williams, R., *Particle Deposition & Aggregation.* Butterworth-Heinemann: Oxford, 1995.

Elwell, D. and Scheel, H.J., *Crystal Growth from High-Temperature Solutions.* Academic Press: London, 1975.

Garside, J., "Industrial Crystallization from Solution". *Chem. Eng. Sci.*, 40, 3-26 (1985).

Gibbs, J.W., *Collected Works,* Longmans, Green: New York, 1928.

Graser, F. and Wickenhauser, G., *Patent US 4,451,654,* September 20, 1982.

Hanna, M. and York, P., *Patent WO 95/01221*, 1994.

Harstad, K. and Bellan, J., "The Lewis number under supercritical conditions". *Int. J. Heat Mass Trans.*, 42, 961-970 (1999).
Hounslow, M.J., Mumtaz, H.S., Collier, A.P., Barrick, J.P. and Bramsley, A.S., "A micro-mechanical model for the rate of aggregation during precipitation from solutions". *Chem. Eng. Sci.*, 56, 2543-2552 (2001).
Jung, J. and Perrut, M., "Particle design using supercritical fluids: Literature and patent survey – review". *J. of Supercritical Fluids*, 20, 179-219 (2001).
Kruis, F.E. and Kusters, K.A., "The collision rate of particles in turbulent flow". *Chem. Eng. Comm.*, 158, 201-230 (1997).
Lou, H. and Miller, R.S., "On the scalar probability density function transport equation for binary mixing in isotropic turbulence at supercritical pressure". *Physics of Fluids.*, 13, 3386-3399 (2001).
Marr, R. and Gamse, T., "Use of supercritical fluids for different processes including new developments – a review". *Chem. Eng. Proc.*, 39, 19-28 (2000).
Mersmann, A., *Crystallization Technology Handbook*. Marcel Dekker Inc., New York, 1995.
Miller, R.S., "Long time mass fraction statistics in stationary compressible isotropic turbulence at supercritical pressure". *Physics of Fluids.*, 12, 2020-2032 (2000).
Mullin, J.W., *Crystallization*. Butterworth-Heinemann Ltd.: Oxford, 1993.
Nielsen, A.E., *Kinetics of Precipitation*. Pergamon: Oxford, 1964.
Perrut, M., Supercritical Fluid Applications: Industrial Developments and Economic Issues, The 5th Int. Symp. on Supercritical Fluids, Atlanta, Georgia, USA (2000).
Podgórska, W., *Ph. D. Thesis*, Warsaw University of Technology, 1993.
Randolph, A.D. and Larson, M.A., *Theory of Particulate Processes*. Academic Press: New York and London, 1971.
Reverchon, E., Della Porta G. and Pallado, P., "Supercritical antisolvent precipitation of salbutamol particles". *Powder Technol.*, 114, 17-22 (2001).
Saffmann, P.G. and Turner, J.S., "On the Collision of Drops in Turbulent Clouds". *J. Fluid Mech.*, 1, 16-30 (1956).
Shekunov, B.Yu., Bałdyga, J. and York, P., "Particle formation by mixing with supercritical antisolvent at high Reynolds numbers". *Chem. Eng. Sci.*, 56, 2421-2434 (2001).
Smoluchowski, M., "Versuch einer Mathematischen Theorie der Koagulations-Kinetik Kolloider Lösungen". *Z. Phys. Chem.*, 92, 129-168 (1917).
Söhnel, O. and Garside, J., *Precipitation*. Butterworth-Heinemann Ltd.: Oxford, 1992.
Spielmann, L.A., „Viscous Interactions in Brownian Coagulation". *J. Coll. Int. Sci.*, 33, 562-571 (1970).
Subra, P. and Jestin, P., "Powders elaboration in supercritical media: comparison with conventional routes". *Powder Technol.* 103, 2-9 (1999).
Težak, B., "Methoric of the Precipitation from Electrolytic Solutions". *Croatica Chemica Acta*. 40, 63-78 (1968).
Valeton, J.J.P., "Wachstum and Auflösung der Kristalle".*Z. Kristallogr.* ,60 1 (1924).
Weidner, E., *Proceedings of the International Meeting of the GVC-Fachausshuß "Hochdruckverfahrenstechnik" on High Pressure Chemical Engineering*, Karlsruhe, p. 217 (1999).

6. AFFILIATIONS

Prof. Jerzy Bałdyga and Dr. Marek Henczka work at the Department of Chemical and Process Engineering, Warsaw University of Technology, Waryńskiego 1, 00-645 Warsaw, Poland.

7. ACKNOWLEDGMENT

Part of this work was supported by the Polish State Committee for Scientific Research (KBN) under the project No. 1547 T09C 2001 21.

K.B. GEERSE AND J.C.M. MARIJNISSEN

ELECTROSPRAY AS MEANS TO PRODUCE MONODISPERSE DRUG PARTICLES

1. INTRODUCTION

In the treatment of asthma and other chronic obstructive lung diseases, inhalation therapy is the most frequently applied method to administer drugs. Direct local administration into the lungs leads to an immediate effect, and smaller doses are needed compared with oral administration. However, with the use of conventional inhalation devices, only a fraction of the inhaled drug reaches the lower airways, where it has its therapeutic effect (Biddiscombe et al. 1993). A large part is deposited in the mouth and the throat, after which it is swallowed and subsequently may be absorbed in the gastrointestinal tract. The low efficiency of the inhalation equipment is related to the less optimal size distribution of the particles released, although recently developed inhalers can show improved distributions (Kunkel et al. 2000, Zierenberg 1999). Research has shown that, in adults, monodisperse 2.8 µm bronchodilator particles were optimal in terms of efficacy (Zanen et al. 1995, Zanen et al. 1996). These experiments point to the fact that the range of optimal aerosol particle sizes might be much smaller (2.0-3.5 µm) than the currently assumed 1-5 µm. It was also shown that administration of these monodisperse aerosols opened the way to reduce the dose emitted from metered or dry powder inhalers by approximately 80% without losing any clinical effect (Zanen et al. 1998). Large particles are less efficacious; therefore, they may be eradicated from the emitted dose without reducing the therapeutic effect. However, if they enter the body they still have the potential to elicit side effects. Smaller particles are exhaled directly or deposit in the alveoli, where they are not effective due to the lack of smooth muscle. They are most probably absorbed rapidly into the systemic circulation and might be responsible for fast-emerging side effects such as hypokaliemia (Zanen et al. 1999). As a result of these findings, an interest in (producing) monodisperse aerosols has emerged as a means to improve the therapeutic quality (Thompson 1998).

To apply bronchodilator data to inhaled corticosteroids would be speculative, but the physics governing deposition of such aerosols are, of course, identical to those for bronchodilators. Therefore, administration of monodisperse steroid aerosols also has the potential to reduce total dose and side effects, but the optimal formulation needs to be defined.

L. Gradoń and J. Marijnissen (eds.), Optimization of Aerosol Drug Delivery, 75–90.

The means to produce monodisperse or narrow size-ranged steroid aerosols are, however, limited. Present systems, like the spinning-top generator, are cumbersome in their use and are, therefore, confined to a laboratory environment (Cheah and Davies 1984). The aim of the work presented in this chapter was to find out if the electrospray technique could be used to generate corticosteroid aerosols in an aerodynamic size range between 1 and 5 μm with a low geometric standard deviation (GSD) and in quantities sufficiently high to make administration to patients feasible.

2. THEORY

The production of particles using electrospray involves the production of droplets containing the dissolved material. Consequently evaporation of those droplets yields the solid particles. This process consists of two steps, droplet production and evaporation. Changing the initial droplet size or the amount of solid material dissolved can be used to control the resulting particle size.

2.1 Electrospray

An electrospray refers to the atomization of a liquid through the Coulombic interaction of charges on the liquid and the applied electric field. The result of this interaction includes both the acceleration of the liquid and subsequent disruption into droplets as well as the buildup of charge. This phenomenon is also referred to as *ElectroHydroDynamic Atomization* (EHDA), to avoid confusion with other spraying techniques using an electric field. This name covers the process of the atomization of a moving (*dynamic*) liquid (*hydro*) in an electric (*electro*) field.

Depending on different parameters like applied flow rate and electric field strength, different modes of electrospray can be obtained (Cloupeau and Prunet-Foch 1994, Grace and Marijnissen 1994). The mode of most interest is the so called cone-jet mode, in which monodisperse droplets in the micrometer range can be produced.

The atomization of the liquid and the resulting droplet size in the cone jet mode is a function of both external (experimental configuration, flow rate and atmosphere) and internal (liquid) properties. The fluid properties receiving the most attention are: the electrical conductivity, the surface tension, the viscosity, the density and the permittivity. Figure 1 shows an electrospray in the cone jet mode.

Figure 1. Photograph of electrospray in the cone-jet mode.

2.2 Controlling the initial droplet size in the cone-jet mode

Scaling laws developed by Hartman (Hartman et al. 1999) can be used to calculate the droplet size as function of intrinsic liquid properties and of the flow rate applied using the cone-jet mode. Changing the intrinsic liquid properties is hard to do, while this involves the addition of some additive to change the conductivity or the surface tension for example. Furthermore, in the case where solid particles have to be produced for inhalation, this addition is unwanted as this complicates the use of the particles. Therefore the applied liquid flow rate is chosen as control parameter. In case of the varicose break up regime, the droplet size scales with the flow rate as given by equation 1:

$$d_d \propto Q^{\frac{1}{2}} \tag{1}$$

in which d_d is the obtained droplet diameter and Q is the applied flow rate.

The varicose break up regime is the wanted regime as this can yield a monodisperse droplet population. To avoid whipping break up, with a wider size distribution, the flow rate range is limited. Although the transformation from varicose to whipping break up can be predicted by calculating the ratio on the jet of the electric stress over the surface tension stress, the complexity of determining this ratio makes it not a practical tool. Therefore on-line monitoring of the final particle size is used to verify whether varicose break up takes place or not.

2.3 Changing the weigth fraction of dissolved material

Evaporation of the droplets produced leads to the formation of solid residues. Depending on the circumstances and the material used particles of different morphology can be obtained (Vercoulen 1995). As the main goal after evaporation is the yield of particles with a certain aerodynamic diameter, the assumption is made that the final particles are solid. Whether this assumption is valid will be proven at the end of this chapter.

The relation between final solid particle size, initial droplet size and weight fraction is given by equation 2:

$$d_p = \left(\frac{\rho_s w}{\rho_s w + \rho_p (1-w)} d_d^3 \right)^{\frac{1}{3}} \tag{2}$$

in where d_p is the particle size, w is the weight fraction of dissolved material and ρ_s and ρ_p are the solvent and solid density, respectively.

Figure 2 shows the aerodynamic particle size dependency on both the weight fraction and the flow rate applied in the cone jet mode. The values are calculated for ethanol as solvent and a solid with density of 1285 kg/m^3, resembling the medicine used.

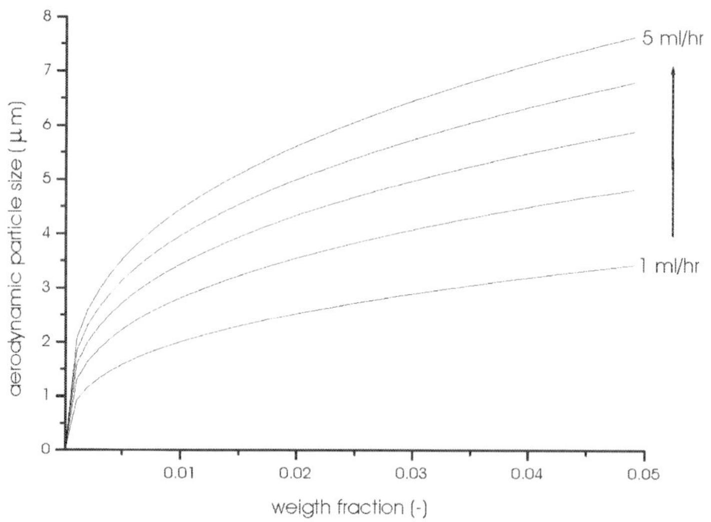

Figure 2. Aerodynamic particle size as function of weight fraction solid material for different flow rates in the cone jet mode. (The increase in flow rate is 1 ml/hr)

2.4 *Evaporation of the droplets*

To make the transition from droplets to solid particles, enough residence time for evaporation is necessary. The evaporation time is highly dependent on the liquid properties, the relative humidity and the surrounding temperature. In the experiments discussed here, ethanol was used as solvent, because it is volatile and dissolves the used medicine easily.

For pure liquid droplets having a diameter larger than 10 μm and reaching the steady state of evaporation, the diminution rate is given by equation 3 (Hinds 1999):

$$\frac{d(d_d)}{dt} = \frac{4DM}{R\rho_s d_d}\left(\frac{p_\infty}{T_\infty} - \frac{p_s}{T_s}\right) \tag{3}$$

in where D is the vapor diffusion coefficient, M is the liquid molar mass and R is the gas constant. T_s and p_s are the temperature and the vapor pressure at the droplet surface, while T_∞ and p_∞ are the same far away from the droplet.

This equation can be solved explicitly to give the droplet life time or total time of evaporation, but in the case of volatile liquids a correction is necessary to account for the cooling of the droplet due to the latent heat of evaporation. The equilibrium droplet temperature depression can be obtained by balancing the heat required for evaporation and the heat gained by conduction from the surrounding air. This results in a steady state depression given by equation 4:

$$T_\infty - T_s = \frac{DMH}{Rk_v}\left(\frac{p_s}{T_s} - \frac{p_\infty}{T_\infty}\right) \tag{4}$$

in where H is the latent heat of evaporation and k_v is the thermal conductivity of the gas.

Solving equation 3 and equation 4 by iteration yields the droplet life time. Figure 3 shows the droplet life time of ethanol droplets with an initial size of 20 μm as function of the ethanol partial pressure at different temperatures.

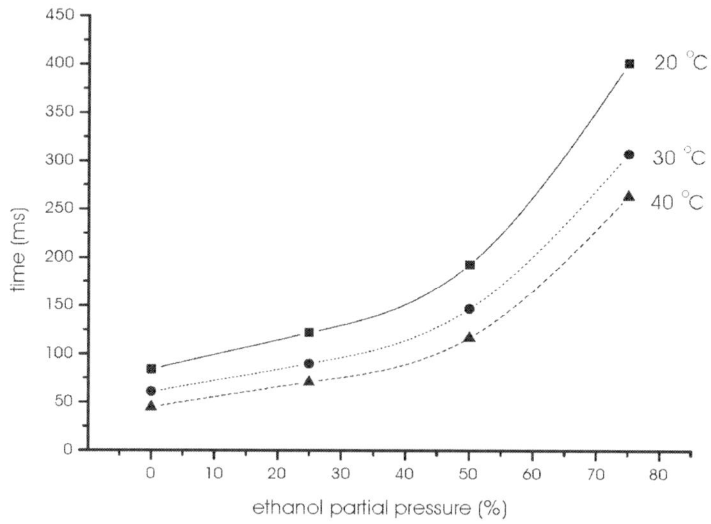

Figure 3. Ethanol droplet lifetime as function of the ethanol partial pressure for three different temperatures. (initial droplet size is 20 μm)

Note that because we do not have pure liquids but solutions, the evaporation time will be different from the calculated results. We assume here that the difference for the total drying time is not too big. However, this needs more research.

3. THE INHALER SET UP

Based on the described theory on electrospray and the possibility to produce monodisperse particles, an inhaler apparatus is developed to be used in the Utrecht Medical Center (UMC) to test the efficacy of monodisperse corticosteroid particles compared to normally used particle populations. The particles will be inhaled directly from the device. The device therefore has to meet the following requirements:

- Particles of sizes between 1 and 5 μm have to be produced with a small GSD.
- The residence time must be long enough to form solid particles.
- The reactor volume must be large enough to contain one inhalation dose.
- A sampling system must be present to verify online the particle size distribution.

The set up is comprised of four sections, a spray section, a neutralization section, a buffer section, and a sampling section. The sections will be discussed below.

3.1 *The spray section*

The spray section of the device consists of a nozzle, a ring and liquid supply entrance put together in a glass tube that is connected to the reactor. Using this nozzle-ring configuration makes it possible to create airborne droplets. The cylindrical nozzle is made of high quality stainless steel to prevent corrosion. It has an inner diameter of 0.2 mm and an outer diameter of 3 mm. The ring has an inner diameter of 15 mm and an outer diameter of 19 mm with a connector for high voltage through the glass tube. The distance between the nozzle and the ring can be adjusted between 0 mm and 19 mm. The nozzle is connected to a metal tube with connections for the high voltage and the liquid supply. The high voltage of the nozzle is put at 13.8 kV, the ring at 10.9 kV. The liquid supply is performed by a common syringe pump having a flow rate between 0.1 and 99.9 ml/hr. Figure 4 shows a detailed photograph of the spray section.

Figure 4. Photograph of the spray section.

3.2 *The neutralization section*

Because the by EHDA generated droplets are highly charged, the need to be neutralized to be used. This is done in the following way. Opposite of the spray section a similar glass tube is mounted at the reactor to be used for droplet discharging. The discharging is performed by small corona discharges using a sharp metallic needle. This needle is grounded. Due to the voltage difference between the ring and the needle and the needle, sharpness a strong electric field is present at the needle tip. This leads to the release of negative charges from its tip. The negative charges neutralize the positively charged droplets to prevent the occurrence of droplet Rayleigh explosions. During evaporation of the droplets the surface charge increases until the Rayleigh charge limit is reached, leading to the disruption of the drops. This process leads to undesirable final particle sizes and can decrease the monodispersity of the droplet population dramatically.

3.3 The evaporation and buffer section

Figure 5 shows a schematic picture of the total reactor as used in the UMC. An earlier version used for experiments at Delft was smaller, having a buffer volume of 0.5 L instead of 5.7 L for this one. This larger buffer volume is necessary to have enough particulate mass present while used for direct inhalation. During the administration of a certain dose of medicine only a few inhalations are necessary. The air flow input was 6.5 L/min, which assures a laminar flow in the reactor.
The experiments show that the evaporation of the produced droplets takes place within the residence time in the reactor. The main part of this section is used as a steady state buffer volume. As the air flow for transportation is perpendicular with respect to the initial droplet velocity, time is needed to reach the terminal velocity in the transport direction. This time can be estimated and is in the order of milliseconds for evaporating 20 μm droplets.

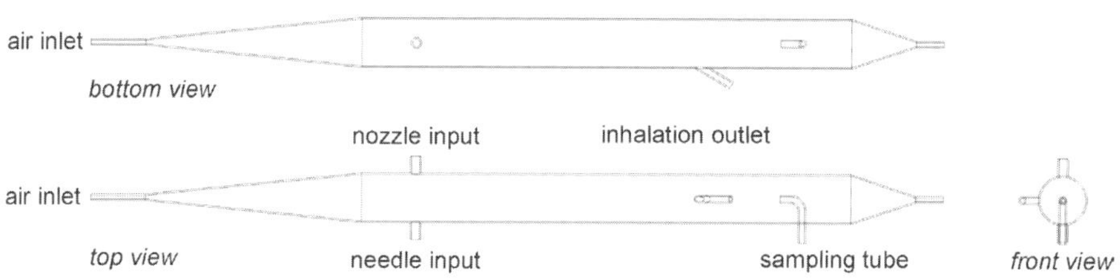

Figure 5. Reactor used for production of inhalation drug particles.

The reactor has the following dimensions: the air flow entrance is a 0.4 m long venturi tube widening to the 0.09 m reactor diameter. This venturi tube prevents the presence of turbulence in the reactor. The evaporation and buffer section have a diameter of 0.09 m and a total length of 1.10 m. The spray and neutralization section are located at the beginning of this tube while the sampling and inhalation section are located at the end. The sampling tube has a diameter of 0.02 m.

3.4 The sampling section

The particle size distribution and particle concentration are monitored using a time-of-flight measurement device. For the experiments at TU Delft an Aerosizer LD (Amherst) was used, in the Utrecht Medical Center (UMC) an Aerodynamic Particle Sizer (APS, TSI) with a 1:100 diluter (for the APS). To sample an aerosol in a laminar flow, three criteria for ideal sampling can be given:

- Isokinetic sampling, sampling with a sampling velocity equal to the velocity in the laminar flow.
- Isoaxial sampling, placing the sample tube in the same direction as the laminar flow duct.
- Sampling with a thin walled sampling probe, to prevent the sampling probe effecting the flow field.

The sampling tube meets the second and third condition. The first one is hard to meet because this implies very high flow rates in the reactor. According to Belyaev and Levin (Belyaev and Levin 1974) the error made by anisokinetic sampling is given in equation 5:

$$\frac{c}{c_0} = 1 + \left(\frac{u_0}{u} - 1\right)\left(1 - \frac{1}{1 + \left(2 + 0.62\frac{u}{u_0}\right)St}\right) \tag{5}$$

In where c is the measured concentration, c_0 is the concentration in the reactor, u_0 is the gas velocity in the reactor, u is the gas velocity in the sampling tube and St is the Stokes number, which is a function of the velocity in the reactor, the sampling tube diameter and the particle relaxation time. The Stokes number indicates the ability of the particles to follow the gas streamlines when an object is present. When this number decreases to zero, the particle will follow the gas streamlines. Equation 6 gives the Stokes number, equation 7 the particle relaxation time in the Stokes regime.

$$St = \frac{\tau u_0}{D_{probe}} \tag{6}$$

$$\tau = \frac{\rho_p d_p^2}{18\eta} \tag{7}$$

in where D_{probe} is the sampling tube diameter and η is the gas viscosity.

The importance of isokinetic sampling is shown in Figure 6. For small Stokes numbers only small (and acceptable) deviations occur, but for higher Stokes number and velocity ratios differing from unity unacceptable errors occur.

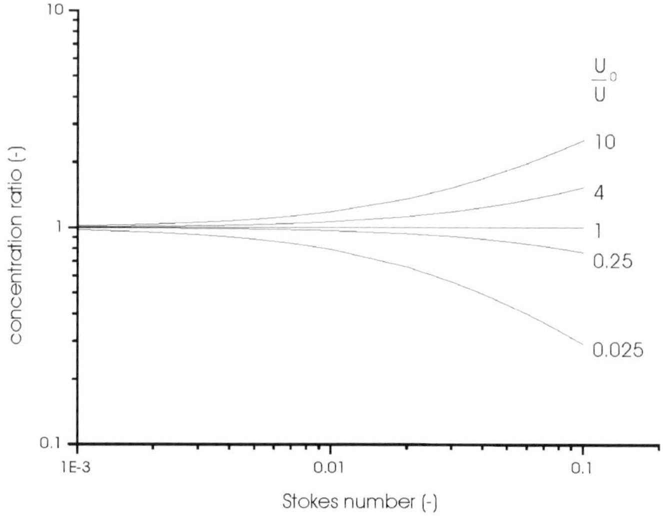

Figure 6. Deviation of ideal sampling as function of the Stokes number and different velocity ratios.

For both the set ups in Delft and the UMC, where solid particles between 1 and 5 µm are produced, the particle relaxation time lies between 4 µs and 100 µs seconds. Table 1 shows the relative sampling velocity and the Stokes numbers for the two particle sizes. It can be easily seen that for both the set ups, the sample efficiency is higher than 95 %, which is acceptable for the measurements.

Table 1. Sample parameters for the two set ups used.

	Aerosizer (Delft set up)	APS (AZU set up)
u_0/u	0.07	0.07
$St_{1\,\mu m}$	$4.5\ 10^{-5}$	$7.4\ 10^{-6}$
$St_{5\,\mu m}$	$1.1\ 10^{-3}$	$1.8\ 10^{-4}$

3.5 Production of monodisperse particles

The solutions used for experiments were based on ethanol as the solvent. Ethanol was chosen because it has excellent properties to be used for electrospray, it is already present in inhalation preparations, and, lastly, the steroid intended to be used in human experiments, beclomethasone diproprionate (BDP), is soluble in ethanol. Because BDP is expensive, the initial experiments were performed with a solution of methylparahydroxybenzoate (MPHB) in ethanol. MPHB was chosen because it is ethanol soluble, and the physicochemical properties closely resemble those of BDP.

3.5.1 Properties of the solutions and settings of the device

Table 2 shows the conductivity of MPHB and BDP solutions used. Because this influence is neglectable (small variations are due different ethanol bottles used), the initial droplet size does not depend on the concentration of the solution. The density of ethanol, 790 kg/m^3, is assumed to be not influenced by the small concentrations used. The surface tension of ethanol, 22.8 mN/m was measured to be constant for all concentrations.

Table 2. Conductivity of the solutions used.

solution	weigth fraction (-)	conductivity µS/cm
pure ethanol	-	0.62
MPHB in ethanol	0.5	0.60-0.70
MPHB in ethanol	2.0	0.52-0.70
MPHB in ethanol	3.0	0.58-1.10
BDP in ethanol	2.0	0.49

To obtain the desired particle range between 1 and 5 µm, the effect of increasing liquid flows and drug concentrations on the final particle size was performed with solutions of 0.5, 2.0 and 3.0 wt% MPHB in ethanol. The applied liquid flow rates for experiments were varied between 1 mL/hr and 4 mL/hr. During the experiments, the potential on the nozzle and the ring, the distance between the ring and the corona needle, and the airflow rate were kept constant (nozzle: 13.75 kV, ring: 10.9 kV, distance: 90 mm, and airflow: 6.5 L/min). The stability of the system was tested by repeating size distribution measurements over 10 consecutive 1 minute intervals. These settings are the result of a lot of initial experiments in which was shown that these settings are optimal for the electrospray system.

4. RESULTS AND DISCUSSION

4.1 Results obtained with the MPHB solutions

Table 3 shows the results obtained for different MPHB solutions at different flow rates. The results from MPHB solutions show a high degree of monodispersity, especially for the small flow rates and concentrations. The increase in the GSD can be due to a number of reasons. An increase in the liquid flow rate can lead to a wider initial droplet size distribution. Higher concentration of MPHB can result in a wider particle size distribution due to a non-constant drying process. As the size distribution still can be considered as being narrow, this is not a problem for the direct inhalation tests.

Table 3. Influence of the MPHB concentration in ethanol on the MMAD and GSD as a function of the applied flow rate.

MBPH conc. *(wt%)*	flow rate					
	1 mL/hr		2 mL/hr		3 mL/hr	
	MMAD *(μm)*	GSD	MMAD *(μm)*	GSD	MMAD *(μm)*	GSD
0.5	1.58±0.05	1.18±0.03	2.07±0.07	1.17±0.02	2.51±0.07	1.23±0.03
2.0	2.22±0.02	1.18±0.03	3.04±0.03	1.22±0.03	3.43±0.17	1.28±0.01
3.0	2.69±0.05	1.37±0.02	3.98±0.05	1.37±0.02	4.55±0.05	1.29±0.03

(Values are the means ± SD of ten consecutive measurements at 1-min. intervals. Each measurement shows a mass median aerodynamic diameter (MMAD) and geometric SD (GSD), and they were considered as independend aerosol particle parameters. The mean with its SD was calculated for both.)

As the device was build for direct inhalation of drug particles, the mass concentration present in the buffer volume is an important value to be known. The concentration is measured with the same instrument as the MMAD (Aerosizer or APS). Table 4 shows the mass concentrations for the measurements shown in Table 3. These concentrations are high enough to be used for direct inhalation as a dose is obtained after several inhalations.

Table 4. Influence of the MPHB concentration in ethanol on the aerosol particle mass(μg/l) as a function of the applied flow rate.

MBPH conc. (wt%)	flow rate		
	1 mL/hr	2 mL/hr	3 mL/hr
0.5	1.42 (0.23) μg/L	2.61 (0.32) μg/L	2.90 (0.65) μg/L
2.0	4.75 (0.48) μg/L	7.27 (0.64) μg/L	5.83 (0.44) μg/L
3.0	5.9 (0.39) μg/L	9.5 (0.39) μg/L	6.04 (0.43) μg/L

(Values are the means and (SD) of ten consecutive measurements at 1-min. intervals.)

4.2 Results obtained with BDP solutions

The results from the initial MPHB experiments were used to select and confirm the electrospray settings to obtain aerosols with the same MMAD and small GSD with BDP. Based on the outcome of these experiments a 0.5% BDP solution at 1 ml/hr, a 2.0% solution at 1.5 ml/hr and a 3.0% solution at 3 ml/hr were used. All the other device settings remain the same. Table 5 shows the results obtained with BDP. The results are in good agreement with the experimental data obtained from the MPHB experiments.

Table 5. Results for experiments with BDP solutions.

BDP conc. (wt%)	flow rate (mL/hr)	MMAD (μm)	GSD (-)	aerosol mass (μg/L)
0.5	1.0	1.58±0.05	1.18±0.03	1.42±0.25
2.0	1.5	2.51±0.03	1.20±0.03	6.30±0.05
3.0	3.0	4.60±0.06	1.25±0.06	6.00±0.40

(Values are the means ± SD of ten consecutive measurements at 1-min. intervals.)

4.3 *Particle morphology*

Besides the particle characteristics in terms of the aerodynamic diameter, which can be considered as the most important parameter in inhalation drug therapy, the particle morphology also can reveal important information. Especially for the transformation from aerodynamic diameter into mass concentration, where the final particle density and form has to be used, we have to know if we really have to do with solid spherical particles. As pointed out by Vercoulen (Vercoulen 1995), the formation of solid particles by evaporation of a solution can yield different final solid particles, depending on the evaporation characteristics and the kind of precursors. Considering Figure 7, it can be concluded that the medicine particles produced are almost solid particles. For our work we have assumed that they are solid.

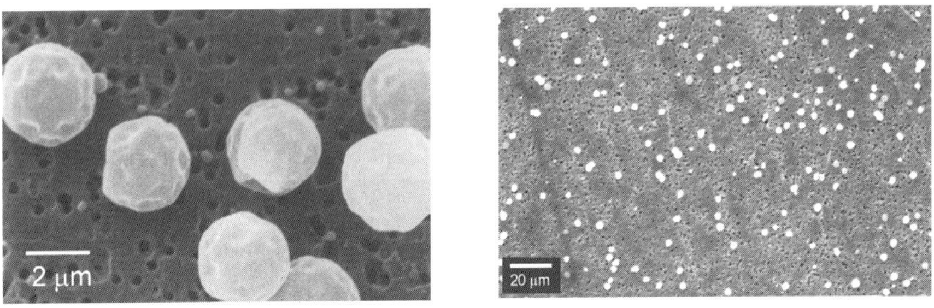

Figure 7. SEM pictures of medicine particles produced with electrospray.

4.4 *Comparison between calculated and measured particle sizes*

In section 2 the control parameters for the final particle size have been discussed. Figure 8 shows the measured values for the MMAD from Table 3 together with the calculated (expected) values for the different parameter settings. It can be said the used model shows good agreement with the measured data, so the model can be used to predict particle sizes that will be obtained depending on the initial droplet size, controlled by the flow rate, and the weight fraction of material dissolved.

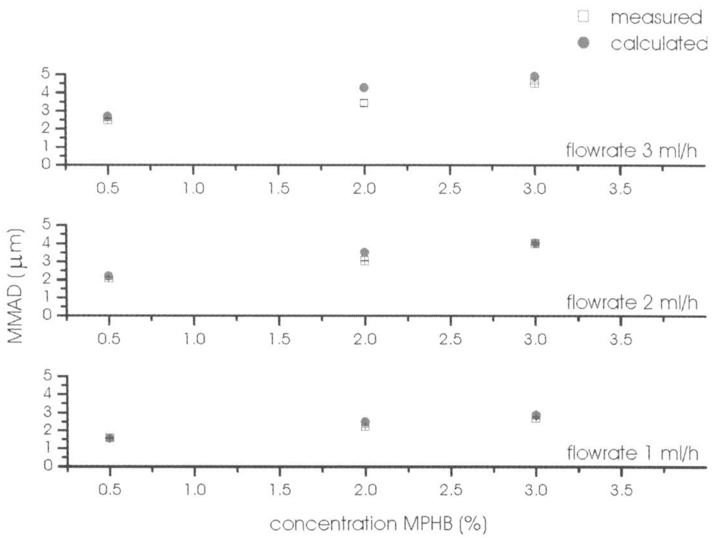

Figure 8. Comparison between calculated and measured values for the used MPHB solutions.

5. CONCLUSIONS

Electrospray in the cone-jet mode was used as a method to produce monodisperse medicine particles between 1 and 5 μm. With both a placebo material (MPBH) and a real medicine (BDP), particles of the required size were obtained, having a small geometric standard deviation. The mass concentration obtained from a single electro-spray nozzle is high enough to be used for in vivo tests.

Using the model of Hartman to predict the initial droplet size the final particle size can be predicted. The calculated aerodynamic diameters show good agreement with the results obtained from the experiments.

6. REFERENCES

Belyaev SP and Levin LM. Techniques for collection of representative aerosol samples. *J. Aerosol Sci* 5: 325-338 (1974)

Cheah PKP and Davies CN. The spinning top aerosol generator: improving the performance. *J Aerosol Sci* 15: 741-751 (1984)

Biddiscombe MF, Melchor R, Mak VHF, Marriot RJ, Taylor AJ, Short MD and Spiro SG. The lung deposition of salbutamol, directly labelled with technetium-99m, delivered by pressurised metered and dry powder inhalers. *Int J Pharm* 91: 111-121 (1993)

Grace JM and Marijnissen JCM. A review of liquid atomization by electric means. *J Aerosol Sci* 25: 1005-1019 (1994)

Cloupeau M and Prunet-Foch B. Electrohydrodynamic spraying functioning modes: a critical review. *J Aersol Sci* 25: 1021-1036 (1994)

Vercoulen PHW. Electrostatic processing of particles. *PhD thesis ISBN 90-9008193-3*; Delft 1995.

Zanen P, Go LT, Lammers J-WJ. The optimal particle size for parasympathicolytic aerosols in mild asthmatics. *Int J Pharm* 114:111-115 (1995)

Zanen P, Go TL and Lammers J-WJ. The optimal particle size for ß$_2$-mimetic and parasympathicolytic aerosols in patients with severe airway constrictions. *Thorax* 51: 977-980 (1996)
Thompson PJ. Drug delivery to the small airways *Am J Respir Crit Care Med* 157: S199-S202 (1998)
Zanen P, Go TL and Lammers J-WJ. The efficacy of a low dosed, monodisperse parasympathicolytic aerosol compared to a standard aerosol from a metered dose inhaler. *Eur J Clin Pharmacol* 54: 27-30 (1998)
Hartman R., Brunner DJ, Camelot DMA, Marijnissen JCM and Scarlett B, Electrohydrodynamic atomization in the cone-jet mode physical modeling of the liquid cone and jet. *J Aerosol Sci* 30: 823-850 (1999)
Hinds W.C. Aerosol Technology. New York, Wiley (1999)
Zanen P and Lammers J-W.J. Reducing adverse effects of inhaled fenoterol through optimization of the aerosol formulation. *J Aerosol Med* 12: 241-247 (1999)
Zierenberg B. Optimising the in vitro performance of Respimat. *J Aerosol Med.*12 S19-24 (1999)
Kunkel G, Magnussen H, Bergmann K, Juergens UR, de Mey C, Freund E,Hinzmann R and Beckers B. Respimat (a new soft mist inhaler) delivering fenoterol plus ipratropium bromide provides equivalent bronchodilation at half the cumulative dose compared with a conventional metered dose inhaler in asthmatic patients. *Respiration* 67:306-14 (2000)

7. AFFILIATIONS

Prof.dr.ir Jan C.M. Marijnissen and Dr.ir. Kees B. Geerse work at the Particle Technology Group of Delft University of Technology, Julianalaan 136, 2628 BL Delft, The Netherlands

RITA VANBEVER

OPTIMIZATION OF DRY POWDER AEROSOLS FOR SYSTEMIC DRUG DELIVERY

1. INTRODUCTION

Since insulin was discovered by Banting and Best in 1921, daily subcutaneous injections have been the only way to administer it to diabetic patients. However, conventional regular insulin does not perfectly mimic the physiological insulin secretion pattern, which results in suboptimal pharmacodynamic actions. Moreover, injections involve inconvenience, poor patient acceptability and adherence, as well as manufacturing constraints related to the sterility requirement of the formulation and the cold chain necessary for storage. Therefore, a non-invasive alternative to injected insulin has continued absorb pharmaceutical scientists in recent decades (Heinemann et al, 2001).

Inhalation aerosols present a particularly attractive alternative to injections for systemic administration of drugs because a number of therapeutic molecules are more efficiently absorbed from the lung than from the oral, nasal or transdermal routes. Efficient systemic absorption results from the unique physiological features of the lungs. i) The alveoli present a large surface for absorption (100 m^2), immediately accessible to drugs; ii) a very thin diffusion path separates the airspace from the bloodstream, i.e., the alveolar epithelium, the vascular endothelium and their respective basal membranes are less than 0.5 μm thick; iii) the high blood flow (5 L/min) of the pulmonary circulation rapidly distributes molecules throughout the body, without first-pass hepatic metabolism; iv) metabolic activity locally in the lung is relatively low (Adjei and Gupta, 1997).

Although inhaled insulin was first attempted in 1925, it is only today that it has reached the successful clinical trial stage (Gansslen, 1925; Heinemann et al, 2001). Insulin is an expensive protein drug with a narrow therapeutic window that requires administration with high efficiency and reproducibility and it is only recently that these performance goals have been attained through the development of high technology inhalers or engineered particles (Heinemann et al, 2001; Owens, 2002).

This chapter will outline some of the obstacles that have been or still have to be overcome in order to achieve an efficient and/or reproducible administration of drugs to the bloodstream using inhalation aerosols. These include the limitations of conventional inhalers, interaction between the drug and formulation excipients, and

L. Gradoń and J. Marijnissen (eds.), Optimization of Aerosol Drug Delivery, 91–103.

biological losses of the therapeutic molecule that can occur locally in the lung. Answers to the drawbacks of conventional inhalers have been provided by a variety of technologies and will be discussed in the particular case of inhalation dry powders. Although insulin is the therapeutic protein with a systemic action for which clinical trials of an inhaled aerosol are at their most advanced stage and for which information is the most abundant, examples in this chapter will not be limited to this molecule.

2. DRY POWDER INHALERS AND PARTICLES

Dry powder inhalers (DPIs) are portable, inexpensive, propellant-free, easy to operate (breath-activated), deliver a broad range of drug doses and show improved stability of the formulation as a result of the dry state (Timsina et al, 1994; Niven, 1997). The advantages of DPIs over nebulizers and/or metered-dose inhalers (MDIs) make them particularly suited to chronic intensive therapy as required by insulin-dependent diabetes mellitus. Discussion in this chapter has been limited to DPIs.

Efficient systemic delivery using inhalation requires aerosol particles to be designed with an aerodynamic diameter of between 1 and 5 μm in order to maximize deposition in the alveolar region (Lippmann, 1977). The aerodynamic diameter of a particle, d_{aer}, is equivalent to the diameter of a unit density (ρ_0) sphere that has the same terminal velocity in still air as the particle :

$$d_{aer} = d\sqrt{\frac{\rho}{\rho_0 \chi}}$$

where d is the geometric diameter of the particle, ρ is the particle density and χ is the particle dynamic shape factor denoting deviation of shape from sphericity (Hinds, 1999). Filtering of large particles ($d_{aer} > 5$ μm) occurs in the upper airways (mouth, trachea and main bronchi) by inertial impaction. 1 to 5 μm d_{aer} particles deposit by gravitational settling in the central and distal tract. Particles with $d_{aer} < 1$ μm are mostly exhaled, or deposit by random Brownian motion in distal regions if d, as d_{aer}, is < 1 μm (Heyder et al, 1986).

Although dry powder aerosol particles have been designed with primary aerodynamic diameters suited to peripheral deposition, the incomplete and variable fluidization and deaggregation generated by standard DPIs lower the efficiency and reproducibility of aerosol delivery to the deep lungs - around 10-20 % of the nominal dose classically deposit in the lungs (Dunbar et al, 1998). Passive conventional DPIs rely solely on the inspiratory flow as a source of energy for powder fluidization and particle deaggregation. Since the inspiratory flow generates insufficient energy to fully overcome the van der Waals forces that exist between 1-5 μm microparticles, particles behave as particle aggregates with aerodynamic sizes > 5 μm and deposit in large upper airways. Additionally, variability in inspiratory flow creates variability in the degree of deaggregation and thereby in pulmonary deposition.

The drawbacks to aggregation have been circumvented using various strategies. An example of how composition of the formulation can be optimized in order to increase delivery efficiency and reproducibility is given below. Active dry powder inhalers and large porous particles that emerged from high tech development are also presented.

2.1. Aggregation and Excipients

Inhalation dry powders were produced in our laboratory by spray-drying using excipients that are FDA-approved for inhalation as lactose, materials that are endogenous to the lungs as albumin and dipalmitoylphosphatidylcholine (DPPC), and/or protein stabilizers as trehalose or mannitol (Bosquillon et al, 2001). This type of powder composition, based on the use of GRAS (Generally Recognized As Safe) excipients, had previously been selected in former studies (Vanbever et al, 1999a-b; Wang et al, 1999; Ben-Jebria et al, 1999). Our dry powders presented primary aerodynamic particle diameters suitable for deep lung deposition (< 3 μm), very low bulk powder tap densities (0.04-0.25 g/cm^3), 3-5 μm geometric particle sizes and excellent aerosolization properties in vitro in the Andersen cascade impactor. Doses emitted from a Spinhaler inhaler device operated at 28.3 L/min ranged from 55 to 90 % and fine particle fractions, defined as the fraction of aerosolized particles (and particle aggregates) with an aerodynamic diameter ≤ 5 μm, ranged between 5 and 50 % (Bosquillon et al, 2001).

As shown in Figure 1, the composition of the particles, excipient type and proportion, significantly influenced aerosol performance and powders prepared with lactose behaved best in the Andersen impactor. The incorporation of albumin improved in vitro deposition as well (Bosquillon et al, 2001).

The superior aerosolization properties of the powder formed of albumin, lactose and DPPC over compositions based on trehalose and mannitol could not be fully explained by differences in single particle physical characteristics (Figure 1). In fact, replacing lactose by trehalose or mannitol did not create any major changes in primary particle geometric size, bulk powder tap density and overall particle morphology, as visualized by electron microscopy (Figs. 1 and 2). However, the experimental mass median aerodynamic diameters (MMADs) of the dry powders were 4.7 (lactose) to 5.3 (trehalose), to 6.9 (mannitol) times larger than the theoretical estimates of primary aerodynamic diameter, d_{aer} (Figure. 1), suggesting that differences in fine particle fractions were rather the result of differences in powder cohesiveness with the nature of the excipient. The Hamaker constants – a material property that defines the strength of the van der Waals interparticulate forces – of sugars and polyols are likely to be similar, as they are made of identical atoms. Differences in hygroscopicity of the materials and thereby alterations of capillary forces and/or differences in surface properties might explain the differences in powder cohesiveness observed (Bosquillon et al, 2001).

Removing albumin from the formulation dramatically decreased fine particle fractions (Figure 1). This effect could be partly explained by an increase in density and therefore in d_{aer} and MMAD. However, it is likely that albumin reduces inter-particle cohesion forces as well, since proteins possess surfactant properties and may limit point-to-point contacts due to highly indented surface geometries (Fig. 2b).

The ratio of the experimental MMAD on the computed d_{aer} is an estimate of the degree of powder aggregation. It was generally smaller for powders formed by albumin than for powders prepared without albumin, in support to this assumption (Figure 1; Bosquillon et al, 2001).

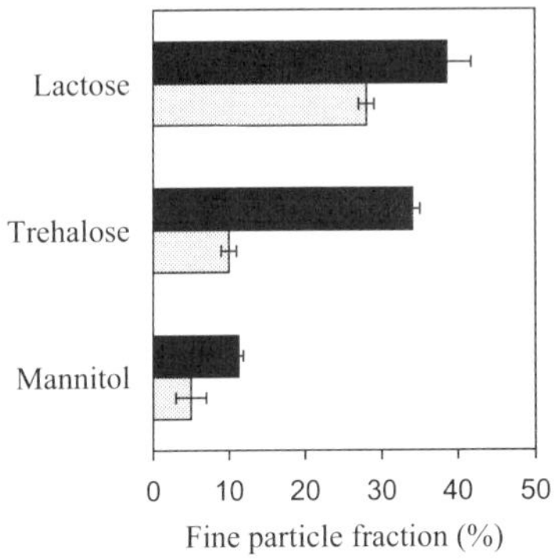

ED (%)	d (µm)	ρ (g/cm3)	daer (µm)	MMAD (µm)
81	4.2	0.08	1.2	5.6
72	4.0	0.12	1.4	7.8
79	4.8	0.05	1.1	5.8
55	4.0	0.10	1.3	8.6
74	4.6	0.11	1.5	10.4
64	4.5	0.18	1.9	10.8

Figure 1 : Influence of composition on aerosolization properties of dry powders. Powders were prepared by spray-drying and formed of 60 % DPPC, 20 % albumin and 20 % lactose, trehalose or mannitol (■); or of 60 % DPPC, 40 % lactose, trehalose or mannitol and no albumin (⁙). Aerodynamic behavior of the aerosols was measured in an Andersen cascade impactor at 28.3 L/min using a Spinhaler inhaler device. ED, emitted dose; d, primary geometric particle diameter; ρ, bulk powder tap density; d_{aer}, theoretical estimate of primary particle aerodynamic diameter; MMAD, experimental mass median aerodynamic diameter of aerosolized powder. From Bosquillon et al, 2001, with permission.

Although the optimal dry powder made of albumin/lactose/DPPC was not fully dispersed by the Spinhaler inhaler device and behaved as particle aggregates in the Andersen cascade impactor, emitted doses and fine particle fractions were high as a result of low bulk powder density which led to small primary aerodynamic diameters and to a large fraction of particle aggregates with an aerodynamic size < 5 µm (Figure 1). In addition, varying the airflow rate from 30 to 100 L/min resulted in limited alteration to the aerosolization characteristics of the dry powder: the emitted dose of a powder made of albumin/lactose/DPPC 30/10/60 w/w/w was worth

83 and 98 %, the fine particle fraction 53 and 59 % and MMAD 4.7 and 4.2 µm using the Spinhaler in the Multi-Stage Liquid Impinger operated at 30 and 100 L/min, respectively (Codrons et al, 2003). This indicates that increasing the airflow rate from

30 to 100 L/min created insufficient differences in mechanical fluidization and shear force to profoundly impact on the degree of dispersion of the particle aggregates.

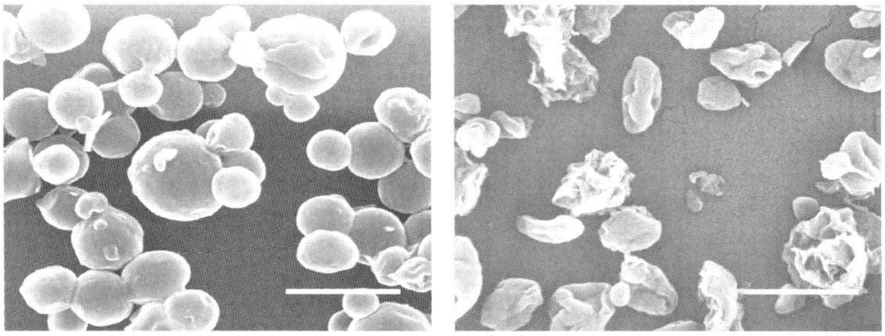

Figure 2. Scanning electron microscope images of particles prepared without (a) or with (b) albumin. The powders were formed of lactose/DPPC (40/60, w/w; a), or albumin/trehalose/DPPC (20/20/60, w/w/w; b). Scale bars are 5 μm. From Bosquillon et al, 2001, with permission.

These data exemplify how a proper choice of formulation composition together with a very low bulk powder density could alleviate the drawbacks of aggregation and make it possible to achieve high and reproducible pulmonary deposition of dry powders in vitro, even though aggregates were present. Other ways to facilitate dispersion and produce respirable particles by modifying the composition and/or physical state of the formulation have been investigated. Physical blending of the drug with a coarse inert carrier such as lactose is a standard technique employed in inhalation products that improve flow and dispersion (Dunbar et al, 1998). In an ideal drug-carrier system, the adhesion of the drug to the carrier is strong enough to prevent demixing during filling, handling and storage, but not so strong as to prevent detachment during inhalation. Another approach involves the agglomeration of micronized particles into soft pellets using spheronization (Wetterlin, 1988). These pellets have improved flow properties and disintegrate into primary respirable particles during inhalation. Co-spray-drying rhDNase with NaCl has been shown to significantly increase fine particle fractions of the dry powder aerosol due to increased crystallinity of the powder (Chan et al, 1997). Supercritical fluid technologies offer a single-step process of engineering respirable drug particles with improved control over particle size, shape and morphology (Shekunov and York, 2000). Supercritical powders have reduced particle surface energy as compared to micronized powders and the decreased cohesion has been shown to benefit aerosolization (Tong et al, 2001; Feeley et al, 2000).

2.2. Active Dry Powder Inhalers

Another strategy that has been developed to overcome the cohesiveness of 1-5 μm particles consists in the use of an additional source of energy to fluidize and disperse the dry powder to that provided by the patient inspiratory flow. The additional energy input increases dispersion of the dry powder in primary respirable particles as well as decreases the dependence of the aerosolisation process on the patient 's breathing pattern and thereby particle size and dose variabilities (Dunbar et al, 1998).

The active dry powder inhaler (Inhance™) of Inhale Therapeutic Systems (recently re-named Nektar Therapeutics) has been the first system used in clinical trials of inhaled insulin, early in the 1990s (Patton and Platz, 1993). It is a gas-assisted dry powder inhaler where fluidisation occurs by passing compressed air through the powder at sonic velocity. Spray-dried insulin is packaged in individual blisters containing 1 or 3 mg of recombinant human insulin in 5 mg of powder made of mannitol, glycine and sodium citrate (Skyler et al, 2001). A blister is inserted and punctured within the inhaler and the powder is withdrawn from the blister by gas-assist fluidization and further deagglomerated by turbulence and shear force into a discrete aerosol cloud in a 210 mL holding chamber. The patient inhales the aerosol bolus through a mouthpiece. The MMAD of the aerosolized particles is approximately 3.5 μm, making it suitable for alveolar deposition (Cefalu et al, 2001).

Clinical trials of inhaled insulin, using the above hand-held dry powder inhaler or other inhaler devices, suggest that pulmonary insulin is an effective and well-tolerated alternative to subcutaneous regular insulin (Cefalu et al, 2001; Skyler et al, 2001; Brunner et al, 2001; Cappelliri et al, 2002). Inhaled insulin results in a more physiological insulin profile than conventional subcutaneous insulin, that is, an earlier onset of action that provides a more adequate control of peak postprandial glucose levels. The relative biopotency of pulmonary insulin is approximately 10 % and its reproducibility is as good as that of injected insulin. Glycosylated hemoglobin (HbA_{1c}) and the occurrence or severity of hypoglycemia episodes are similar in patients treated by inhalation or subcutaneous injection. In a phase III 6-month multicenter study of the Inhance™ dry powder insulin aerosol, inhalation-treated patients developed increased serum antibodies to insulin without attributable clinical manifestations (Skyler, 2002). Pulmonary function (forced expiratory volume in one second) was stable, except for the carbon monoxide diffusing capacity (DL_{co}) which slightly decreased. Patients using the inhaler expressed greater overall satisfaction than those under injection. The clinical significance and the possible mechanisms of the altered DL_{co} are currently being investigated (Skyler, 2002).

2.3. Large Porous Particles

AIR™ technology, or the engineering of powder particles with large geometric size (> 5 μm), low density (< 0.4 g/cm^3) and aerodynamic diameter between 1 and 5 μm, represents a major innovation among pulmonary delivery systems (Edwards et al, 1997). The large geometric size of large porous particles (LPPs) leads to ease of particle dispersion and their small aerodynamic size, resulting from their low particle density, allows LPPs to effectively escape impaction in upper airways and penetrate deep into the lungs. LPPs are produced by co-spray-drying drug and GRAS excipients (sugars, amino acids, lipids) and contain drug loads between 1 and 90 % (Vanbever et al, 1999a). The particles can be administered using a simple breath-activated dry powder inhaler without the need for an additional power source, giving advantages in terms of inhaler size, ease of use and costs (Edwards, 2002). The ease of dispersibility of LPPs also permits a wider possible dosage range (up to 50 mg of powder delivered) as compared to conventional DPIs or MDIs that typically deliver doses up to a few milligrams only. Moreover, apart from excellent flow properties, LPPs may more easily escape phagocytosis by alveolar macrophages due to large geometric sizes and may therefore render the pulmonary administration of sustained-release formulations attainable because of diminished clearance of drug particles (Edwards et al, 1997; Ben-Jebria et al, 1999; Vanbever et al, 1999b

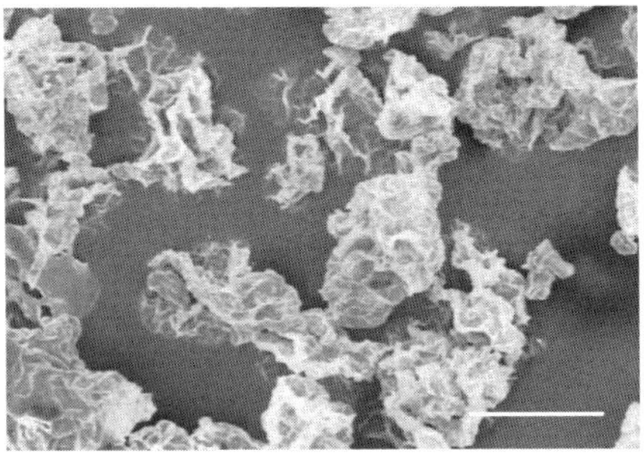

Figure 3. Scanning electron microscope image of large porous particles, made of insulin/albumin/DPPC 5/35/60 w/w/w. Scale bar is 10 μm. From Vanbever et al, 1999b, with permission.

A scanning electron microscopy image of LPPs made of a high protein content is shown in Figure 3 as an example. The particles present a sponge-like shape and a highly indented surface. This morphology is not very different from that presented in Fig. 2b for small porous particles and resulted from the high protein load common to both particle types.

Lung deposition of LPPs was measured both in vitro and in vivo in humans (Dunbar et al, 2002). LPPs with a 3 μm MMAD exhibited approximately 90 % emitted doses and lung deposition of 60 % (expressed as the percentage of nominal dose). These values are strikingly elevated. They were previously attained using active inhalers but here could be achieved using a passive dry powder inhaler of a pen size due to the free-flowing character of LPPs. Aerodynamic behavior in vitro and in vivo did not show any variability with airflow rate, indicating that efficient powder fluidization and deaggregation in primary respirable particles occurred already at low inspiratory flows. Good aerosolization properties of LPPs primarily depend on single particle physical characteristics and not on the nature of the excipients, as is the case for small porous particles (see above).

The time-action profile of LPPs of insulin was investigated in comparison with subcutaneous regular insulin in a phase I clinical trial by means of the euglycaemic clamp method (Edwards et al, 2001). The biopotency of inhaled insulin was worth 16 % relative to subcutaneous injection and inter-subject variability was similar between inhalation and injection.

3. INTERACTION BETWEEN THE DRUG AND EXCIPIENT

We recently demonstrated that a physical interaction between a therapeutic molecule and an excipient could be a significant obstacle to an efficient drug absorption to the systemic circulation following inhalation (Codrons et al, 2003). The 1-34 fragment of human parathyroid hormone (PTH) significantly bound to albumin used to formulate a dry powder of PTH (a binding of 78 % was measured by equilibrium dialysis) and the presence of albumin was shown to decrease the systemic absorption of inhaled PTH in vivo in rats. The absolute PTH bioavailability was 21 % and 34 % following intratracheal administration of a powder formed of PTH/albumin/lactose/DPPC (1/30/10/60 w/w/w/w) and PTH/lactose/DPPC (10/30/60 w/w/w), respectively. Similar effects of albumin were observed following subcutaneous PTH injection with the absolute bioavailability decreasing two-fold when incorporating albumin in the solution. Our hypothesis is that binding to albumin increased the duration of PTH presence in the alveolar and subcutaneous space and thereby the exposure time of the peptide to local enzymes and/or other degradation processes (Codrons et al, 2003).

Drug binding to excipients is likely to influence drug transport to the systemic circulation in general and the presence of endogenous albumin or other compounds in the alveolar region may decrease systemic absorption of PTH or other drugs aerosolized through similar mechanisms. PTH solutions containing albumin or plasma have previously been used for subcutaneous injections in pre-clinical or clinical studies without consideration of binding and the ensuing decrease in systemic bioavailability (Kent et al, 1985; Fox et al, 1997). Albumin has been employed to formulate inhalation dry powders as well (Zeng et al, 1994; Vanbever et al, 1999a-b; Wang et al, 1999; Bosquillon et al, 2001). Our data therefore indicate that additives might need to be screened for possible binding before in vivo use.

The pharmacokinetic profile that follows the pulmonary administration of the PTH powder prepared without albumin was superimposed on that following the intratracheal instillation of a PTH solution. This suggests that the initial physical state of PTH had no influence on its systemic absorption and that the other excipients, lactose and DPPC, did not impede absorption as did albumin, provided that the powder and liquid formulations led to a similar deposition within the respiratory tree (Codrons et al, 2003).

4. BIOLOGICAL LOSSES IN THE LUNG

Although the efficiency and reproducibility of the administration of aerosols to the deep lung have been optimized, systemic bioavailabilities of inhaled therapeutic peptides and proteins remain low due to significant biological losses occurring in the lung (Wall, 1995; Adjei and Gupta, 1997). Drug losses due to passage through the lung can be quantified based on pharmacokinetic studies and pulmonary deposition data collected in vivo. For instance, since inhaled insulin reaches 10 % bioavailability in clinical trials, and given an average pulmonary deposition of 60 % of the nominal dose, it follows that only 1 to 2 molecules of insulin out of 10 deposited in the human lung reach the bloodstream (Heinemann et al, 2001). The local losses of therapeutics that occur in the lungs and decrease systemic absorption have not been ascribed to any precise mechanism up to now (Wall, 1995; Folkesson et al, 1996).

One of the principal objective of the research efforts of our laboratory concentrates specifically on understanding the causes of biological losses of therapeutics from the lungs and on developing formulation strategies to overcome these losses. Potential origins to drug losses include enzymatic degradation, mucociliary clearance, aggregation with surfactant, binding to lung components and/or capture by particular pulmonary cells, for example, macrophages. Although the involvement of some of these losses have been separately demonstrated (Morita et al, 1994; Shen et al, 1999; Pohl et al, 1999), no studies have globally investigated the fate of inhaled molecules or identified the most significant causes of losses.

As a first step towards the identification of the biological obstacles to an efficient systemic absorption, we studied the pulmonary disposition of inhaled fluorescent molecules using confocal imaging of rat lungs (Lombry et al, 2002). The technique allowed us to obtain a global view of pulmonary fate and to localize hot spots, for example, capture by particular cells. Green fluorescent dyes of various solubility and size were insufflated as a dry powder or instilled as a solution in the lungs of rat in vivo. At pre-determined times following delivery, the lungs were colored with sulforhodamine, a red fluorescent dye, fixed, removed and grossly sliced for visualization by confocal laser scanning microscopy. In comparison with other microscopy techniques, our method of tissue preparation was rapid (less than 30 min), did not involve micro-dissection or tissue incubation in various solutions, embedding, freezing and/or dehydration, which might lead to profound disruption of the native milieu of the lungs and significant loss of the probe (Pinkerton et al, 1993; Hastings et al, 1995).

Significant differences in the localization behavior and clearance from the lungs between the diverse fluorescent probes were evident (Lombry et al, 2002). The small molecules, coumarin and calcein, diffused within minutes across the trachea, airways and alveolar tissue, while the model protein, FITC-albumin, remained for hours in the airspaces and displayed only a low penetration in the tissue. Differences appeared also in the interaction of molecules with pulmonary cells. Coumarin (log P = 5.43) accumulated in alveolar type II cells, probably partitioning into lamellar bodies due to its lipophilicity. In contrast, FITC-albumin was preferentially captured by alveolar macrophages, probably as a result of the long residence time of the macromolecule in the alveolar fluid and the continuing pinocytic activity of alveolar macrophages. The small hydrophilic molecule, calcein, did not exhibit any specific affinity for particular cells.

No major differences in the pulmonary fate of the three probes resulted from their physical state, dry powder or solution, and/or from the excipients of the powder, except in the rate of disposition of FITC-albumin (Lombry et al, 2002). Dry powders were made of albumin, lactose and DPPC and contained one of the above green markers. The distribution of coumarin and calcein in the pulmonary tissue was as fast when delivered as a dry powder or as a solution and the same affinity of coumarin for type II cells was evident in the confocal images. However, confocal laser scanning microscopy indicated that delivering the bulky protein, FITC-albumin, as a powder prolonged the step of diffusion within the airspace fluid and delayed the uptake by alveolar macrophages (Lombry et al, 2002).

A few previous studies, in line with the present data, are relevant to note. Pohl et al have explored the disposition of 6-carboxyfluorescein, rhodamine B, FITC-insulin and 10 kDa FITC-dextran in peripheral alveoli of intact inflated rat lungs using confocal imaging (Pohl et al, 1998, 1999). Carboxy-fluorescein and rhodamine B disappeared from the alveolar space within minutes. The green fluorescent dye then diffused to the tissue interstitium while the red lipophilic marker principally located in lamellar bodies of type II cells, similarly as coumarin. FITC-insulin and FITC-dextran were retained for more than one hour in the alveolar space but no uptake by alveolar macrophages was apparent in the confocal images. Hastings et al reported that human albumin instilled in vivo in rabbit lungs was endocytosed by both type I and type II pneumocytes, but that alveolar macrophages took up protein earlier and more avidly than the epithelial cells (Hastings et al, 1995). These latter results were obtained using light and electron microscopy and support the data we collected by confocal imaging.

5. CONCLUSION

In this chapter, obstacles to an efficient systemic absorption of peptides and proteins using inhalation aerosols have been underlined. Aggregation of particles of respirable size lowers the efficiency and reproducibility of aerosol delivery to the deep lungs and various strategies have been invented to overcome these drawbacks. It has been shown that elevated fine particle fractions and limited variability in aerodynamic behavior with airflow rate could be attained by properly choosing the excipients of the dry powder aerosols and by designing powder particles with low mass density. An external source of energy can be added to that provided by the patient inspiratory flow in order to fluidize and disperse dry powders and this approach has been exploited by developing hand-held active dry powder inhalers. Engineering powder particles with large size, low density and 1-5 μm aerodynamic diameter represents another strategy that has been demonstrated to highly increase the efficiency and reproducibility of aerosol delivery to the deep lungs. In this case, excellent powder flow properties result from the large geometric size of the particles and allow the use of a simple passive inhaler of pen size for administration.

The binding of the drug molecule to an excipient can be a significant obstacle to an efficient absorption to the systemic circulation as can be biological losses of therapeutics that occur in the lungs following inhalation. Up to now, these aspects of the optimization of pulmonary drug delivery have been overlooked. However, understanding them might lead to more improved pulmonary therapies, in the same way as understanding aggregation and the mechanisms of pulmonary deposition has lead to the development of efficient dry powder inhalers.

6. ACKNOWLEDGEMENTS

The valuable contributions of my Ph.D. students Cynthia Bosquillon, Catherine Lombry and Valérie Codrons are gratefully acknowledged. The Fonds National de la Recherche Scientifique (FNRS, Belgium), the Special Funds for Research of the Université catholique de Louvain and the Belgian Walloon Region are thanked for their financial support. Rita Vanbever is Chercheur Qualifié of the FNRS.

7. REFERENCES

Adjei AL, Gupta PK, editors. *Inhalation Delivery of Therapeutic Peptides and Proteins*. Marcel Dekker New York-Basel-Hong Kong, 1997.

Ben-Jebria A., Chen D., Eskew M.-L., Vanbever R., Langer R. and Edwards D.A. Large porous particles for sustained protection from carbachol-induced bronchoconstriction in guinea pigs. *Pharm. Res.* 16 (1999): 555-561.

Bosquillon C, Lombry C, Préat V and Vanbever R. Influence of formulation excipients and physical characteristics of inhalation dry powders on their aerosolization performance. *J Control. Release* 70 (2001): 329-339.

Brunner G.A., Balent B., Ellmerer M., Schaupp L., Siebenhofer A., Jendle J.H., Okikawa J., and Pieber T.R. Dose-response relation of liquid aerosol inhaled insulin in Type I diabetic patients, *Diabetologia* 44 (2001): 305-308.

Cappelleri J., Cefalu W.T., Rosenstock J., Kourides I.A., and Gerber R.A.. Treatment satisfaction in type 2 diabetes : Comparison between an inhaled insulin regimen and a subcutaneous insulin regimen. *Clinical Therapeutics* 24, 4 (2002): 552-564.

Cefalu W.T., Skyler J.S., Kourides I.A., Landschulz W.H., Balagtas C.C., Cheng S.L. and Gelfand R.A. Inhaled human insulin treatment in patients with type 2 diabetes mellitus. *Annals of Internal Medicine* 134 (2001): 203-207.

Chan H.K., Clark A. Gonda I. Mumenthaler M. and Hsu C. Spray dried powders and powder blends of recombinant human deoxyribonuclease (rhDNase) for aerosol delivery. *Pharm. Res.* 14, 4 (1997): 431-437.

Codrons V, Vanderbist F., Verbeeck R.K., Arras M., Lison D., Préat V. and Vanbever R. Systemic delivery of parathyroid hormone (1-34) using inhalation dry powders in rats. *J. Pharm. Sc.* 92 (2003): 938-950.

Dunbar C., Hickey A., and Holzner P. Dispersion and characterization of pharmaceutical dry powder aeorosols. *Kona* 16, 7 (1998): 7-45.

Dunbar C., Scheuch G., Sommerer K., DeLong M., Verma A., Batycky R. In vitro and in vivo dose delivery characteristics of large porous particles for inhalation. *Int. J. Pharm.* 245 (2002): 179-189.

Edwards D.A., Hanes J., Caponetti G., Hrkach J., Ben-Jebria A., Eskew M.L., Mintzes J.D., Deaver D., Lotan N., Langer R. Large porous particles for pulmonary drug delivery. *Science* 276 (1997): 1868-1871.

Edwards D.A., Verna A., Heise T., and Batycky R.P. Efficacy, simplicity and control of systemic protein delivery using large porous particle technology. *J. Aerosol Med.* 14 (2001): 384.
Edwards D.A. Delivery of biological agents by aerosols. *AIChE J.* 48 (2002): 2-6.
Feeley J.C., Gilbert D.J., Palakodaty S., Walker S.E. and York P. Engineering of particle size distributions for respiratory drug delivery by supercritical fluid processing. *Respiratory Drug Delivery* VII (2000): 357-360.
Folkesson H.G., Matthay M.A., Weström B.R., Kim K.J., Karlsson B.W. and Hastings R.H. Alveolar epithelial clearance of protein. *J. Appl. Physiol.* 80 (1996): 1431-1445.
Fox J, Miller MA, Stroup GB, Nemeth EF and Miller SC. Plasma levels of parathyroid hormone that induce anabolic effects in bone of ovariectomized rats can be achieved by stimulation of endogenous hormone secretion. *Bone* 21, 2 (1997): 163-169.
Gansslen M.U. Inhalation von insulin. *Klein Wochenschr* 4 (1925): 71.
Hinds WC. Uniform particle motion. In Hinds WC, editor. *Aerosol Technology – Properties, Behavior, and Measurement of Airborne Particles.* 2nd ed., John Wiley & Sons, New York, p 42-74, 1999.
Hastings R.H, Folkesson H.G., Petersen V., Ciriales R. and Matthay M.A. Cellular uptake of albumin from lungs of anesthetized rabbits. *Am. J. Physiol.* 269 (1995): L453-462.
Heinemann L., Pfutzner A., and Heise T. Alternative routes of administration as an approach to improve insulin therapy : Update on dermal, oral, nasal and pulmonary insulin delivery. *Current Pharm. Des.* 7 (2001): 1327-1351.
Heyder J., Gebhart J., Rudolf G., Schiller C. and Stahlhofen. Deposition of particles in the human respiratory tract in the size range 0.005-15 μm. *J. Aerosol Sci* 17 (1986): 811-825.
Kent GN, Loveridge N, Reeve J and Zanelli JM. Pharmacokinetics of synthetic human parathyroid hormone 1-34 in man measured by cytochemical bioassay and radioimmunoassay. *Clin Sci* 68 (1985): 171-177.
Lippmann M., Lee D.H.K., Falk H.L., Murphy S.O. and Geiger S.R. Regional deposition of particles in the human respiratory tract. *Handbook of Physiology, Reaction to Environmental Agents.* American Physiological Society, Bethesda, 1977.
Lombry C., Bosquillon C., Préat V. and Vanbever R. Confocal imaging of rat lungs following intratracheal delivery of dry powders or solutions of fluorescent probes. *J Control Release* 83 (2002): 331-341.
Morita T., Yamamoto A., Takakura Y., Hashida M. and Sezaki H. Improvement of the pulmonary absorption of ($Asu^{1,7}$)-Eel calcitonin by various protease inhibitors in rats. *Pharm. Res.* 11 (1994): 909-913.
Niven R, editor. Dry powder formulations for inhalation. *Adv. Drug Deliv. Rev.* (special issue) 26 (1997): 1-67.
Owens D.R. New horizons - Alternative routes for insulin therapy. *Nat. Rev. Drug Discov.* 1 (2002): 529-540.
Patton J.S. and Platz R.M. Method and device for delivering aerosolized medicaments. International patent WO 93/00951, 1993.
Pinkerton K.E., Gallen J.T., Mercer R.R., Wong V.C., Plopper C. G. and Tarkington B. K. Aerosolized fluorescent microspheres detected in the lung using confocal scanning laser microscopy. *Micr. Res. And Tech.* 26 (1993): 437-443.
Pohl R., Kramer P. A., Thrall R. S. Confocal laser scanning fluorescence microscopy of intact unfixed rat lungs. *Int. J. Pharm.* 168 (1998): 69-77.
Pohl R., Thrall R.S., Rogers R.A. and Kramer P.A. Confocal imaging of peripheral regions of intact lungs following intratracheal administration of 6-carboxyfluorescein, FITC-insulin and FITC-dextran. *Pharm. Res.* 16 (1999): 327-332.
Shekunov B.Y. and York P. Crystallisation processes in pharmaceutical technology and drug delivery design. *J. Crystal Growth* 211 (2000): 122-136.
Shen Z., Zhang Q., Wei S. and Nagai T. Proteolytic enzymes as a limitation for pulmonary absorption of insulin : in vitro and in vivo investigations. *Int. J. Pharm.* 192 (1999): 115-121.
Skyler J.S., Cefalu W.T., Kourides I.A., Landschulz W.H., Balagtas C.C., Cheng S.L. and Gelfand R.A. Efficacy of inhaled human insulin in type 1 diabetes mellitus a radomised proof-of-concept study. *Lancet* 357 (2001): 331-335.

Skyler J.S. Efficacy and safety of inhaled insulin (Exubera ®) compared to subcutaneous insulin therapy in an intensive insulin regimen in patients with type 1 diabetes : results of a 6-month, randomized, comparative trial. *Diabetes* 51 (2002): 540.

Timsina M.P., Martin G.P., Marriott C., Ganderton D. and Yianneskis M. Drug delivery to the respiratory tract using dry powders inhalers. *Int. J. Pharm.* 101 (1994): 1-13.

Tong H.H.Y., Shekunov B.Y., York P. and Chow A.H.L. Characterization of two polymorphs of salmeterol xinafoate crystallized from supercritical fluids. *Pharm. Res.* 18 (2001): 852-858.

Vanbever R., Mintzes J.D., Wang J., Nice J., Chen D., Batycky R., Langer R. and Edwards D.A.. Formulation and physical characterisation of large porous particles for inhalation. *Pharm Res* 16, 11 (1999a): 1735-1742.

Vanbever R., Ben-Jebria A., Mintzes J. D., Langer R. and Edwards D. A. Sustained-release of insulin from insoluble inhaled particles. *Drug Dev. Res.* 48 (1999b) 178-185.

Wall D.A. Pulmonary absorption of peptides and proteins. *Drug Deliv.* 2 (1995): 1-20.

Wang J. Ben-Jebria A. and Edwards D.A. Inhalation of estradiol for sustained systemic delivery. *J. Aerosol Med.*, 12 (1999): 27-36.

Wetterlin K. Turbuhaler : A new powder inhaler for administration of drugs to the airways. *Pharm. Res.* 5, 8 (1988): 506-508.

Zeng XM, Martin GP, Marriot C. Albumin microspheres as a means of drug delivery to the lung : analysis of the effects of process variables on particle sizes using factorial design methodology. *Int J Pharm* 107 (1994): 205-210.

8. AFFILIATIONS

Dr. Rita Vanbever works at the Chercheur Qualifii FNRS, Chargi de cours UCL, Universiti catholique de Louvain, School of Pharmacy, Uniti de Pharmacie Galinique, Industrielle et Officinal, Avenue Emmanuel Mounier, 73 UCL 73.20 1200 Brussels - Belgium

E-mail : vanbever@farg.ucl.ac.be

JOHN N. PRITCHARD

RECENT ADVANCES IN DRUG DELIVERY VIA PRESSURED METERED-DOSE INHALERS

1. INTRODUCTION

Riker Laboratories launched the first commercial pressurised metered-dose inhalers (pMDIs), containing isoprotenerol and epinephrine, in 1956. To this day, it remains the most commonly prescribed device for inhaled asthma medications, with world-wide production exceeding 500 million devices (Tansey, 1997a). The size of these canisters will vary, but assuming that they contain on average 120 doses, this equates to 2000 doses of medication taken every second !

Despite, or perhaps because of, this success, technical advances remained modest over the next 30 years. Even something as fundamental as design of the expansion chamber and spray orifice was not studied theoretically until 1992 (Clark, 1992). However, in the 1970s, Molina and Rowland (1974) raised concerns over the possible detrimental effect of chlorofluorocarbons (CFCs) on levels of stratospheric ozone. Confirmation of ozone depletion was first reported over the Antarctic in 1985 (Farman et al., 1985) and two years later, a great many (now over 150) nations signed the Montreal Protocol on Substances That Deplete the Ozone Layer. Specific exemptions were granted for defined "essential uses" where there were no technologically or economically viable alternatives; at that time this included pMDIs. However, these exemptions are reviewed annually by the Technical and Economic Assessment Panel of the United Nations Environment Programme, with a background that all economically feasible steps have been taken to minimise the "essential use". Faced with this requirement, plus a diminishing supply of CFCs, the pharmaceutical industry has committed large resources to the development of CFC-free pMDIs, plus alternative delivery systems. This impact is reflected in the numbers of related patents granted by the US Patent Office during the 1990s (Figure 1).

L. Gradoń and J. Marijnissen (eds.), Optimization of Aerosol Drug Delivery, 105–121.

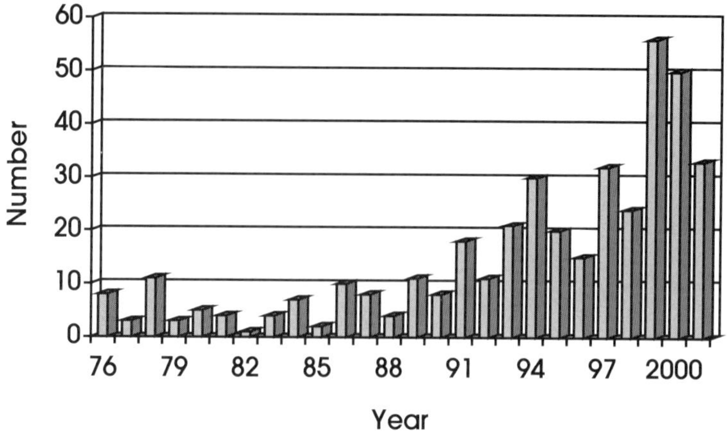

Figure 1 Inhaler-related patents granted each year in the USA

At the outset, the challenge was seen mainly in identifying safe alternative propellants that did not contain chlorine atoms, often referred to as hydrofluoroalkanes (HFAs). A specialist conference hosted by The Mangement Forum (1992) concluded that the easiest products to reformulate would be launched in the USA by 1997, with the most difficult following around 2 years later. Even in 1998, the Commission to the Council and the European Parliament (1998) envisaged that "essential use" status would be lost for most common inhaled drugs by 2002. The reality has proven much harder, with many technical challenges to overcome. Witness the fact that the New Drug Application (NDA) for the original isoprotenerol pMDI contained only 34 pages covering the chemistry, manufacturing and controls, whereas that for 3M's CFC-free salbutamol pMDI was a staggering 1130 pages !

This paper reviews some of the advances that have been made, with examples often drawn from GlaxoSmithKline's (GSK's) development programme. It also briefly considers what the future may hold.

2. CFC REPLACEMENT

Any replacement propellant for the ozone-depleting CFCs needs to be a liquefied gas, so that as the canister empties through use, a small amount of the liquid boils off to maintain the pressure above the liquid. Otherwise, if vapour pressure drops, the size of the spray will increase, leading to a change in product performance (Polli et al., 1969). As may be seen from the key propellant properties, summarised from the review of McDonald and Martin (2000) in Table 1, the CFCs allowed great flexibility in terms of selecting the appropriate vapour pressure. There was also plenty of scope to match the propellant density to that of the drug, thereby minimising any impact of suspended drug particles either sedimenting or floating to the surface.

Replacements have not been easy to find; hydrocarbons, such as propane and butane, have not generally been considered on the grounds of density (< 0.6 g/ml), flammability and taste. Two non-chlorinated HFA propellants (134a and 227ea) were selected that most closely matched the properties of the CFCs. These were subjected to rigorous safety testing by a pharmaceutical industry consortium, before being endorsed in Europe by the Committee for Proprietary Medicinal Products (1994, 1995). Nonetheless, it can be seen from Table 1 that there is much less flexibility in creating a new product. Some of the key considerations are summarised below.

2.1 The formulation

Perhaps the most obvious challenge is the lack of a propellant that is liquid at room temperature. This means that to make a formulation, one must either work with all the constituents at a low enough temperature to keep the propellant liquid, or to add the constituents to a vessel that can have liquid propellant introduced under pressure. This both adds complexity to laboratory experiments, as well as to factory manufacture. Thus, entirely novel processes have had to be introduced (e.g. Burt and Kwok, 1993). However, it does mean the devices will work at lower temperatures than had been possible with CFC-based products.

Most currently marketed CFC products contain drug particles suspended in the fluid. In order to generate a stable homogeneous dispersion, surfactants are normally added. However, none of the surfactants that have been used in CFCs are directly soluble in HFAs (Tansey, 1997a). This has prompted the search for a new class of surfactants suitable for use in HFAs (eg Stefely et al., 2000). None have yet appeared in a marketed product. Another approach has been to add a co-solvent, usually ethanol, for the surfactant (Tansey, 1997b). This may also offer the opportunity to pre-mix part of the formulation before adding propellant. However, the co-solvent will lower the vapour pressure of the product, so may impact efficacy unless other measures are taken.

Furthermore, some drugs are soluble in ethanol, which at low concentrations will lead to instability of a suspension. Therefore, for these products, it may be feasible to add sufficient ethanol to make a solution of the drug. This has been successfully achieved by 3M Pharmaceuticals with the launch of a solution formulation of beclomethasone dipropionate (BDP).

Code Name	Chemical name	Formula	Ozone depletion potential	Atmospheric life (years)	Global warming potential (CFC 11 = 1)	Boiling point (°C)	Vapour pressure (psig @ 20 °C)	Liquid density (g/ml)
CFC 11	Trichlorofluoromethane	CCl_3F	1	60	1	23.7	- 1.8	1.49
CFC 12	Dichlorodifluoromethane	CCl_2F_2	1	125	3.0	-29.8	67.6	1.33
CFC 114	Dichlorotatrafluoroethane	$C_2Cl_2F_4$	0.7	200	3.9	3.6	11.9	1.47
HFA 134a	Tetrafluroethane	$C_2H_2F_4$	0	16	0.3	-26.5	68.4	1.21
HFA 227	Heptafluoropropane	C_3HF_7	0	33	0.7	-17.3	56.0	1.41

Table 1 Key physico-chemical properties of pMDI propellants (McDonald and Martin, 2000)

However, as may be seen from Figure 2, the size of drug particles formed when the propellant and solvent evaporate is very different to that of the micronised drug (Harrison et al., 1997). This in turn leads to changes in lung delivery (Leach et al., 1998), efficacy (Busse et al., 1999) and systemic exposure (Harrison et al., 1999). Thus, although the same drug has been used, the product characteristics will be very different. To overcome this problem, Brambilla et al. (1999) have shown that addition of a non-volatile ternary component to the formulation (e.g. glycerol) means that the particle size can be tailored to match that of the original suspension. Whilst glycerol is hygroscopic, and so could influence lung deposition patterns, the safety and efficacy of the formulation has been endorsed by Regulatory Authorities in France and Italy, amongst others.

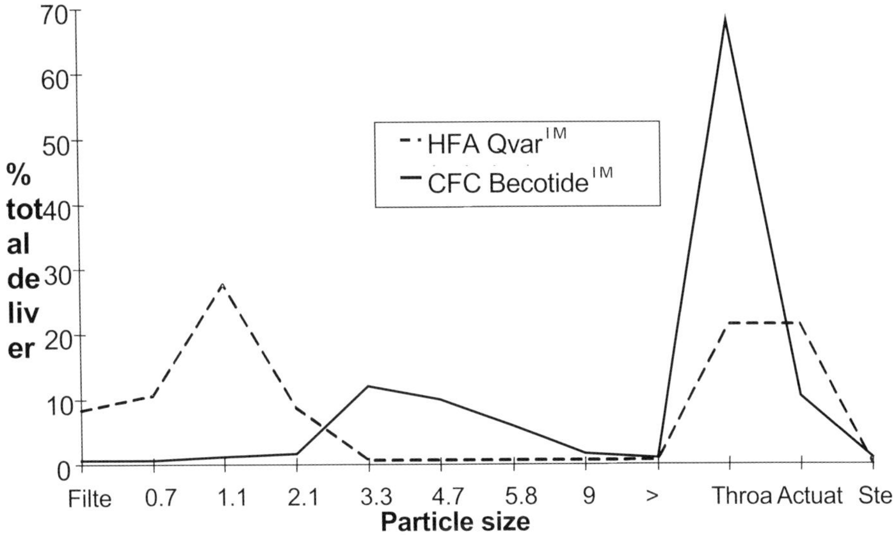

Figure 2 Size distribution of particles form two pMDI formulations of BDP as measured by cascade impaction. (adapated from Harrison et al., 1997)

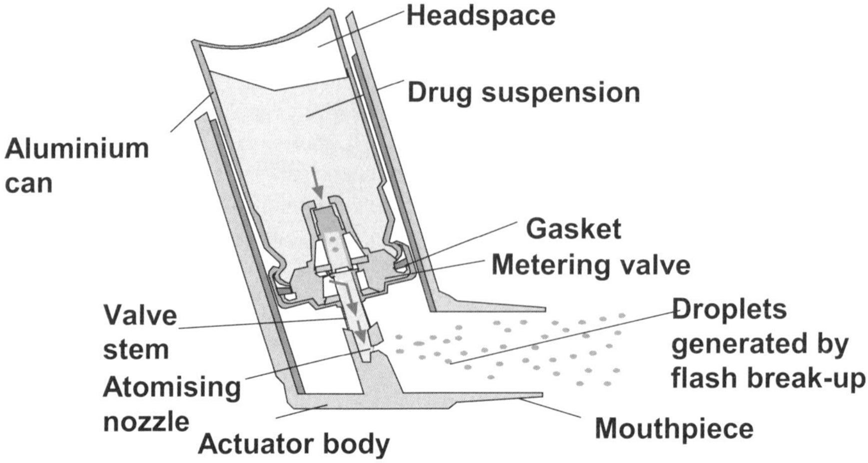

Figure 3 Schematic of a pMDI

2.2 The pack

Less obvious is the way in which propellants interact with the materials of construction of the canister and metering valve (Figure 3). Firstly, a lack of surfactant will tend to cause drug particles to adhere more strongly to these surfaces. As this is a process that is likely to increase both on storage and as the can empties through use, it can lead to unacceptable variation in dose. This has led to the development of coatings, such as fluorinated polymers, in order to meet Regulatory standards (Britto, 2001).

The metering valve is required to retain and protect the contents of the canister at the same time as delivering a fixed volume of the formulation in a reproducible manner. Seals are used to prevent leakage around the site of the crimp between valve assembly and can, together with the moving parts of the valve itself. Surfactants act as lubricants for these moving parts. Without surfactants compatible with HFAs, novel valve designs and materials have had to be developed to prevent valves from sticking during operation, such as the fluorinated components described by Warby (2002). In reformulating GSK's products, it has been possible to use such advances in pack design to achieve high-performing pharmaceutical products without the need for excipients such as surfactants or co-solvents.

The solvency properties of the propellant also affect the degree of swelling (or shrinkage) of elastomeric components which in turn can affect propellant leakage (Tiwari et al., 1998). HFAs also exhibit significant uptake of water, which in turn can influence product performance (Williams and Hu, 2001). Hence, seal quality is equally important in controlling moisture ingress, whilst the choice of material can also influence drug absorption. Elastomer developments required for compatibility with HFAs have included the use of novel materials, removal of sources of polynuclear aromatic compounds and avoidance of sulphur-based curing processes, which also have benefit in reducing the amount of extractable material (Howlett and Colwell, 1997).

The materials used in construction of the pack (and actuator) are also of importance from another aspect. As the spray leaves the actuator, triboelectric processes lead to the build up of electrostatic charge on many surfaces. Choice of material impacts the delivered dose, whilst the same material will impact the performance of different drugs to a greater or lesser extent (Peart et al., 2002).

2.3 Actuator

As well as the orifice diameter, the degree of expansion allowed by the metering valve stem and seating chamber in the actuator will influence the size and velocity of the drug particles and propellant droplets emitted from the mouthpiece (Clark, 1992). The patient is no directly aware of this, but will sense the temperature and impact force of the propellant flume. By using actuators with small orifice diameters, it is possible to produce a relatively slow and warm (> 0 °C) spray from HFA pMDIs compared with the CFC products (Gabrio et al., 1999). Whilst this may make it easier for a patient to coordinate the act of firing the pMDI with onset of inhalation, there will be clear recognition that the product is not the same as their CFC pMDI. Furthermore, changes in orifice dimensions affect delivered dose characteristics (Clark, 1992) and may lead to clogging if too small.

3. DEVELOPING CFC-FREE PRODUCTS

With all of the changes described above, it is clear that the HFA pMDIs require rigorous *in vitro* testing before replacing their CFC equivalent (Summers, 1998). However, the nature of the accompanying clinical programme will depend on whether the HFA product is designed to be equivalent to the CFC, or whether it is treated as an entirely new product. In almost all cases, the pharmaceutical industry adopted the approach that the replacement products should be equivalent, as this would maintain patient and prescriber confidence and promote a seamless transition. In turn, this would facilitate phase-out of CFCs in the most expeditious way. However, bioequivalence testing between inhaled products can include pharmaceutical testing, assessment of lung deposition and, most importantly, appropriately designed robust studies of clinical efficacy and safety. Each are considered in turn below.

3.1 Pharmaceutical testing

Pharmaceutical testing, including tests of stability over time, susceptibility to moisture, consistency of dose delivery as a percentage of label claim and estimates of fine particle dose (i.e. particles < 5 μm and which may represent the lung dose) are all key regulatory requirements. Both European authorities and FDA have stringent specifications in this respect for claims of equivalence (Summers, 1998). There seems to be a good correlation between pharmaceutical performance (such as particle size distribution) and the degree of lung deposition and systemic exposure. However, when two products differ in their pharmaceutical properties (e.g. the particle size distribution) these differences may not necessarily translate to clinically relevant differences in efficacy or safety. This is probably due to the fact that other factors, such as intra and inter-subject variability in inhaler technique, disease process and degree of airway occlusion, mask the more subtle differences that are due to changes in pharmaceutical performance.

3.2 Lung deposition

A number of methods can be used to test lung deposition; some of these methods are more sophisticated than others. Gamma scintigraphy provides a 2D image of deposition of drug in the lung. However, it does not allow for the accurate quantification of regional deposition in relation to airway morphology. Furthermore, the deposition that is seen cannot be directly linked to safety and efficacy; this is strongly dependant on where the product sits on the relevant dose-response curves. For most marketed doses of inhaled corticosteroids, the curve is very flat such that, for example, an 8-fold increase in deposition of beclomethasone dipropionate from an HFA formulation (Leach, 1998) is only associated with an observed 2-fold increase in efficacy (Busse et al., 1999). Therefore, total lung deposition cannot be regarded as a reliable surrogate marker.

Pharmacokinetic evaluations are important in understanding bioequivalence of inhaled drugs, and such studies can provide reliable estimates of lung deposition and systemic exposure. However, the relationship between topical efficacy and systemic exposure remains unclear. For example, Diskhaler™ and MDI are to some extent different in terms of their pharmacokinetic properties, yet in large multi-centre clinical studies involving several hundreds of patients, the Fluticasone Propionate (FP) Diskhaler™ and FP MDI are microgram equivalent in terms of safety and efficacy (Lundback et al., 1994).

Data from pharmacokinetic studies can be confounded by a number of factors; e.g, data in healthy volunteers are not necessarily representative of what happens in asthmatic patients (Brutsche et al., 2000). Time of dosing of a drug, especially corticosteroids, can influence the effects seen when comparing two different drugs due to their different pharmacological properties (Derendorf et al., 1998). Different assay methodologies can also lead to misinterpretation of the data, especially if the sensitivity is inadequate to fully characterise systemic exposure (Derendorf et al., 2002).

3.3 Clinical equivalence

Whilst there is a place for pharmaceutical and pharmacokinetic studies examining deposition, the ultimate test remains clinical efficacy and safety. The types of study that are appropriate in design will depend upon the class of drug being studied, but where feasible, should be conducted on the steep part of the dose-response curve, rather than on the plateau. Regulatory and clinical requirements demand that testing of safety and efficacy is carried out in the appropriate patient population. For example, data in adults cannot be extrapolated to paediatrics; data in asthmatics cannot be extrapolated to COPD and data in mild patients cannot be extrapolated to data in severe patients. The examples below are drawn from the equivalence-testing programme that was recently conducted for GSK non-CFC products against their CFC counterparts, which was designed to meet regulatory guidelines.

3.3.1 Beta-agonists

The GSK clinical programme for transition of CFC to non-CFC Ventolin™ included two single-dose histamine challenge studies: one in paediatrics (Custovic et al., 1995) and one in adults (Taggart et al., 1995). These were crossover in design, included a placebo limb and around 24 patients each. The differences seen between the two groups were well within 2 doubling dilutions at around 1, with a 90% confidence interval of 0.6 – 1.4. Such studies are useful initial guides for equivalence. However, they do not reflect the real life efficacy and safety of β_2 agonist usage in the clinical setting and, therefore, in accordance with regulatory guidelines and also to meet the reality of true clinical practice, further studies were conducted. One such study was a switch design in which patients receiving 200 µg q.i.d. CFC Ventolin™ were assessed for 4 weeks and then continued on that regime, or 200 µg q.i.d. HFA for a treatment period of 12 weeks. All patients then returned to the CFC formulation for a 4-week run-out period. The primary assessment of this safety-based study was heart rate and the study demonstrated no differences between the two treatment groups. (Regular use of □2 agonists in this way is not advocated by international guidelines, but the study was designed in recognition that such practice still occurs.) An additional study looked at a treatment period of 4 weeks in which patients continued their normal therapy but received p.r.n. salbutamol via the HFA or CFC formulation. The median number of puffs that was used during the treatment period was similar between the two treatment groups indicated comparable efficacy. This represents a programme of clinical work that not only meets regulatory needs, but also attempts to more closely reflect the true clinical use of the product. However, the final safety evaluation included a large post-marketing surveillance study, looking at the incidence of adverse events (Craig-McFeely et al., 2001).

3.3.2 Inhaled Steroids

Bronchial hyperresponsiveness studies (BHR), including methacholine or antigen challenge studies, can be a useful starting point when comparing the therapeutic benefit of inhaled steroids. However, if the dose response curve is not steep, the noise in the challenge study technique and the interpretation of data may disguise any subtle differences that may exist between two products. Furthermore, the relationship between effects on BHR and clinical efficacy/safety are not clearly understood. BHR studies therefore need to be complemented with conventional clinical studies using end points, such as lung function and symptoms.

It is important that when testing equivalence of inhaled steroids, that the study design is appropriate. For example, studies conducted at the plateau of the dose response curve for two steroids can easily show no difference between two doses of the same inhaled steroid, or indeed two different doses of two different inhaled steroids. The most appropriate design should include appropriate patients who have room for improvement and are on the steep part of their dose response curve for the two doses or two different preparations being studied. Using such criteria and using different doses of inhaled steroids, a clear dose response relationship may be established for one or both preparations, and equivalence claims can then be made (eg Tonnel et al., 2000; Ayres et al., 2000). However, such studies are complex in design, can require a large number of patients and can take a long time to recruit. An alternative, pragmatic design takes patients who are already well controlled, or whose control is optimised through a course of oral steroids. These patients are then stepped down to achieve their minimal effective dose. A comparison of two products can be made in this way by determining what the minimal effective dose is with two different preparations.

Systemic exposure data, reporting of local side-effects such as candida and effects on cortisol, are useful starting points in evaluating and comparing the safety of inhaled steroids. However, this data may not be predictive of long-term clinical side-effects. In children, knemometry studies and measurements of growth velocity are useful additional measures. Regulatory requirements demand robust comparisons of systemic exposure and monitoring of adverse event profiles for a period of between 3 – 12 months. The ultimate parameter may be final attained height. However, such data are difficult to obtain and even when possible (e.g Pedersen, 2001), are confounded by a number of variables, including the effect of the disease itself or the disease severity or the background intermittent use of oral steroids.

4. ACHIEVING TRANSITION

4.1 Regulatory Guidance

The Parties to the Montreal Protocol required the preparation of national strategies for the transition to non-CFC pMDIs by 31 January 1999. Within the European Union, this was informed by guidance from the European Parliament (1998), as shown in Table 2. When sufficient products containing a particular drug meet the defined criteria, the "essential use" exclusion will be removed and CFC will no longer be permitted for manufacture of pMDI products containing that drug. However, the strategy also stipulates that there needs to be dose ranges to accommodate all patient groups, including the young and the elderly and, in some Member States, sufficient post-marketing surveillance, prior to removal of a particular "essential use" exemption. (It should be noted that this does not preclude manufacture of CFC products within that country for export).

4.2 Progress to date

To date, six companies have successfully launched non-CFC products into a total of 56 countries (see Table 3). As a consequence, the European Union has removed "essential use" status for salbutamol, whilst Germany has also removed it for Category B (steroids). However, launch of a product does not automatically mean that prescribers will immediately switch their patients to the non-CFC variant. The experience of GSK in the United Kingdom was that it was not until CFC salbutamol was removed from the supply chain, that there was a significant switch to the HFA product. Overall, this process took around 12 months to complete.

It is also clear from Table 3 that two companies are much further advanced in the worldwide introduction of replacement products than the rest. By the time transition is complete for GSK, there will have been a global reduction in pMDI CFC usage of 25 % compared to a decade before. Nonetheless, the pace is still some way from those projections of the early 90s.

Product	Guidance to consider CFCs no longer essential
Category A Short-acting □-agonists	
Salbutamol*, Clenbuterol, Fenoterol*, Bitolterol, Orciprenaline, Procaterol, Reproterol, Carbuterol, Hexoprenaline, Pirbuterol	Two non-CFC salbutamol products from different producers PLUS one other product defined as necessary
Category B Inhaled steroids	
Beclomethasone*, Dexamethasone, Flunisolide, Fluticasone*, Budesonide*, Triamcinolone	Two non-CFC beclomethasone products from different producers PLUS two other products defined as necessary
Category C Non-steroidal anti-inflammatories	One alternative product except where both products are defined as necessary
Cromoglicic acid*, Nedocromil*	
Category D Anticholinergic bronchodilators	
Ipratropium bromide, Oxiptropium bromide	One alternative product to replace either of the two
Category E Long-acting □-agonists	
Salmeterol*, Formoterol**	One alternative product to replace either of the two
Category F Combination Products	Both of the components are available as stand-alone non-CFC products

Table 2 Replacement of CFC containing pMDIs (European Parliament, 1998)

Moiety	Company	Countries launched
Beclomethasone dipropionate	3M Pharmaceuticals	19
	Chiesi	1
	IVAX	2
Fenoterol	Boehringer Ingelheim	5
Fluticasone propionate	GSK	24
Nedocromil sodium	Aventis Pharmaceuticals	1
Salbutamol	3M Pharmaceuticals	37
	IVAX	1
	GSK	43
Salmeterol/fluticasone propionate	GSK	1
Sodium cromoglycate	Aventis Pharmaceuticals	5

[*] Denotes products deemed necessary under this strategy in one or more Member States of the European Union

Table 3 Worldwide availability of non-CFC pMDIs in Spring 2001 (International Pharmaceutical Aerosols Consortium, 2001)

5. FUTURE TRENDS

There has been much development around this delivery system, as evidenced by the complicated patent landscape that now exists (Bowman and Greenleaf, 1999). Whilst many patents are being opposed this has promoted greater interest in alternative delivery systems. Nonetheless, the investment in developing the pMDI to meet the requirements of new propellants will ensure that there will continue to be new developments.

5.1 Delivery system

It is well known that patients may suffer problems with coordination or excessive throat deposition when using pMDIs (Terzano, 2001). To overcome this, breath-activated pMDIs that were developed for CFC pMDIs have been adapted to accommodate the HFA products. However, if products are developed that produce a slower spray, there may be much less need for this type of device in the future. Nonetheless, we can expect to see added complexity as counters become an established feature of pMDIs, just as they have recently become for multi-dose dry powder inhalers (for 3 examples, see Julius et al., 2002). The most sophisticated may even act as compliance monitors to aid physicians in understanding why a particular form of treatment may be less than optimal.

Advances are also likely in the formulation. The search for novel surfactants was referred to above (e.g. Stefely et al, 2000). However, as most excipients have not previously been approved for use in inhaled products, extensive safety testing programmes will be required. It is unlikely, therefore, that these will be introduced until at least the second half of this decade. An alternative is to engineer the physical form of the suspended particles in order to make more stable formulations; improved efficiency of lung delivery can then also result (Hirst et al, 2002).

5.2 New applications

There has been tremendous interest in recent years in using the lung as a portal of entry for the systemic delivery of drugs (Agu et al., 2001). Much of this has been because of the difficulty of delivering proteins and peptides by any route other than injection. The large surface area, good vascularity, immense capacity for solute exchange and the thinness of the alveolar epithelium are unique features of the lung that facilitate systemic delivery. However, it should be noted that the particle size that is required to optimise alveolar delivery (≤ 1 □m) is different to that of most asthma medications.

pMDIs have already been used to demonstrate potential for the systemic delivery of peptides such as leuprolide (Wood and Knowles, 1994), proteins (Quinn et al., 1999), vaccines and DNA-plasmids (Brown and Pickrell, 1995). In particular, if solution formulations can be developed, then the pMDI has the potential to be a very efficient delivery system, a key consideration when working with molecules that are very expensive to produce. Other applications include those where an injection-like pharmacokinetic profile is needed, such as the use of opioids (eg Derschwitz et al., 2000). Topical uses in other therapy areas are now being considered as well, such as the development of inhaled drugs in the treatment of cancer (Sharma et al., 2001).

5.3 Regulatory changes

HFAs are not without environmental impact, potentially contributing to the greenhouse effect and global warming (Smith, 1995). Although it has been estimated that HFA from pMDIs will contribute less than 0.1 % of worldwide greenhouse gas emission by 2005 (McDonald and Martin, 2000), they are substances that are proposed for control under the Kyoto Protocol of 1997 (United Nations, 1997). Therefore, the prospects of another reformulation exercise akin to the transition from CFC to HFA propellants cannot be ruled out. Should this ever come about, the knowledge gained during the switch to non-CFC propellants should ensure that the process will not be as long.

It should also be remembered that advances are being made in other forms of inhaled delivery systems. If these result in the ability to give more reproducible dosing, or even better targeting within the respiratory tract, then regulators and prescribers will expect similar performance from a pMDI. This puts pressure to continue to research and evolve this product well into the future. However, ever-increasing pressure on the price of prescription medicines means that the costs of these new developments must be managed prudently by the industry.

6. SUMMARY

There have been a great many challenges in replacing CFC propellants in pMDIs, with the result that the transition, whilst now approaching its midway point, has taken much longer that originally envisaged. However, a great many advances in the understanding and design of pMDIs have been brought about. Whilst many have led to improved dosing consistency and product stability, these changes will have gone largely unnoticed by patients and prescribers. For the most part, this has been a deliberate strategy to make the transition as seamless as possible. However, as new drugs are developed, both for topical and systemic delivery, there will be plenty of scope to introduce more advances in the future.

7. REFERENCES

Agu RU, MI Ugwoke, M Armand, R Kinget and N Verbeke. "The lung as a route for systemic delivery of therapeutic proteins and peptides." *Respiratory Research* (2001) 2: 198-209.

Ayres JG, AB Millar and AP Sykes. "Clinical efficacy and safety of fluticasone propionate 1 mg twice daily administered via a HFA 134a pressurised metered dose inhaler to patients with severe asthma." *Respiratory Medicine* (2000) 94 (Supplement B): s42-s50.

Bowman PA and D Greenleaf. "Non-CFC metered dose inhalers: the patent landscape." *International Journal of Pharmaceutics* (1999) 186: 91-94.

Brambilla G, D Ganderton, R Garzia, D Lewis, B Meakin, P Ventura. "Modulation of aerosol clouds produced by pressurised inhalation aerosols." *International Journal of Pharmaceutics* (1999) 186: 53-61.

Britto IL. "Metered dose inhaler for fluticasone propionate." United States Patent No. US6253762 (2001)

Brown AR and JA Pickrell. "Propellant-driven aerosols for delivery of proteins in the respiratory tract." *Journal of Aerosol Medicine* (1995) 8:43-58.

Brutsche MH, IC Brutsche, M Munavvar, SJ Langley, CM Masterson, PT Daley-Yates, R Brown and A Woodcock. "Pharmacokinetics and systemic effects of fluticasone propionate are different in asthmatics and normal volunteers". *Lancet* 356 (August 2000):556-561.

Burt PCW and LKH Kwok. "Method and apparatus for introducing into a container a suspension or solution of a material in a propellant held under pressure. European Patent No. 0419261 (1993)

Busse WW, S Brazinsky, K Jacobson, W Stricker, K Schmitt, J Vanden Burgt, D Donnell, S Hannon and GL Colice. "Efficacy response of inhaled beclomethasone dipropionate in asthma is proportional to dose and is improved by formulation with a new propellant." *Journal of Allergy and Clinical Immunology* 104 (1999): 1215-1222.

Clark A. "The physics of aerosol formation by MDIs – limitations of the current approach." *Journal of Biopharmaceutical Sciences* (1992) 3: 69-76.

Commission to the Council and the European Parliament. "Strategy for the phase-out of CFCs in metered-dose inhalers. *Official Journal of the European Communities* (1998) COM 603.

Committee for Proprietary Medicinal Products. "Results of the co-ordinated review of 1,1,1,2-tetrafluroethane HFC-134a." European Commission, Brussels 1994.

Committee for Proprietary Medicinal Products. "Results of the co-ordinated review of 1,1,1,2,3,3,3-heptafluropropane (HFC-227)." *CPMP/503/95* .European Agency for the Evaluation of Medicinal Products, London 1995.

Craig-McFeely PM, JB Soriano, WC Maier, J Riley and SAW Shakir. "Prospective observational cohort safety study to monitor the introduction of a non-CFC formulation of salbutamol with HFA134a in England: Final results." *American Journal of Respiratory and Critical Care Medicine* (2001) 163: a857.

Custovic A, SCO Taggart, A Stuart, A Robinson and A Woodcock. "Efficacy of a new non-ozone depleting formulation for salbutamol." *Journal of Pharmaceutical Medicine* 5 (1995): 161-168.

Derendorf H, G Hochhaus, B Meibohm, H Mollmann and J Barth. "Pharmacokinetics and pharmacodynamics of inhaled corticosteroids." *Journal of Allergy and Clinincal Immunology* 101 (Suppl S) (1998): S440-S446.

Derendorf H. PT Daley-Yates, LN Pierre, J Efthimiou. "Bioavailability and metabolism of mometasone furoate: Pharmacology versus methodology." *Journal of Clinical Pharmacology*. (2002) 42: 383-387.

Dershwitz M, JL Walsh, RJ Morishige, PM Conners, RM Rubsamen, SL Shafer and CE Rosow. "Pharmacokinetics and pharmacodynamics of inhaled versus intravenous morphine in healthy volunteers." *Anesthesiology*. (2000) 93:619-628.

Farman JC, BG Gardiner and JD Shanklin. "Large losses of total ozone over Antarctica reveal serious ClOx/Nox interaction." *Nature* (325): 207-210.

Gabrio BJ, SW Stein and DJ Velasquez. "A new method to evaluate plume characteristics of hydrofluoroalkane and chlorofluorocarbon metered dose inhalers." *International Journal of Pharmaceutics* (1999) 186: 3-12.

Harrison LI, I Soria, J Machacek et al. "Pharmacokinetic profile of improved lung delivery with CFC-free ultrafine aerosol steroid inhaler [Abstract]". *Am J Respir Crit Care Med* (1997) 155: A666.

Harrison LI, I Soria, AC Cline and BP Ekholm. "Pharmacokinetic differences between chlorofluorocarbon and chlorofluorocarbon-free metered dose inhalers of beclomethasone dipropionate in adult asthmatics." *Journal of Pharmacy and Pharmacology*. (1999) 51: 1235-1240.

Hirst PH, GR Pitcairn, JG Weers, TE Tarara, AR Clark, LA Dellamary, G Hall, J Shorr J and SP Newman. "In vivo lung deposition of hollow porous particles from a pressurized metered dose inhaler." *Pharmaceutical Research* (2002) 19:258-264.

Howlett D and J Colwell. "Improvements in extractables from pMDI elastomer systems." Pp 36-38 in *Drug Delivery to the Lungs VIII*. Portishead, UK: The Aerosol Society, 1997.

International Pharmaceutical Aerosols Consortium. "World-wide availability on CFC-free MDIs". www.ipacmdi.com (2001)

Julius SM, JM Sherman and L Hendeles. "Accuracy of three electronic monitors for metered-dose inhalers." *Chest* (2002) 121: 871-876.

Leach CL. "Improved delivery of inhaled steroids to the large and small airways." *Respiratory Medicine* 92 (Suppl A) (1998): 3-8.

Leach, CL, PJ Davidson and RJ Boudreau. "Improved airway targeting with CFC-free HFA-beclomethasone metered-dose inhaler compared with CFC-beclomethasone." *European Respiratory Journal*. (1998) 12: 1346-1353.

Lundback B, R Dahl, M De Jonghe, N Hyldebrandt, R Valta and SL Payne. "A comparison of fluticasone propionate when delivered by either the metered-dose inhaler or the Diskhaler inhaler in the treatment of mild-to-moderate asthma." *European Journal of Clinical Research* 5 (1994): 11-19.

McDonald KJ and GP Martin. "Transition to CFC-metered dose inhalers – into the new millennium." *International Journal of Pharmaceutics* (2000) 201:89-107.

Molina MJ and FS Rowlands. "Stratospheric sink for chlorofluoromethane: chlorine atom catalysed destruction of ozone." *Nature* (1974): 810-812.

Peart J, JC Orban, P McGlynn, MP Redmon, CM Sargeant and PR Byron "MDI electrostatics: valve and formulation interactions that really make a difference." Pp 223-230 in *Respiratory Drug Delivery VIII*: Raleigh, NC, Davis Horwood International, 2002.

Pedersen S. "Do inhaled corticosteroids inhibit growth in children?" *American Journal of Respiratory & Critical Care Medicine*. (2001) 164: 521-535.

Polli GP, WM Grim, FA Bacher and MH Yunker. "Influence of formulation on aerosol particle size." *Journal of Pharmaceutical Sciences* (1969) 58:484-486.

Quinn EA, RT Forbes, AC Williams, MJ Oliver, L McKenzie and TS Purewal. "Protein conformational stability in the hydrofluoroalkane propellants tetrafluoroethane and heptafluoropropane analysed by Fourier transform Raman spectroscopy." *International Journal of Pharmaceutics* (1999) 186:31-41.

Sharma S, D White, AR Imondi, ME Placke, DM Vail and MG Kris. "Development of inhalational agents for oncologic use." *Journal of Clinical Oncology* (2001) 19:1839-1847.

Smith IJ. "The challenge of reformulation." *Journal of Aerosol Medicine* (1995) 8 (Supplement): S19-S27.

Stefely JS, D Duan, P Myrdal, D Ross, DW Schultz and CL Leach "Design and utility of a novel class of biocompatible excipients for HFA-based MDIs." Pp 82-90 in *Respiratory Drug Delivery VII*: Raleigh, NC, Serentec Press 2000.

Summers M. "Control and regulation of HFA pMDIs". Pp 119-22 in *Respiratory Drug Delivery VI:* Buffalo, Il, Interpharm Press 1998.

Taggart SCO, A Custovic, DH Richards, AA Woodcock. "GR106642X: A new, non-ozone depleting propellant for inhalers." *British Medical Journal* 310 (1995): 1639-1640.

Tansey I. "Changing to CFC-free inhalers: the technical and clinical challenges." *Pharmaceutical Journal* (1997) 259: 896-898.

Tansey I "The technical transtion to CFC free inhalers." *British Journal of Clinical Practice* (1997) 89 (Suppl): 22-27.

The Management Forum "The CFC crisis and portable therapy." *Reported in Manufacturing Chemist* (July 1992): 22-23.

Terzano C. "Pressurized metered dose inhalers and add-on devices." *Pulmonary Pharmacology and Therapeutics* (2001) 14: 351-366.

Tiwari D, D Goldman, S Dixit, WA Malick and PL Madan. "Compatibility evaluation of metered dose inhalers with tetrafluoroethane (P134a), a non-CFC propellant." *Drug Development and Industrial Pharmacy* (1998) 24:345-352.

Tonnel AB, J Bons, M Legendre, A Prud'Homme, B Bugnas, I Evano-Celli and AM Stuart. "Clinical efficacy and safety of fluticasone propionate 250 □g twice daily administered via a HFA 134a pressurised metered dose inhaler to patients with mild to moderate asthma." *Respiratory Medicine* (2000) 94 (Supplement B): s29-s34.

United Nations. "Kyoto Protocol to the United Nations Framework Convention on Climate Change". *UNEP/INC/98/2*. Geneva: Information Unit for Conventions, UNEP (1998).

Warby RJ "Improvements in drug delivery devices" European Patent Application No. 1208864 (2002).

Williams RO and C Hu. "Influence of water on the solubility of two steroid drugs in hydrofluoroalkane (HFA) propellants." *Drug Development and Industrial Pharmacy* (2001) 27:71-79.

Wood RE and MR Knowles. "Recent advances in aerosol therapy." *Journal of Aerosol Medicine* (1994) 7:1-11.

8.0 AFFILIATIONS

John N Pritchard works at GlaxoSmithKline
881-895 Greenford Road, Greenford, Middlesex, UB6 0HE
United Kingdom

L. GRADOŃ, T.R. SOSNOWSKI AND A. MOSKAL

RESUSPENSION OF POWDERS AND DEPOSITION OF AEROSOL PARTICLES IN THE UPPER HUMAN AIRWAYS

1. INTRODUCTION

Aerosols for therapy are produced by ultrasonic nebulizers, pneumatic nebulizers, pressurized metered dose inhalers (pMDI) and dry powder inhalers (DPI). Each method of generation has its own characteristic with specific advantages and disadvantages related to aerosol properties and technical usefulness. A particular system is chosen based on the ease of use for patient and the intent to provide the required therapeutic dose with minimal adverse effects. The most widely used delivery system is the pMDI, which creates therapeutic aerosols from pressurized solution or suspension (Schultz, 1995). Properties of this method of generation are described by J. Pritchard in another chapter of this book.

More than 80 countries have signed the Montreal Protocol to ban the production of chlorofluorocarbons in the next few years, which necessitates the replacement of chlorofluorocarbon-propelled pMDI's with other aerosol delivery devices. Among the developing alternative means of delivering therapeutic aerosols, dry powder inhalers are gaining in popularity (Ganderton and Kassem, 1992).

Aerosols of dry powder are created by directing air through an aliquot of loose powder. As DPI are breath-actuated, the need to synchronize inhalation with actuation is eliminated. The powder contains either micronized drug particles bound into loose aggregates or micronized drug particles that are loosely bound to large lactose or glucose particles (Ganderton, 1992). The release of respirable particles of the drug requires inspiration at relatively high flow rates, which creates a turbulence in the region of powder layer in the device's structure. Aerosol delivery from DPI's is influenced by the materials used in their manufacture, resistance to airflow, proper assembly of the components, and by accumulation of the powder within the device.

There is a wide range of different types of DPI device already on the market or under development. Some of them dispense individual doses of drug from punctured gelatin capsules, while others - from individual blister packets of drug, or a tape-system containing multiple individually sealed doses. The structure of the powder exposed to the inhaled air is the most sensitive element of the performance of DPI.

The main goal of this study is the measurement of the influence of air flowrates on the reentrainment of particles from powders, and the modeling of the influence of the shape of breathing curves on the deposition of aerosolized powder particles in the model of upper human airways.

L. Gradoń and J. Marijnissen (eds.), Optimization of Aerosol Drug Delivery, 123–137.

EXPERIMENTAL TECHNIQUES

Resuspension of Particles from Powders

The structural stability of the powder layer depends on the relation between adhesive and separation forces acting on each particle of structure. This relation has its specific spatial and temporary distributions. By taking into account the adhesive strength distribution in the powder layer, the surface renewal of the powder surface due to the resuspension of single particles or their aggregates, and the time dependence of the process should be recognized. The experimental analysis of the problem described below is capable to predict the rate of resuspension and size distribution of resuspended particles at the breathing condition.

Figure 1 shows a schematic diagram of the experimental system. The most essential part of the experimental system consisted of the specially designed resuspension channel (RC – part 6 in Figure 1) with a rectangular cross-section 5 mm x 14 mm and the total length of 480 mm. The powder was periodically loaded via the filling orifice in the cover of the RC, and it was placed at the face of the feeding piston driven by a micrometrically threaded shaft (part 7 in Figure 1). Such system allowed for the precise adjustment of the initial level of the powder sample. The RC was embedded into the system for the preparation and distribution of the air streams used for the experiments.

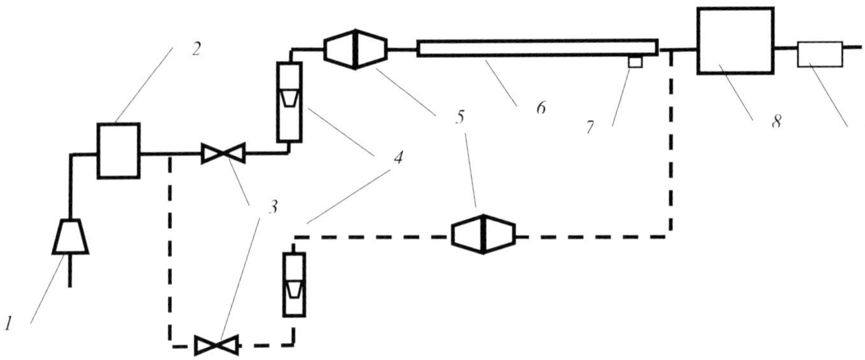

Figure 1. Schematic diagram of experimental system.
1 – air compressor, 2 – air dryer/humidifier and the measuring system of air temperature and relative humidity, 3 – valves, 4 – flowmeters, 5 – absolute filters, 6 – resuspension channel (RC), 7 – micro-feeder, 8 – aerosol collector, 9 – airflow controller.

The resuspension process was studied in two arrangements:
to obtain the total mass of particles eluted from the RC, the highly efficient filter was placed in the holder mounted at the outflow from RC (part 8 in Figure 1). Filtering material used was a low flow-resistance fibrous structure obtained in our

laboratory by the melt-blown technique (Gradoń and Ciach, 1996). Its filtering properties are characterized in Figure 2;

to obtain the aerosol mass distribution of the resuspended particles, the Andersen Cascade Impactor (ACI – Copley Scientific, UK) was connected at the RC outlet (part 8 in Figure 1). The desired airflow rate in the ACI (1 cfm) independently of the flow via the RC, was assured by additional air distribution system (marked as dashed lines in the Figure 1) had to be used together with the airflow controller (part 9 in Figure 1).

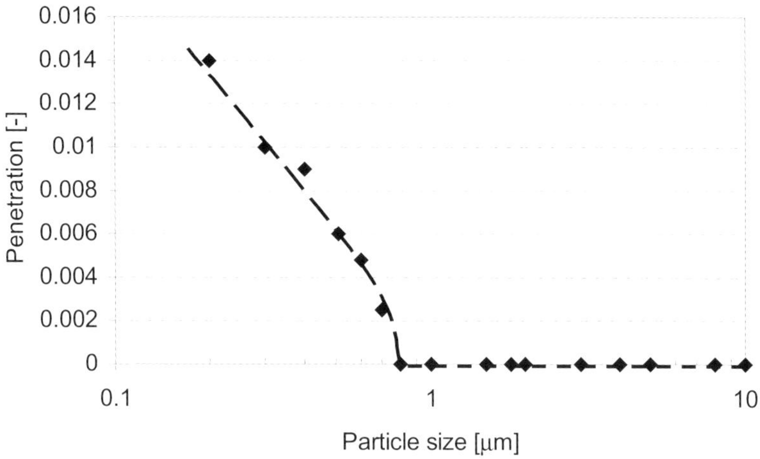

Figure 2. Penetration of aerosol through the material used for collection of particles elutriated from RC.

The very initial part of the study was conducted for SAE Fine Test Dust (Powder Technology Inc., USA) at room temperature and low relative air humidity (15-17%). The primary particles size distribution of the powder used in these experiments as measured with Coulter counter (manufacturer's data) is shown in Figure 3. In order to study the overall resuspension of the powder from a randomly prepared sample, the collection filter was cut, weighed and mounted in the holder at the RC outflow. Then the known mass of the powder was dropped on the surface of the feeding piston inside the RC. After the filling orifice was sealed, the air stream of desired flow rate was directed into the channel and the aerosol generated from the powder was collected on the filter for 30 seconds. Then the airflow was switched off and the procedure was repeated 5 times with the same filter, because the mass collected in the single run was too small to be determined gravimetrically. From the differences in the filter mass after and before experiments, the overall resuspension mass efficiency was found as a function of the airflow rate or - the air velocity.

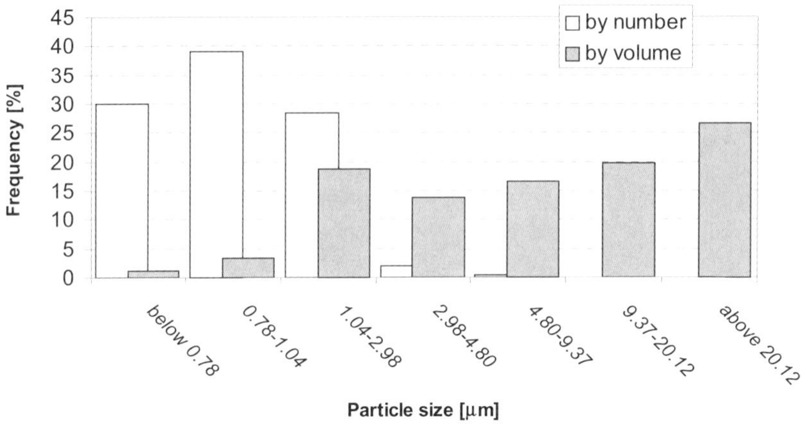

Figure 3. Size distribution of SAE Fine Test Dust determined with the Coulter particle counter (manufacturer's data).

More informative data are provided by the analysis of the particle size distribution of the aerosol formed from the powder. This part of the study was done using the ACI operating at the flow rate 28.3 ± 0.5 L/min. In the cases when flow of air directed to the RC was below that value, additional clean air was mixed with aerosol stream before the impactor to ensure that the mass collected on each stage of the ACI reflected particles of the well-defined size ranges. Each stage of the impactor and the ACI inlet was weighed before the experiment. The procedure of powder loading was identical to the one in the previous part of the study described earlier. The collection of aerosol particles was done for 30 seconds, and after that, the airflow was switched off. A new portion of powder was loaded to the RC, and such experiment was repeated without disassembling the ACI in order to collect enough mass on each stage. After 10 runs the impactor was disassembled and each stage was weighed again. From the mass difference after and before the experiments, the size distribution of particles elutriated from the channel could be found as a function of air velocity in the RC.

2.2. Deposition of Aerosol in the Model of Trachea and Main Bronchi.

Similar experimental conditions were used for the second part of experimental studies, which was focused on aerosol deposition in the model of the trachea and main bronchi. The geometrical model of the system was created based on the data by Horsfield et al. (1971). In the model of upper airways the bifurcation of trachea into left and right lobes and then - one more bifurcation in each lobe were considered. The model (cast) was made at natural scale from transparent polycarbonate - the material, which allows observing particles deposition during the measurements. Moreover, the model could be disassembled into two parts (top and bottom) what allows for better characterization of deposition pattern and for easy access for cleaning. To avoid reentrainment of deposited particles, the internal walls of the cast were coated with Vaseline. Figure 4 shows the photograph of the model and its scheme with designation of dimensions and angles of each element. The detailed numeric data are listed in Table 1.

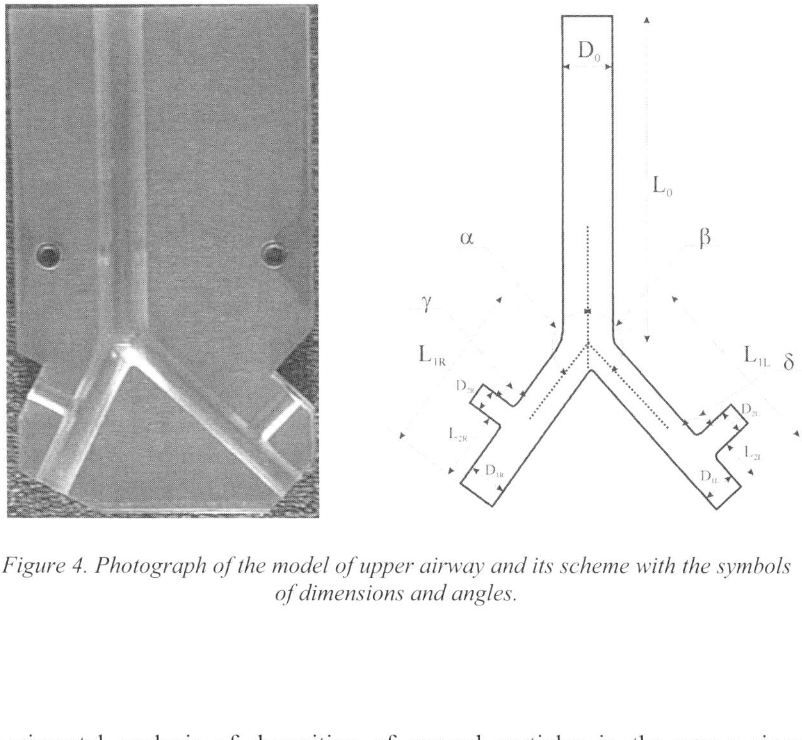

Figure 4. Photograph of the model of upper airway and its scheme with the symbols of dimensions and angles.

Experimental analysis of deposition of aerosol particles in the upper airways allows to determine the regions with highest rate of particle deposition ("hot spots") for different air velocity, and to evaluate the total mass efficiency of deposition in the model system. Figure 5 shows the experimental system used during the measurements.

Table.1. Dimensions of elements of the model of upper airways.

Symbol of element	Length [mm] or angle [°]
D0	18
D1L	12
D1R	14
D2L	9
D2R	8
L0	115
L1L	68
L1R	62
L2R	25
L2L	15
α	150°
β	130°
γ	90°
δ	90°

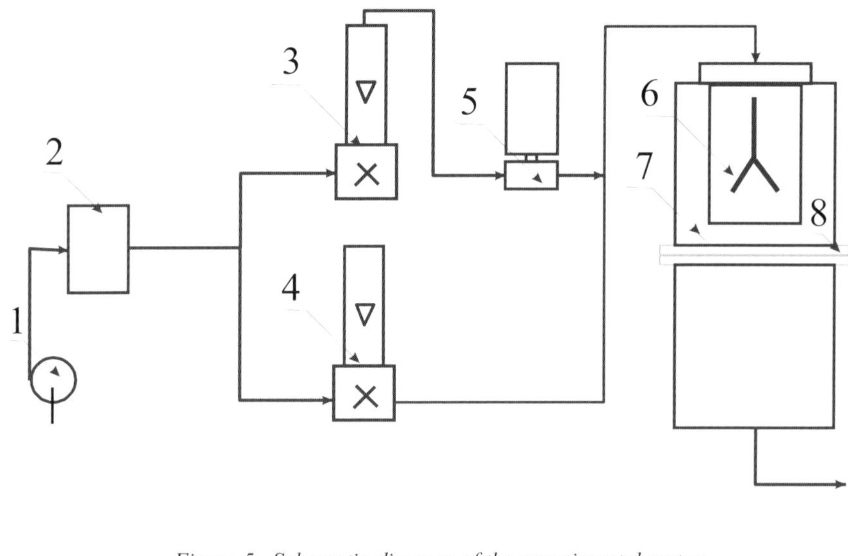

Figure 5. Schematic diagram of the experimental system.
1-air blower; 2-air dryer/humidifier; 3,4 – flowmeters with valves; 5- powder dispenser; 6- model of upper airways; 7- container; 8- filter.

The measurements were carried out for different values of air velocity associated to the airflow conditions in the upper airways during the natural breathing cycle, starting from 0.5 m/s (which corresponds to the start of inspiration) up to 9.8 m/s (which corresponds to middle part of inspiration). This part of the study was conducted with SAE Fine Test Dust, which was also used in the first part of experimental works described earlier. The deposition in the model of upper airways was studied in two arrangements:

1. to obtain the location of "hot spots" of the aerosol particles deposition, the stream of aerosol produced in the powder dispenser (part 5 in Figure 5) was directed to the inlet of the model (6 in Figure 5). The exposure of the aerosol was carried out for 20 min, and after that the photographs of the deposition pattern in the model were made in order to determine the "hot spots";
2. to obtain the total mass efficiency of deposition in the system, the highly efficient filter was used (part 8 in Figure 5). The total mass of aerosol particles deposited on the filter was determined gravimetrically after each measurement. By comparing it with the total mass of aerosol produced in the powder dispenser, the total mass efficiency of deposition of aerosol particles in the model of the upper airways was calculated.

1. RESULTS OF THE EXPERIMENTS

3.1. Total Efficiency of Powder Resuspension.

The results demonstrating the dependence of the relative mass of the aerosol formed from the powder and collected on the final filter are presented in Figure 6. The percent of resuspended mass (the resuspension efficiency) is defined here as:

$$E\% = \frac{m_C}{m_0} \cdot 100\% \qquad (1),$$

where m_C stands for mass collected on the final filter and m_0 the mass loaded into the resuspension channel.

It can be seen that there is a strong dependence of the mass that is reentrained in the form of an aerosol on the velocity, i.e., on the flow intensity in the channel. The aerodynamic forces acting on the particles, which form the powder are too low comparing to inter-particle cohesion forces when the air velocity is lower than 2 m/s, so only a minute percent of initially loaded powder can be elutriated with the air stream. With the increasing air velocity, the mass of the resuspended powder increases sharply, and reaches the value of about 80 percent. Further increase of flow intensity does not elevate the relative mass of aerosol produced what suggests that there is always an amount of the powder that either strongly adheres to the surface of the channel or remains in the zones of insufficient flow intensity (e.g., in the corners of the rectangular cross-section of the channel).

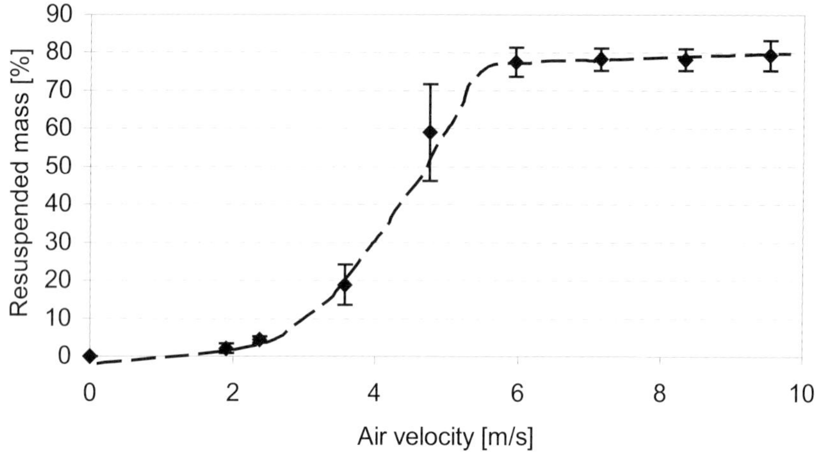

Figure 6. Percent of mass resuspended from the powder as a function of the air velocity through the channel (average of 10 experiments ± std deviation).

3.2. Particle Size Distribution of the Resuspended Powder.

Large amount of powder was resuspended to aerosol particles larger than 10 micrometers, which were captured before the impactor and in the ACI throat. From their mass and the impactor data, the fractions of resuspended particles in the given size intervals were evaluated as a function of air velocity in the channel. The percentage values were determined as follows:

$$p_m = \frac{m_i}{\sum m_i} \cdot 100\% \qquad (2),$$

$$p_n = \frac{n_i}{\sum n_i} \cdot 100\% \qquad (3),$$

where m_i and n_i denote, respectively, the mass and the number of particles in the size interval *i*. The results are presented in Figures 7 and 8.

It can be seen from Figure 8 that higher air velocities allow for entertainment of higher amounts of larger (i.e. heavier) particles. This explains, why the mass fractions of particles smaller than 5 μm decrease at higher airflow rates (Figure 7). The experimental data presented do not indicate any deaggregation of resuspended particles with the increasing airflow.

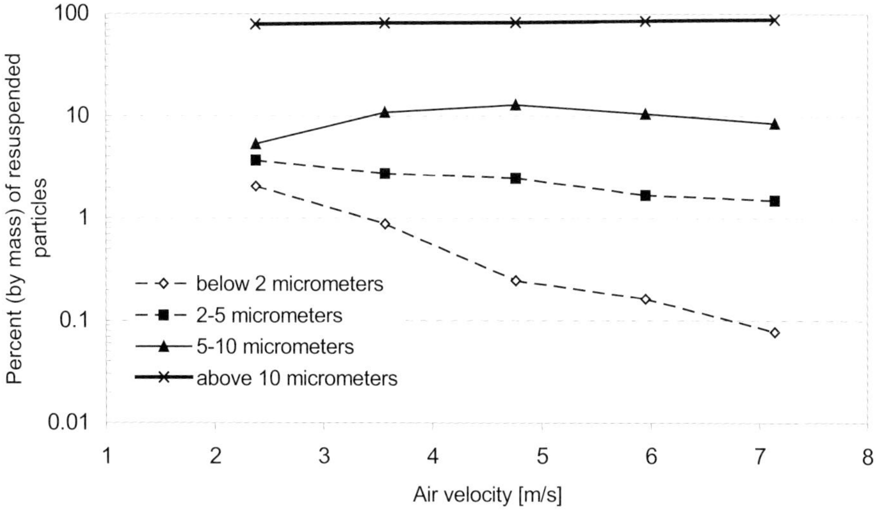

Figure 7. Changes in particle mass distribution of resuspended particles with the air velocity in the RC.

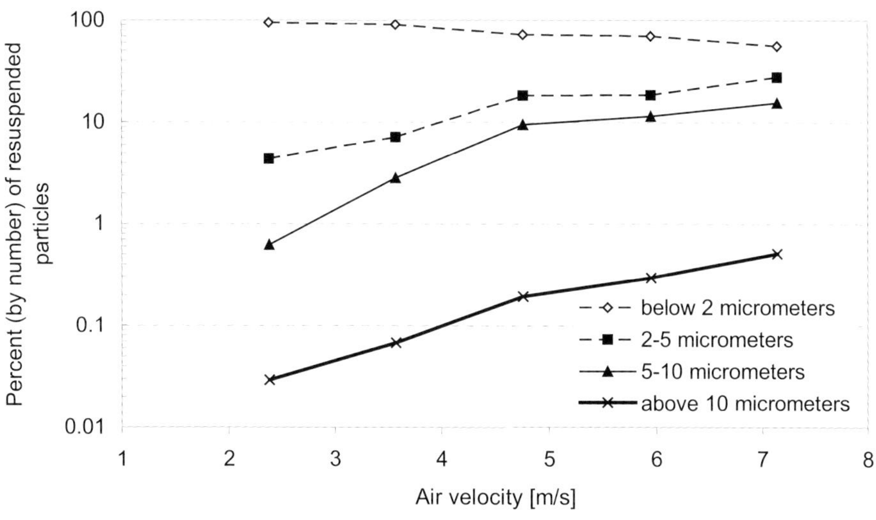

Figure 8. Changes in particle number distribution of resuspended particles with the air velocity in the RC.

Probably, the dry powder studied here do not agglomerate significantly, so simply single particles are reentrained from the sample of powder. Another explanation of the above observation is that the particles may form agglomerates, which are, however, not disintegrated under the shear forces created by the airflow in the range investigated in this work. This view is supporte by comparison of the coarse particles (> 10 μm) fraction in the Coulter analysis (~45%, Figure 3) with the one determined during the resuspension study (~90%, Figure 7), which suggests an increased amount of larger particles in the aerosol. This issue certainly requires further research.

3.3. Particle Deposition in the Model of Human Upper Airways.

Typically aerosol deposition in the cascade impactor is directly related to the deposition in the respiratory tract (US Pharmacopeial Convention, 1996). It is believed that first stages of the ACI operating at 28.3 L/min correspond to laryngeal region, the third stage - to the trachea and primary bronchi, stages 4 and 5 – to bronchial tree, and stages 6 to F – to alveoli. To validate this approach it seems reasonably to measure the deposition of resuspended powder in the physical model of the upper airways (cast) described in the earlier part of this paper.

The total mass efficiency of aerosol particle deposition was measured for a wide range of the volumetric flow rates of aerosol through the cast. Because of polydispersity of the aerosol particles, it is difficult to define a functional relationship between the mass efficiency of deposition and the flow rate. For higher values of local velocity, larger particles are deposited with higher efficiency due to inertial impaction, and smaller particles, for which turbulent diffusion is a main mechanism of deposition, pass the cast easier because the residence time of these particles in the cast is reduced. For smaller values of the flow rates, the contributions of both mechanism are different. Picture is even more complicated for particles, for which the influence of both mechanisms is comparable. The competition between a contribution of particles of different dimensions and predominance of particular mechanism of deposition for different flow rates causes a complicated relation between efficiency of deposition and the flow rate.

The experimental system described before gives the possibility of observation of distribution of deposits in the considered region. The "hot spots" of deposition of aerosol particles was observed in the series of experiments. Figure 9 shows an example picture obtained after measurements.

It is evident from our observations that larger particles deposit mostly at the carinal region of bifurcations. Smaller particles more or less uniformly cover the surface of airways.

For explanation of results of these qualitative observations, the modeling of particle deposition in the wide range of their diameters was undertaken.

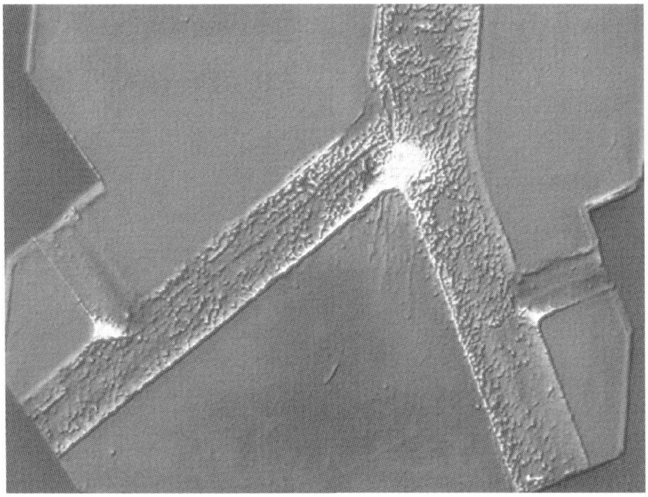

Figure 9. Picture of "hot spots" of deposited aerosol particles in the model.

4. MODELING OF DEPOSITION OF AEROSOL PARTICLES IN THE UPPER HUMAN AIRWAYS DURING BREATHING CYCLE.

4.1. Description of the Mathematical Modeling Technique.

In the experimental part of these investigations, the process of deposition in the model of human upper airways was examined. Knowledge of local deposition rates of aerosol particles in the respiratory system together with information on particle size, shape, mass and clearance rates at the deposition sites are required for complete evaluation of the dose of a drug delivered with inhaled particles. The efficiency of deposition of particles along the respiratory tract depends significantly on the local gas flow structure and particle properties. An important parameter, which influences the deposition pattern, is the local geometry of the respiratory system. Using the experimental set-up described above it is not possible to measure the spatial and temporal deposition of aerosol in the system. To get this information, the mathematical modeling is needed.

The mathematical model of particle deposition should consist of the following elements: morphology and geometry of the respiratory system, fluid dynamics of the respiratory air flow and the deposition mechanism of the aerosol particles of given physicochemical properties. Based on the geometrical data presented in the experimental chapter, the numerical model of the upper airways was created. The conditions of the airflow in the tracheobronchial tree depend on the pattern of the breathing cycle characterized by a volume-time (V-t) curve. The representative breathing curve, which was used in the calculations, has a shape determined under following assumption: duration of the breathing cycle = 4 s, frequency of breathing =15 cycles/min, volumetric flow rate of the air = 45 dm^3/min. Proposed curve corresponds to the symmetrical hyperventilation with spontaneous breathing. The shape of such breathing curve is presented in Figure 10.

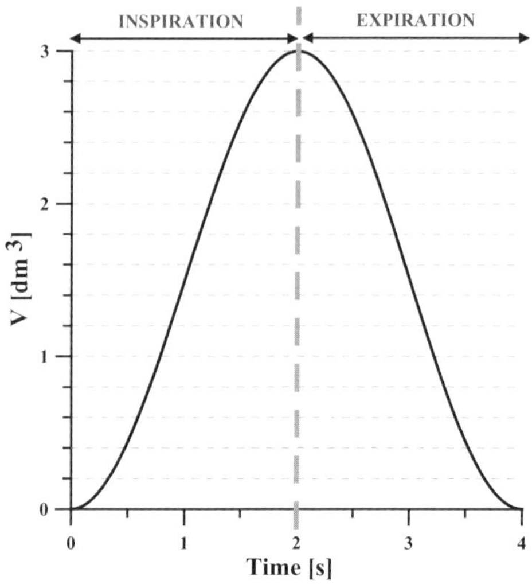

Figure 10. Shape of breathing curve used in the calculations.

The computational problem of the air flow structure during cyclic breathing was solved through integration of Navier –Stokes equations with the concept of k-ε model for description of turbulent flow, using CFD Fluent software package of version 6.0. Fluent used in this work was applied for calculation of displacement of solid spherical particles dispersed in the gas. The particle trajectory was calculated through an integration of balance equation of forces acting on the particle. For the purpose of this investigation, gravitational and thermal forces causing Brownian motion of particles are considered. For submicron particles an effect of Brownian displacement is optionally added to the component of external forces and is used in the Fluent software package. The mass stream of particles entering the trachea was assumed to equal $4 \cdot 10^{-6}$ kg/s. Monodispersed particles of diameters 10, 1, 0.1 and 0.01 μm of density 2.65 g/cm^3 were used for simulation of deposition efficiency. Those ranges of diameters correspond to particles, which are able to penetrate to the tracheobronchial tree in reality.

4.2. Results of Computations.

The airflow structure within upper airways of the respiratory system, for any instant of breathing cycle has been calculated. The computations show a strong instability of the flow in the model of upper airways. The cyclic pattern of flow forcing causes circulation of air observed in any cross-sectional planes of the airways. The deposition efficiency of aerosol particles was also calculated. The efficiency of deposition within the considered system during whole breathing cycle is presented in Figure 11.

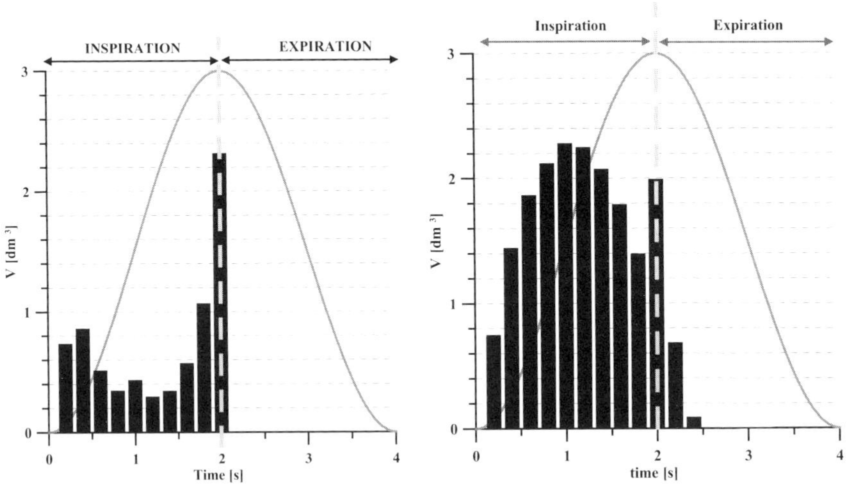

Figure 11. Deposition efficiency of aerosol particles with diameter 0.01 μm (left) and 10 μm (right).

For particles of high inertia (10 μm), the highest deposition efficiency is reached at the middle phase of the inspiration, when the average air velocity has the maximum value during the whole breathing cycle. Significantly high deposition for these particles is observed at the transition between inspiration and expiration. For particles in the diffusion range (0.01 μm), strong convection near the airways wall causes reduction of the thickness of the boundary layer and increasing of the mass transfer coefficient. Characteristic peak of the momentary deposition at the transition between inspiration and expiration is observed. It is because of intensification of airflow circulation at this moment. The "hot spots" of particle deposition was found using modeling technique. Figure 12 shows the picture of "hot spots" of deposition of particles of diameter 10 μm.

5. DISCUSSION AND CONCLUSIONS

The broad scope of current research allowed for examination of several aspects of powder resuspension and deposition in the respiratory tract. All those issues are essential for proper drug delivery from powder inhalers. As expected, air velocity through the resuspension channel is essential for the dose of aerosol formed from the powder aliquot.

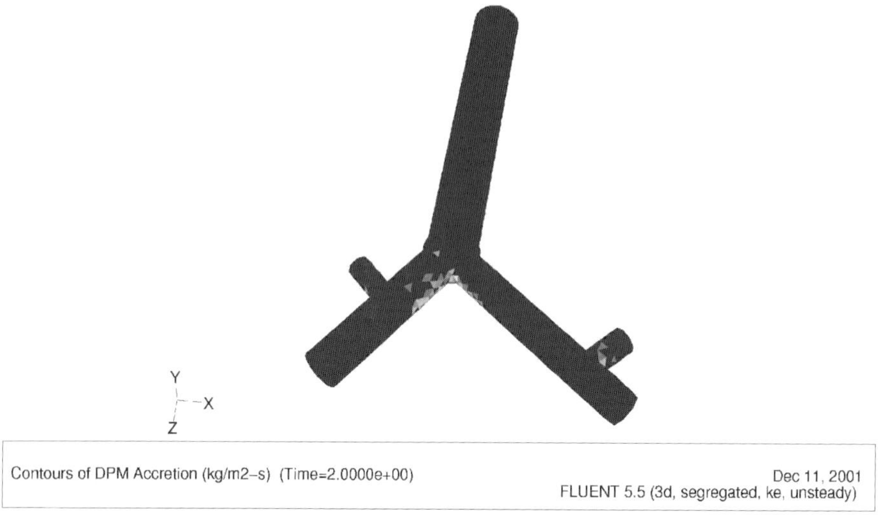

Figure 12. Computed "hot spots" of deposition of aerosol particles with diameter 10 μm.

Below the critical value of the velocity (about 2-3 m/s in these studies), no significant mass of the powder can be transformed into aerosol, which can be transported to the respiratory tract. This observation indicates the importance of the geometry of powder inhalers It is expected that several factors will influence the value of the critical air velocity – first of all the intrinsic properties of the powder (particle size distribution, hygroscopic properties, electrostatic charge, etc.) which influence the inter-particle cohesion forces. Beside the velocity, the properties of air used for resuspension, especially its relative humidity, should also influence the aerosol formation. The impact of all these factors needs detailed study what will be done in the near future.

In the case of the powder used in these preliminary experiments, it was demonstrated that increase in the airflow through the resuspension zone allowed for higher numbers of large inhalable particles entrained into the air stream. No deagglomeration of such particles was observed with the influence of air velocity in the studied range. Such experiments could be compared to the deposition of the same aerosol particles in our model of the upper airways. It was found that large particles are deposited preferentially at regions of bifurcations due to their high inertia, forming the characteristic "hot spots". Smaller particles better penetrate through the upper airways, but some of them are deposited along the airways due to turbulent diffusion.

Both experimental studies of resuspension and deposition were conducted for different steady-state flow conditions (constant airflow rate). Their results may be related to different phases of inspiration - for the given instant of the inspiration the airflow rate is given by the slope of the tangent to the breathing curve.

The constraints of the current experimental technique can be overcome with mathematical modeling, which is able to consider realistic breathing curves. In the example presented in this paper the results of computations of the airflow pattern and particle deposition rates are in agreement with experimental observations, and they also predict additional effects which intensify particle deposition in the airways. These effects are related to air circulation in the phases of the change of airflow direction, i.e., for conversions: inspiration-to-expiration, and expiration-to-inspiration. More results of the influence of the shape of the respiratory curve on the deposition rate will be publishedsoon (Gradoń and Moskal, 2002). In the near future we will be also able to measure these effects experimentally for different breathing curves with the programmed oscillatory-flow air pump.

The better understanding of mechanisms of powder resuspension and its associations with the subsequent deposition of aerosol particles in the respiratory tract is the important factor for optimization of dry powder inhalers used in inhalation drug delivery.

6. REFERENCES

Shultz, R.K., 1995. Drug delivery characteristics of metered dose inhalers., *J. Allergy Clin. Immunol.*, 96, 284-287.

Ganderton, D., N.M. Kassem, 1992. Dry powder inhalers., *Adv. Pharmaceutical Sci.*, Academic Press, London.

Ganderton, D., 1992. The generation of respirable clouds from coarse powder aggregates. *J. Biopharm. Sci.*, 3, 101-105.

Gradoń, L., Ciach, T., 1996. Formation of fibrous filters with melt blown technique., *J. Aerosol Sci.*, 27, 613-614.

Gradoń, L., Moskal, A., 2002. Temporary and spatial deposition of aerosol particles in the upper human airways during breathing cycle. *J. Aerosol Sci.*, 33, 1525-1539.

Horsfield, K., Dart, G., Olsen, D.E., Filley, G.F., 1971. Models of the human bronchial tree., *J. Appl. Physiol.* 31, 207-217.

The United States Pharmacopeial Convention, 1996. *Pharmacopeial Forum*, 22, 3048-3098.

7. AFFILIATIONS

Prof. Leon Gradoń, Dr Tomasz R. Sosnowski and Dr Arkadiusz Moskal work at the Department of Chemical and Process Engineering,
Warsaw University of Technology,
Waryńskiego 1 , 00-645 Warsaw, Poland.

8. ACKNOWLEDGMENTS

This work was supported by the State Committee for Scientific Research (KBN) under the grant No. 7 T09C 054 21.

JOACHIM HEYDER, JOSEF GEBHART, CHRISTA ROTH, GEORGE A. FERRON

TRANSPORT AND DEPOSITION OF HYDROPHILIC DRUG PARTICLES IN THE LUNGS - EXPERIMENT AND THEORY -

1. INTRODUCTION

A hydrophilic material absorbs water vapour from its surroundings. If this material forms an airborne particle with a density larger than unit density the particle increases in size and decreases in density due to the water uptake. The larger the relative humidity rh in the air surrounding the particle the larger becomes the particle. It grows in size with increasing rh. If the particle is a droplet at zero rh and composed - for instance - of sulphuric acid or ethanol the particle grows monotonously with increasing rh. In case of a crystalline material the particle grows only above a certain value of rh, the deliquescence point. At this rh the crystalline particle becomes a droplet.

For sodium chloride the deliquescence point is 0.76 (Figure 1). When rh exceeds a value of 0.98 a sodium chloride particle grows considerably in size. A small change in rh results in a huge change in particle diameter. Sodium chloride particles are thus ideal tools for probing with great precision the relative humidity of air close to water vapour saturation like the rh in the human lung.

At the deliquescence point the density of a sodium chloride particle decreases from 2.16 to 1.19 g cm^{-3}. Above this point it decreases further until it approaches the density of water for unit rh.

Modelling deposition patterns in the respiratory tract of inspired drug particles with hydrophilic surfaces must be based on the longitudinal pattern of rh in the respiratory tract and on the growth of the drug particles with rh. In this study sodium chloride particles were therefore used to determine in-vivo the longitudinal pattern of rh in the human respiratory tract. The dependency of the size of drug particles on rh was determined in-vitro. These data were incorporated in the mathematical deposition model developed by Ferron et al. (1988b) and, finally, deposition patterns of the drug particles in the human respiratory tract were estimated .

2. PARTICLE GENERATION

Compact 0.67 μm and 0.25 μm sodium chloride particles and 1.10 μm polystyrene particles were used for the inhalation studies.

L. Gradoń and J. Marijnissen (eds.), Optimization of Aerosol Drug Delivery, 139–147.

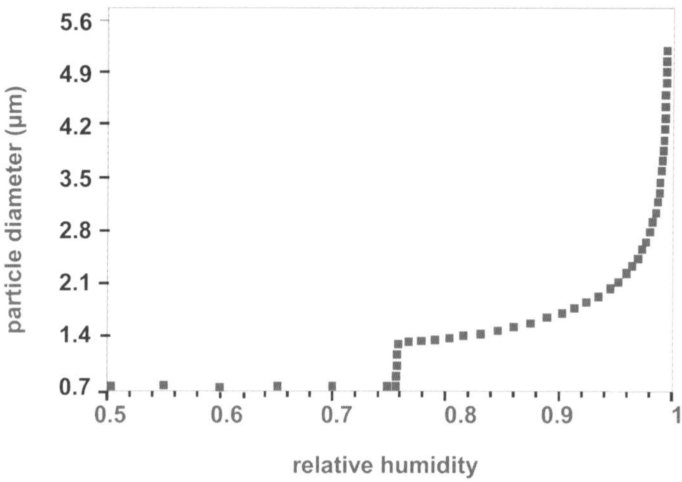

Figure 1. Growth of an airborne 0.7 µm sodium chloride particles with increasing relative humidity of the air surrounding the particle.

2.1. Sodium Chloride Particle

Monodisperse sodium chloride droplets were produced by nebulization of an aqueous sodium chloride solution with a vibrating orifice generator (Berglund and Liu 1973). The droplets contained sodium chloride equivalent to 0.67 µm sphere. After complete water evaporation the diameter distribution of the remaining solid sodium chloride particles was determined with a non-commercial low-angle light scattering spectrometer (Gebhart et al. 1976). Its modal diameter was 0.70 µm and its geometric standard deviation 1.03.

When the settling velocities of the solid particles were determined in a home-made sedimentation channel their densities turned out to be slightly lower (1.89 ± 0.17 g cm^{-3}) than the bulk density of sodium chloride (2.165 g cm^{-3}). The sodium chloride content of the 0.7 µm particles of 1.89 g cm^{-3} density is identical with that of the 0.67 µm particles of 2.165 g cm^{-3} density. The same procedure was applied to generate 0.25 µm sodium chloride particles.

2.2. Polystyrene Particles

Monodisperse polystyrene particles of 1.05 g cm^{-3} density were generated by nebulization of aqueous polystyrene suspensions. After complete water evaporation the diameter distribution of the remaining polystyrene particles was calculated from their settling velocities determined with a home-made sedimentation channel. Its modal diameter was 1.10 µm and its geometric standard deviation 1.04.

3. IN-VIVO DETERMINATION OF INTRAPULMONARY RELATIVE HUMIDITY

3.1. Determination of the Relative Humidity in the Lung Periphery

Since rh in the upper and central airways increases during inspiration and decreases during expiration an inspired sodium cloride particle will grow in size during inspiration due to water vapour uptake and may shrink in size during expiration due to water evaporation. Therefore, the size a particle assumes in the lung periphery has to be determined on the spot. For this purpose, aerosol-derived airspace morphometry was applied (Palmes et al. 1973, Heyder 1983, Rosenthal 1989). An aerosol bolus carries monodisperse particles in the lung periphery. During a subsequent respiratory pause the particles settle onto airspace surfaces. The longer the pause the more particles are deposited on these surfaces and the less particles remain airborne. The probability RC that a particle is still airborne in the lung periphery at the end of the pause and can be recovered from the lung with the expired air can be approximated by

$$RC = \exp(-1.27\, vt / x)$$

where t is the time of breath-holding, v the settling velocity of the particles, and x the mean linear intercept in the lung parenchyma. If the settling velocity of the particles in the lung periphery is known the mean linear intercept can be calculated. This settling velocity is known for particles composed of hydrophobic material.

In this study monodisperse 1.10 μm polystyrene particles with a density of 1.05 g cm^{-3} were used to determine the mean linear intercept of three volunteers (Anselm 1989). The particles were carried over an average period of 3 seconds into a mean lung depth of 700 cm^3. Their recovery was determined by aerosol photometry (Gebhart et al.1988).

In Figure 2 the recovery function RC(t) obtained for one subject is shown. Without breath-holding 76 % of the inspired polystyrene particles were expired; only 25 % after 8 seconds breath-holding. The exponential decline in the number of expired particles indicated that the particles settled during breath-holding in the periphery of a lung characterized by a mean linear intercept of 400 μm. This mean linear intercept was then used to evaluate from the recovery function of 0.25 μm sodium chloride particles (Figure 2) their settling velocity in the lung periphery after bolus delivery of these particles to the same mean lung depth of the subject.

Since the recovery function of the sodium particles also turned out to be an exponential function as that of the hydrophobic polystyrene particles, it can be concluded that all sodium chloride particles assumed their equilibrium size with the moist air in the lung periphery at the onset of breath-holding. However, they deposited faster in the lung periphery during breath-holding than the polystyrene particles. Without breath-holding 60 % of the inspired sodium chloride particles were expired; only 12 % after 8 seconds breath-holding. They assumed therefore a larger size in the lung periphery than the polystyrene particles. They grew from 0.25

to 1.52 μm. This increase in size takes place when sodium chloride particles are exposed to air at 0.9948 rh. This value is identical with the anticipated value of (0.99473 $\pm$ 0.00009) which is the relative humidity of air in thermodynamic equilibrium with blood serum (Ciba Geigy 1979).

The peripheral rh of the second volunteer was found to be identical with that of the first volunteer (0.9946). However, the third volunteer had a lower peripheral rh (0.9928).

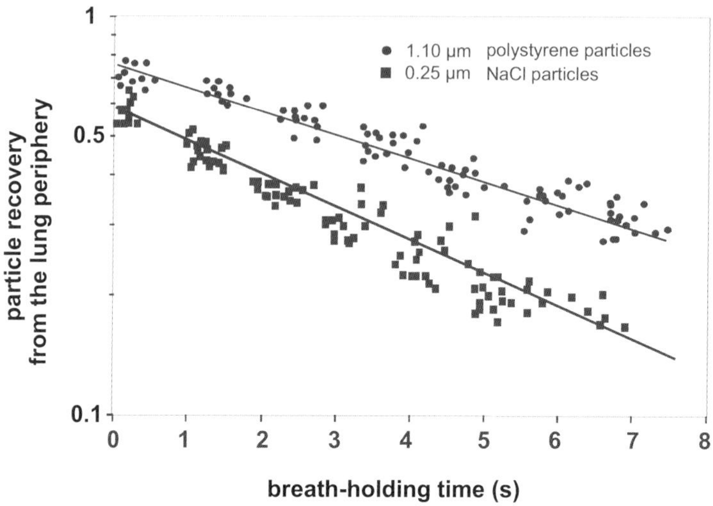

Figure 2. Recovery of polystyrene and sodium chloride particles after breath-holding from the lung periphery of a volunteer.

3.2. Determination of the Longitudinal Distribution of Relative Humidity in the Respiratory Tract

For this purpose compact 0.67 sodium chloride particles were carried with boluses at 250 cm^3 s^{-1} flow rate into varies volumetric depths in the respiratory tract of the two volunteers with 0.9947 peripheral rh. Without breath-holding they were then expired at the same flow rate. When leaving the respiratory tract their size was measured with an on-line particle size spectrometer (Gebhart et al. 1976). The results are shown in Figure 3.

Assuming that the particles were exposed to 0.995 rh when entering the respiratory tract their growth function is also shown in Figure 3. It turned out that measured and calculated growth functions were very similar beyond a lung depth of about 200 cm^3; the particles grew to a size in equilibrium with the water vapour in the peripheral lung generated by the blood serum.

Since less time was available for the expiration of particles penetrating less into the respiratory tract than for those particles penetrating into the lung periphery these particles also maintained the size they assumed during inspiration. Consequently, the size of all sodium chloride particles leaving the respiratory tract is the size to which they grew during inspiration regardless of how deep they penetrated into the respiratory tract. On-line particle size spectrometry could therefore be applied for the evaluation of particle growth as a function of the volumetric depth to which the particles were inspired.

In depths proximal to 200 cm^3 particles grew much less than those penetrating into the lung periphery, indicating that rh in upper airways is less than that in the lung periphery. However, intrapulmonary particle growth calculated for the longitudinal rh profile shown in Figure 4 agrees well with the experimental data at all lung depth (Figure 3). This rh profile was therefore experimentally confirmed with this study.

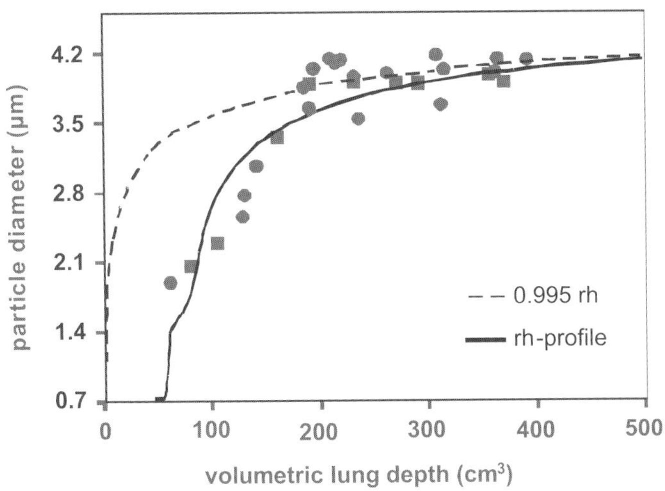

Figure 3. Growth of compact 0.67 µm sodium chloride particles in the respiratory tract of two volunteers as a function of the volumetric depth to which the particles were carried with aerosol boluses. The broken curve characterizes intrapulmonary particle growth assuming the particles were exposed to air at 37 °C and 0.995 rh when entering the respiratory tract, the unbroken curve intrapulmonary particle growth predicted for the intrapulmonary longitudinal rh profile shown in Figure 4.

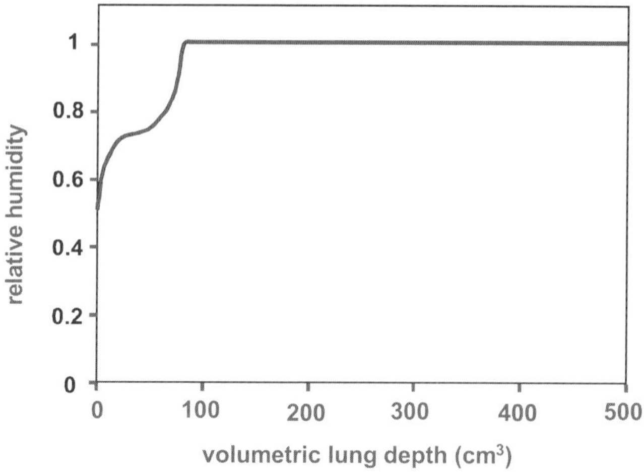

Figure 4. Intrapulmonary longitudinal rh profile (Ferron et al. 1988a).

4. IN-VITRO DETERMINATION OF GROWTH OF HYDROPHILIC DRUG PARTICLES IN MOIST AIR

To simulate transport through and deposition in the respiratory tract of hydrophilic particles the dependency of their size on rh has to be known. This dependency was determined in this study with two differential mobility analysers. The first analyser was used to collect 0.165 μm particles from polydisperse aerosols at rh below the deliquescence point. This fraction was then exposed to air of increasing rh. The particle size distribution of the humidified particles was determined with a second analyser.

In Figure 5 the growth of terbutaline sulphate particles is shown (Seemann et al. 1995). At 0.995 rh the drug particles generated from the liquid formulation grew to larger sizes (0.64 μm) than those generated from the powder formulation (0.36 μm). The powder formulation was simulated by nebulization of a 1% aqueous solution of pure terbutaline sulphate and subsequent drying of the generated droplets. For simulating the liquid formulation the aqueous terbutaline sulphate solution was replaced by a 1% Bricanyl solution.

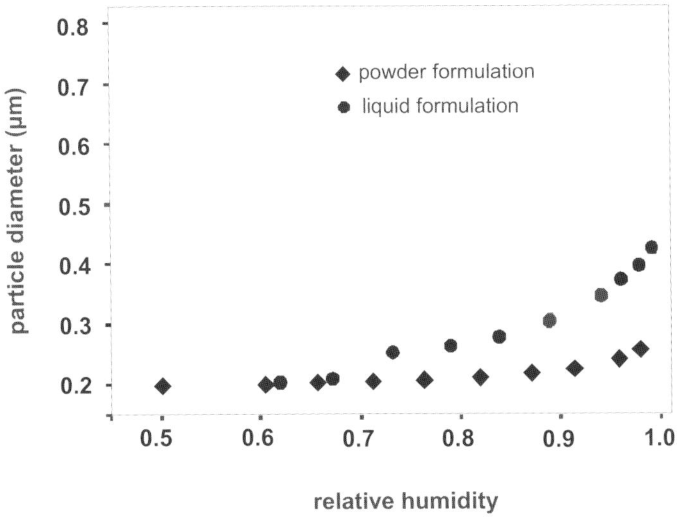

Figure 5. Growth of airborne 0.16 µm terbutaline sulphate particles with increasing relative humidity of the air.

5. MATHEMATICAL SIMULATION OF DEPOSITION OF HYDROPHILIC DRUG PARTICLES IN THE HUMAN RESPIRATORY TRACT

The mathematical model developed by Ferron et al. (1988b) was used to simulate for mouth-breathing at rest the effect of intrapulmonary growth of polydisperse terbutaline sulphate particles on total and alveolar deposition (Figure 6). The growth has a huge effect deposition. In case particle growth is neglected (simulated by a fictitious hydrophobic terbutaline sulfate formulation) alveolar deposition is underestimated by up to a factor of 2.4 in the range of 0.3 – 3 µm median diameter and overestimated by up to a factor of 1.8 in the range of 0.03 – 0.1 µm median diameter. However, the intersubject variability of intrapulmonary rh (0.995 vs. 0.993) has only a marginal effect on deposition. Therefore, in the size range of current pharmaceutical aerosols (0.3 – 3 µm) alveolar deposition can be enhanced when particles with hydrophilic rather than hydrophobic surfaces are used.

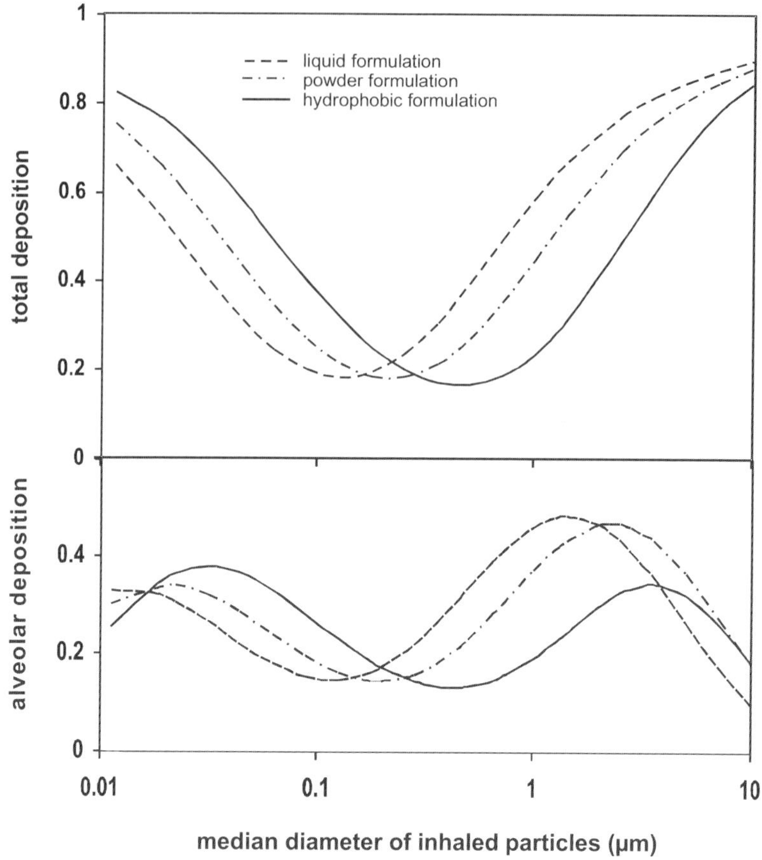

Figure 6. Mathematical simulation of total and alveolar deposition for breathing terbutaline sulphate particles through the mouth at the reference resting pattern for an adult Caucasian male sitting awake (300 cm^3 s^{-1} flow rate and 5 s breathing cycle period, Cotes 1979). The sizes of the particles in the inhaled aerosol were assumed to be logarithmic normally distributed with a geometric standard deviation of 2. The simulations were performed for particles of 1.59 g cm^{-3} density generated from the hydrophilic liquid formulation, for particles of 1.34 g cm^{-3} density generated from the hydrophilic powder formulation, and for particles of 1.34 g cm^{-3} density genera-ted from a fictitious hydrophobic formulation.

6. CONCLUSIONS

The longitudinal profile of intrapulmonary relative humidity has been measured in vivo. In the lung periphery it is 0.9948. This value is identical with the value of 0.99473 $\pm$ 0.00009 which is the relative humidity of air in thermodynamic equilibrium with blood serum. The relative humidity in the upper airways is lower than this equilibrium value and in agreement with values estimated by Ferron et al. (1988a).

Deposition of pharmaceutical particles can considerably be enhanced when particles with hydrophilic rather than hydrophobic surfaces are used.

7. REFERENCES

Anselm, A. Untersuchungen zum Verhalten hygroskopischer Aerosolteilchen im Atemtrakt des Menschen. Ph. D. Thesis, University of Frankfurt am Main, 1989.

Berglund, R. N. and B. Y. H. Liu. Generation of Monodisperse Aerosol Standards. *Env. Sci. Techn.* 7: 147-152, 1973.

Cotes J. E. Lung Function Assessment and Application in Medicine. *Blackwell Scientific Publications*, Oxford 1979.

Ciba-Geigy. *Wissenschaftliche Tabellen Geigy.* Teilband Hämatologie und Humangenetik. Ciba-Geigy Limited, Basle, CH, 1979.

Ferron, G. A., B. Haider, and W. G. Kreyling. Inhalation of salt aerosol particles - I. Estimation of the temperature and relative humidity of the air in the human upper airways. *J. Aerosol Sci.* 19: 343-363, 1988a.

Ferron, G. A., W. G. Kreyling, and B. Haider. Inhalation of salt aerosol particles - II. Growth and deposition in the human respiratory tract. *J. Aerosol Sci.* 19: 611-631, 1988b.

Gebhart, J., J. Heyder, C. Roth, and W. Stahlhofen. Optical Size Spectrometry Below and Above the Wavelength of Light. *Fine Particles* (editor: B.Y.H.Liu), Academic Press, New York: 794-815, 1976.

Gebhart, J., G. Heigwer, J. Heyder, C. Roth, and W. Stahlhofen. The Use of Light Scattering Photometry in Aerosol Medicine. *J. Aerosol Med.* 1: 89-112, 1988.

Heyder, J. Charting Human Thoracic Airways by Aerosols. *Clin. Phys. Physiol. Meas.* 4: 29-37, 1983.

Palmes, E .D., C. Wang, M. Goldring, and B. Altshuler. Effect of Depth of Inhalation on Aerosol Persistance during Breathholding. *J. Appl. Physiol.* 34: 356-362, 1973.

Rosenthal, F. S. Aerosol Recovery following Breathholding derived from Distribution of Chord Lengths in Pulmonary Tissue. *J. Aerosol Sci.* 20: 267-277, 1989.

Seemann, S., B. Busch, G. A. Ferron, A. Silberg, and J. Heyder. Measurements of the Hygroscopicity of Pharmaceutical Aerosols in situ. *J. Aerosol Sci.* 26: S537-538, 1995.

8. AFFILIATIONS

GSF – National Research Centre for Environment and Health, Institute for Inhalation Biology, D – 85758 Neuherberg/Munich, Germany.

FRANS DE JONGH AND RUPINO GRIFFIOEN

VENTILATION MODELLING AND DRUG DEPOSITION IN VERY YOUNG ASTHMATIC CHILDREN

1. INTRODUCTION

In clinical practice, the drug deposits in the lungs of a young child using an aerosol are much less than in those of an adult. However, the reasons for this are not well understood. We will discuss the necessity of administering drugs in aerosol form for this category of patients. Some recent in vivo studies are reviewed. These show that lung deposits are, indeed, very low in children. Also an in vitro study of an upper airway cast of a child is discussed. An explanation for the low deposition is given using a mathematical model that focuses on the upper airways. Finally, own model calculations are performed in the lung of an adult to investigate if the desired deposition site can be targeted with specific flow rates and tidal volumes.

The wide spectrum medication inhaled by children can be divided into three groups. The two main groups are corticosteroids to decrease inflammatory processes and beta-2-agents and anti-cholinergics to achieve bronchodilatation. The third group covers specific medications like recombinant human DNase for Cystic Fibrosis (CF). More substances are being investigated as being potentially beneficial to patients when inhaled in an aerosolised form. An example is surfactant, a surface tension reducing agent covering the alveoli that is being studied especially in pre-term neonates suffering from respiratory distress syndrome. Drugs like insulin and morphine are now being investigated to see if aerosolised forms are useful in treating diabetes and pain respectively.

Aerosolised medication is much preferred when the drug has to act in the lung. It avoids the oral tract which may affect the bio-activity of that drug, and also avoids the invasiveness of injections. The inhaled drug will be available to the airways and lungs in higher local concentrations than is toxicologically possible when it is applied systemically. When a drug is given orally or intravenously it will reach all the tissues in the body. Because it acts everywhere it will yield systemic effects. Side effects are often seen when oral corticosteroids are used. The consequences of too much corticosteroids include the suppression of the adrenal cortex, leading to reductions in growth velocity and height and also changes in bone structure, bone mass and bone turnover rate, weight gain and thin skin and Cushing's syndrome.

L. Gradoń and J. Marijnissen (eds.), Optimization of Aerosol Drug Delivery, 149–164.

Carlsen reported that inhaled corticosteroids could cause local side effects like perorial dermatitis, oral candidiasis hoarseness and dysphonia, but systemic side-caused the most concern (Carlsen et al).

Price reviewed eighteen studies where the effects of more than one year's treatment with inhaled corticosteroids on growth were investigated (Price et al). Patients with asthma were divided into four groups: a group on inhaled corticosteroids, a group on placebo, a group where different inhaled corticosteroids were compared and so-called "real life" groups where the corticosteroid dose was adjusted. The main message was that after beclomethason and budesonide had been inhaled, a small degree of growth suppression was seen over a period of one to two years. Fluticason had no effect on growth. However, all studies showed no effect on final height. These results suggest that even high doses of inhaled corticosteroids in children with asthma will not result in lasting side effects and precise optimisation of drug dose delivery is not obligatory.

However, Shinwell conducted a randomized, double blind, placebo controlled study with 132 pre-term neonates (29 weeks of gestational age, weight 1240 grams). The patients received a three-day course of a systemically given corticosteroid (dexamethasone) or placebo within the first twelve hours after birth with the aim of improving the condition of their lungs (Shinwell et al). They found no favourable effects, but reported considerable differences in severe side effects between the two groups. Cerebral palsy (spastic diplegia) was seen in 49% (39/80) of the dexamethasone group compared to 15% (12/79) of the placebo group (Odds Ratio: OR=4.62). Also delays in development were seen in 55% (44/80) of the group receiving dexamethasone compared to 29% (23/79) in the placebo group (OR 2.87). This study showed that for this group of very small patients it may be very important to investigate whether corticosteroids could be inhaled. Here, it should be taken into account that the reason why medication is still given systemically is because in aerosolised form almost no medication reaches the lungs of these patients.

There is an abundance of in vivo studies comparing different drugs, in different doses, with different delivering systems, in different patient groups and with different objectives and results. These studies do not present unanimous results. Recently, for instance, much higher lung doses in adults as well as in children where reported with drugs with a smaller mass median aerodynamic diameter (MMAD) than before (as a consequence from the transition from chlorfluorcarbon (CFC) pressurized metered dose inhalers (pMDI's) to some hydrofluoralkane (HFA) pMDI's). However, the problem with lung dose is that it does not specify in which area and in what generations of the airways the particles deposit. Drugs act at specific receptors and optimal occurs when the particles reach the receptors. Do these small particles mainly deposit in regions like the alveolar lining where there are probably fewer receptors. In the alveoli these particles can easily be absorbed – as quick as an intravenous injection - and pass into the bloodstream causing systemic effects.

In (very) young and (very) small patients inhalation is dictated by obligatory nasal and tidal volume respiration when they are co-operative. Between the ages of four and six years children can change, when asked, from transnasal to transoral inhalation, thereby changing from an obligatory, airtight fitting nose-mouth-facemask to a mouthpiece. In the same period they can change on command from tidal volume respiration to slow, deep inhalation, which by itself renders inhalation more effective.

2. IN VIVO STUDIES

Reliable results of aerosol studies on small children are scarce. The number of very young children needing aerosol medication is small compared to the adult population. Studies of this group are expensive and cumbersome. One has to bear in mind that studies involving children are more difficult than studies of adults and that the results of studies in schoolchildren cannot be used for very small children like pre term babies. Schoolchildren can cooperate in a study, but for pre-school children special ways of investigation are needed that do not depend on the patient co-operation. Ethical considerations raise other problems because young children are not allowed to decide whether they want to co-operate in investigations. Results of studies on animal models are of limited value because airways have species-specific properties. And this is especially so where studies concerning the inhalation of drugs are concerned.

There are only a few trials studying drug deposition in small children. Anhoy investigated three groups of mild asthmatic patients: 2-3 years (n=8), 4-6 years (n=8) and adults 20-41 years (n=10) (Anhoy et al). All persons received the same dose, two times 200 microgram budesonide (an anti-inflammatory corticosteroid) pMDI, with a holding chamber (the metal Nebuchamber). The striking result was that the plasma-concentrations of the drug in the three groups did not differ. If the lung dose and thereby the fraction of drug entering the cardiovascular circulation would be equal, the younger age group should have higher plasma-concentrations since the children have less blood volume. Since the plasma concentrations of all groups were equal, they concluded that this was an indirect indication that the lung dose for an adult must be higher than that for children.

Chua studied 20 patients with Cystic Fibrosis, twelve from 0.8 years of age and eight of almost 11 years (Chua et al). Radio labelled saline (99mTc-DTPA) was nebulised with a Turret nebuliser at 9 l/min. This produced an aerosol with a MMAD of 3.0 μm. The twelve infants received the drug while sleeping (in supine position and via nasal route), while the older age group inhaled in sitting position both orally and nasally. The lung deposition by the nasal route was 1.3% (0.3-1.6%) in the infants, 2.7% (1.6-4.4%) in the older children through the nose and via the oral route 6.3 % (4.9 - 9.1%). Again a lower lung dose is found in both young and older children, even when breathing through their mouth – than in adults.

Diot studied lung deposition as a function of age in patients with Cystic Fibrosis in the age range of five to thirty years. Radio labelled saline was used (99mTc-HAS) (Diot et al). When upper airway (mouth and throat) deposits were plotted as a function of age, the graph revealed a high deposition fraction (about 50%) for the youngest children while the adult upper airway deposition was around 5%. Recently Amirav studied aerosol deposition in twelve patients between the ages of 0.8 years with Respiratory Virus (RSV) bronchiolitis (Amirav et al). A radio labelled aerosol of a bronchodilating drug (99mTc-albuterol) from a micromist nebuliser (Hudson) at 8 l/min with a MMAD of 4.2 (1.8) μm was used. From the applied dose 3% deposited in the lungs (1.2% in the peripheral parts of the lungs), 7.8% deposited in the upper airways (nose, mouth, throat), and 11% on the face. So again a low lung dose is seen in young children.
All these studies show that the amount of aerosol deposited in the lungs of children is considerably less - 2 to 8% of the prescribed dose – compared to that being deposited in the lungs of adults, where it varies between 20 and 50% of the prescribed dose. It is therefore common clinical practice to prescribe the same dose of drugs for all age groups.

3. IN VITRO STUDIES

Recent in vitro studies focus on aerosol deposition as a function of age. Janssens did aerosol research with a model made of a 3D-CT scan of a nine-month-old child of 10 kilogram, breathing through the nose (Janssens et al). To mimic normal breathing while measuring with the standard Andersen impactor operating at 28.3 l/min the setup shown in Figure 1 was used.

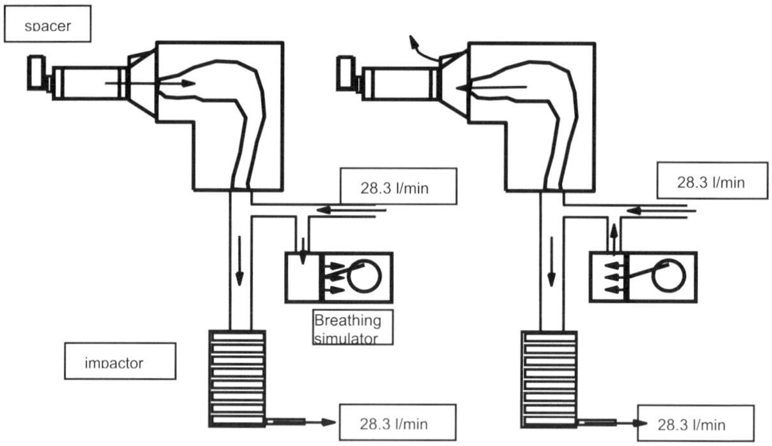

Figure 1. Setup of measuring MMAD with an Andersen impactor working with a constant flow rate while simulating normal breathing through a cast of the upper airways (left side inspiration, right side expiration).

The flow of a breathing simulator, which replayed pre-recorded breathing patterns, was superimposed on a bias flow of 28.3 l/min. Two hundred μgram budesonide with a metal Nebuchamber was administered while the model breathed with a respiratory rate of 25 per minute with an inspiration/expiration time ratio of 1:2. The amount of budesonide passing through the cast, called "lung dose'' in this study, as percent of nominal dose was measured as function of the tidal volume. Also the respiratory dose (particles <4.7 μm) and the extra fine particle dose (particles < 2.1 μm) were measured with the impactor. With a tidal volume of 50 ml the lung dose was 14% (12% particles < 4.7μm, 1.5% particles <2.1 μm). With a tidal volume of 100 ml the lung dose decreased: 9% (8% < 4.7 μm, 1.5% <2.1 μm), and with a tidal volume of 200 ml the lung dose was 4 % (3.8% < 4.7 μm, 1.5% < 1.5 μm). In contrast to Diot's study all deposition fractions where lower for higher tidal volumes. However, in the experiments conducted by Janssens et al the respiratory rate was fixed which implies that doubling the tidal volume leads a doubling of the flow-rate through the cast. Higher flow rates will produce increased inertial impaction in the upper airways.

Also the difference between HFA-beclamethason (QVAR^R) and CFC-beclomethason – an anti-inflammatory corticosteroid - was investigated with an Aerochamber^R as spacer (Janssens 2001). For HFA-beclomethason the lung dose was about 25% of the nominal dose for a tidal volume (TV) of 50 ml, increasing to 30% at a TV of 200 ml. Almost all particles reaching the "lung" where less than 2.1 μm. For the CFC-beclomethason the 'lung dose' was only 7% for a TV of 50 ml decreasing to 3% with a TV of 200ml. Only 1% was less then 2.1 μm. This huge difference is due to the composition of the aerosol. The HFA formulation has a MMAD of 1.1 μm while the CFC formulation had a MMAD of about 3.0 μm.

Again it is clear from these model studies that the amount of drug reaching the lung of a small child is less than that of the expected lung dose in an adult, except perhaps for aerosols with a small MMAD. Applying aerosols with a small MMAD will considerably enlarge the lung dose - also in small children. However, it is not clear whether these particles will target their receptors.

4. MATHEMATICAL MODEL STUDIES

Finally, mathematical model studies are investigated. The three main variables in these studies are the aerosol characteristics (MMAD, GSD), the respiratory pattern, and the upper airway and lung model.

Martonen and Hofmann calculated deposition as a function of age at 7, 22, 48, 98 and 360 months (Martonen and Hofmann). For sedentary breathing he calculated a tracheobronchial deposition fraction of around 30-20% for seven-months old children for small particles (less then 2 μm). For 10 μm particles the deposition fraction increased to 80%. With increasing age the tracheobronchial deposition fraction curve kept exactly the same shape although it was about 10% lower for the adults. The pulmonary deposition curve was more complicated. It varied between 10 and 30%, with the highest fraction for 48 months old children and the lowest fraction for adults and 7-month-old children. Total deposition - the sum of the two factors – mainly followed the tracheobronchial curve, since this contributed the highest fraction. In a recent model study Phalen found similar results (Phalen et al). Bronchial and total deposition is the highest in children for each particle size, while pulmonary deposition is the highest in adults.

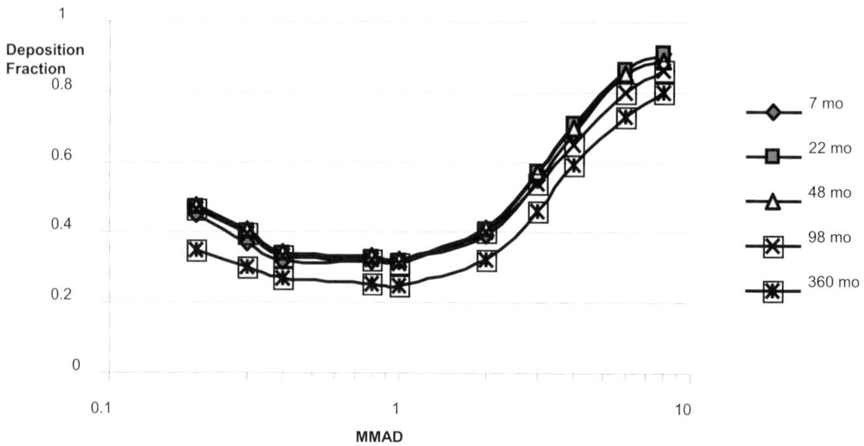

Figure 2. Total (tracheobronchial and pulmonary) aerosol lung deposition during sedentary breathing as function of age (7, 22, 48, 98 and 360 months) after Martonen (Martonen et al, 1989).

There seems to be a contrast between these model studies and the vivo and in vitro studies where considerably lower lung depositions were found in children. The explanation for this discrepancy is that the model studies calculate which fraction of the aerosol entering the trachea deposits in the lung. The in vivo and in vitro studies consider the fraction of aerosols depositing in the lung when delivered at the mouth. Although often neglected, the upper airways - comprising nose, mouth and throat - take the major part in aerosol deposition for children. Schulz showed the empirically found relation between aerosol deposition fraction (as inspiratory collection efficiency) in the upper airways and particle size and mass flow rate (see Fig 3) in human adults breathing through their nose as well as breathing through their mouth (Schulz et al).

Considering normal breathing, so with a known mass flow rate (Q) vertical lines can be drawn for a specific particle size and deposition fraction can easily be determined from the figure. For instance, when inhaling 5 μm particles through the nose, around 63% will deposit in the upper airways. During mouth breathing this fraction is reduced to 7%. For this reason aerosols are preferably inhaled through the mouth and

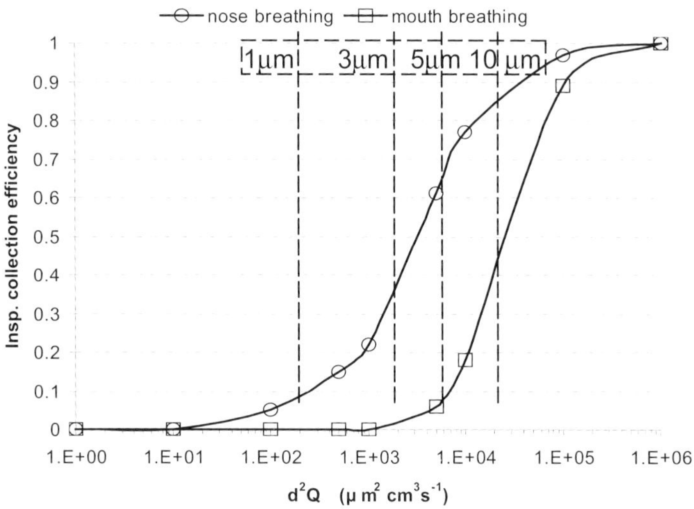

Figure3. Aerosol deposition in the upper airways by nose or by mouth breathing in an adult human (after Schulz et al). The vertical lines compare results for particles of 1, 3, 5 and 10 µm during normal respiration of an adult human, so with an average mass flow rate Q of 250 ml/s (d= particle diameter).

particles less than 5 µm are named respirable, indicating that these particles normally have a great chance to deposit in the lungs.

Phalen showed a similar graph for a one-month year old child (see Fig 4). When the mass flow rate of a child is known, again vertical lines for different particle sizes can be drawn. Since children prefer to breath through their noses the left curve has to be chosen. From the 5 µm particles 80% will deposit in the child's upper airway and still 60% of the 3 µm particles will not reach the trachea. This explains why the lung deposition found in the vivo and in vitro studies is considerably less in children than in adults. For children, particle sizes of less than 3 µm will be more favourable, while that size is now considered to be optimal in adults. Smaller particles will deposit deeper in the lungs, so in the small and more peripheral airways at higher generation numbers. When the receptors for the delivered drug are more evident in the central airways or lower generation numbers – then the therapeutic effects will be small and the chances of undesired systemic effects will be high.

In Table 1 characteristic numbers are shown for bodyweight (BW), height, respiratory rate (RR), tidal volume (TV), minute volume (Minvol; the volume that is moved while breathing during a minute), diameter of the trachea and flow velocity in the trachea (Vtrach) as function of age.

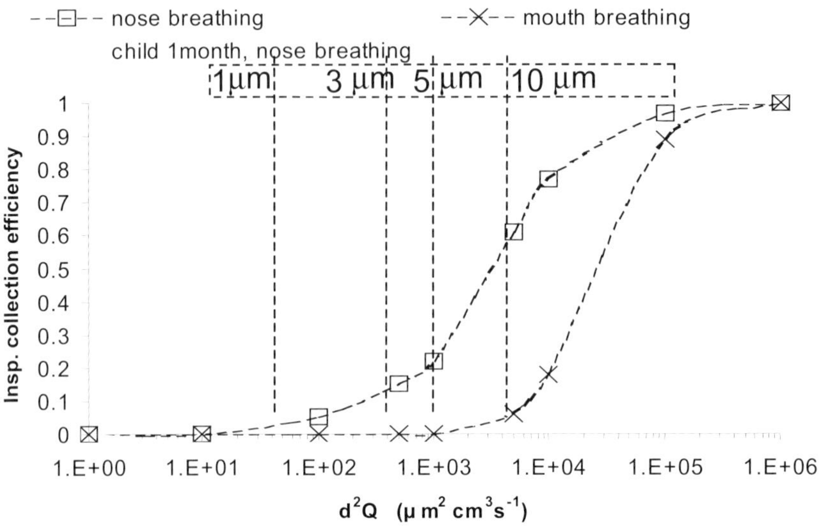

*Figure 4. Aerosol deposition in the upper airways during breathing through the nose or through the mouth in a one month old child (afterPhalen et al), the vertical lines are for normal respiration of such a child, Q =42ml/*s

Although for the young age groups in particular no consensus exists in numbers between different sources, it is clear that the average velocity in the trachea of a child during normal breathing is higher than in the trachea of an adult. Inertial impaction might, therefore, be a more important deposition mechanism in small children and to overcome this particle size and/or flow rate must be reduced to limit upper airway deposition.

Table 1 Bodyweight, length, respiratory rate, tidal volume, minute volume, diameter of the trachea and average velocity in the trachea as function of age (Hughes et al, Martonen et al, Stocks et al. and own measurements). All values refer to a resting state. The values between brackets are for small children who often will increase their respiratory rate instead of decreasing their tidal volume as soon as they are touched by inhalation devices or are 'stimulated' by an aerosol.

	Preterm	Newborn	1 year	7 year	Adult
BW (kg)	1	3.5	10	25	70
Height (cm)	35	50	75	120	175
RR (min-1)	60(90)	45 (60)	32 (49)	20	12
TV (ml)	7 (6)	28 (24)	97 (83)	180	500
Minvol (l/min)	0.42 (0.54)	1.26 (1.44)	3.1 (4.1)	3.6	6
Trachea ∅ (mm)	2.5	4.5	8.0	13	18
Vtrach (m/s)	2.9	2.6	2.1	0.9	0.8

To investigate if a change in breathing pattern can influence the site of deposition of particles that pass the upper airways we performed own model calculations. The standard 1-dimensional model is used with L as deposition term.

$$\frac{\partial C}{\partial t}=\frac{a}{A}D\frac{\partial^2 C}{\partial x^2}+\frac{1}{A}\frac{\partial(aD)}{\partial x}\frac{\partial C}{\partial x}-U\frac{\partial C}{\partial x}-\frac{L}{A} \qquad (1)$$

The deposition term $L=L_i+L_s+L_d$ consists of a term modelling inertial impaction (Li), a term modelling deposition by sedimentation (Ls) and a term for deposition by molecular diffusion.

The commonly used semi-empirically found relations have been chosen.

For L_i the equation after Taulbee and Yu is used (Taulbee et al)

$$\begin{aligned} L_i &= 1.3\frac{CAU}{l}(Stk-0.001) && Stk \geq 0.0001 \\ L_i &= 0 && Stk < 0.0001 \end{aligned} \qquad (2)$$

With $$Stk=\frac{\rho_p d_p^2 U}{18\mu\, d} \qquad (3)$$

ρ_p = density of the aerosol particle
d_p = diameter of the aerosol particle
μ = dynamic viscosity of air
d = characteristic length (airway diameter)

For sedimentation $$L_s=\frac{cAu}{l}(1-\alpha)\eta_s+\frac{N_a}{l}\phi_s \qquad (4)$$

With $$\eta_s = 1-\exp\left(-\frac{2v_s l}{Q}\sqrt{\frac{a}{\pi}}\right) \qquad \phi_s = c\iint \vec{v}_s \cdot d\vec{s}_a \qquad (5)$$

And the terminal settling velocity as $$v_s=\frac{\rho_p d_p^2 g C_c}{18\mu} \qquad (6)$$

The deposition term for diffusion is $$L_d=\frac{cAu}{l}(1-\alpha)\eta_d+\frac{N_a}{l}\phi_d \qquad (7)$$

With $$\eta_d=1-\exp\left(-\frac{36D_B l}{Q}\right) \quad \text{and} \quad \phi_d = cs_a\sqrt{\frac{D_B}{\pi t}} \qquad (8)$$

With the Brownian diffusion coefficient : $D_B = \dfrac{kTC_c}{3\pi\mu d_p}$ (9)

All equations are discretised and solved on a Iris (Unix) system. The adult Weibel lung model is chosen to compare results with existing models. In Figure 5 results are shown for the aerosol deposition fraction of particles entering the trachea with different constant mass flow rates while breathing the same tidal volume of 500 ml. Results proved to be very similar (most within 5%) to the results of the existing model calculations made by various authors including for example, Martonen 1996 and Schulz 2000. The highest deposition was found for the lowest mass flow rate (Q= 50 ml/sec, inhalation and exhalation time are 5 seconds each) due to the high residence time available for sedimentation and Brownian diffusion. The higher the flow rate the lower the deposition fraction for all particle sizes in the lung. Sinusoidal breathing was also studied with a frequency of 15 breaths/min. This can be compared with the line for Q=250 ml/sec for which the inspiration time and expiration time is 2 seconds in each case, so the respiratory rate is again 15 breaths/min). The graph shows that the deposition fractions with a constant and with a sinusoidal mass flow rate shows negligible difference for each MMAD.

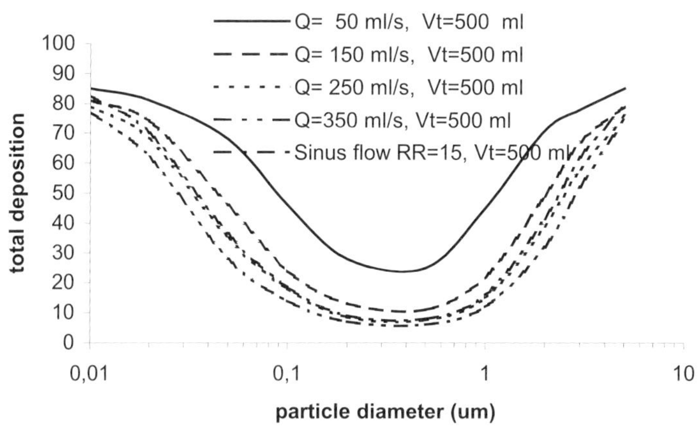

Figure 5 Aerosol deposition in the lung as a function of flow rate with a constant tidal volume of 500 ml and with varying respiratory rates .The constant flow rate pattern and that for a sinusoidal breathing pattern are plotted.

In Figure 6 results for the deposition fraction for normal respiration (RR=12. TV=500 ml) are plotted as a function of the generation number. The deposition fraction is calculated as the number of particles deposited in the lungs, where 100% account for all the particles that have passed the upper airways.

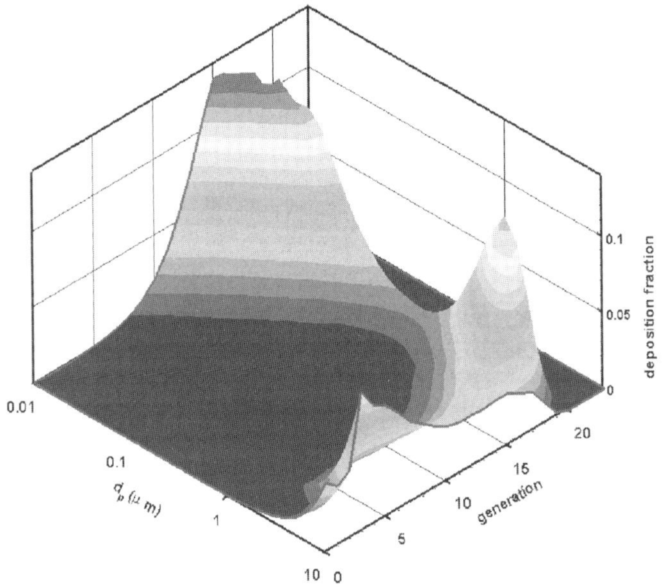

Figure 6. Aerosol deposition in the lung as function of MMAD and generation number for sedentary breathing (TV=500 ml RR= 15/min),(results are comparable with Martonen 1996).

Three peaks can be distinguished. One peak is around generation number 5 for the large particles (5-10 μm) and is the result of inertial impaction. Both the high flow velocities in the first generations as result of a small cross-sectional area and a large MMAD determine this deposition site. A second peak is deep in the lung, at about generation 18-20 for particles of 1-3 μm and is the result of sedimentation. These particles are small and light enough to avoid inertial impaction in the first generations, but have enough size and mass to deposit deep in the lungs where airway dimensions are small and residence time is still sufficient. The third peak is again deep in the lung, around generation 15-18 for particles of 0.01-0.1 μm and is the result of deposition by Brownian diffusion. These small particles deposit where residence time is long enough and airway dimensions are small enough.
It is clear from this graph that with this normal breathing pattern, and independent of the chosen particle size, a predominant deposition in airway generation 5-15 is impossible. For some drugs, most receptors are thought to be in these generations.

Deposition in the first generation is possible by choosing large particles but these particles will also leave the maximum deposit in the upper airways by the same deposition mechanism. Therefore, the major fraction of these particles will be

swallowed, enter the gastro-intestinal system possibly causing unwanted side effects.
Deep deposition in the last lung generations is possible for small and very small particles, so for drugs which have to deposit in the alveolar region. There they might easily pass the alveolar membrane and enter the blood stream. There is no peak for particles between 0.1 and 1.0 μm since none of the three deposition mechanisms affect the course of the particles strongly enough to cause deposition. They are too light for inertial impact and gravitation, but also too heavy to be diffused and reach an airway wall in the residence time that is available during normal respiration.
To see if targeting at specific airway generations is possible with other breathing patterns, two more calculations have been performed with 3 μm particles since this particle size is the most desirable particle size available for medication. The reason why this particle size is produced a lot by pharmaceutical industries is the fact that these 3 μm particles are small enough to escape deposition in the upper airways of adults, and are heavy enough to deposit in the lung (mainly due to sedimentation). A breath hold of 10 seconds after an inhalation of these particles is recommended to increase residence time and therewith the deposition fraction.
In Figure 7 results are plotted in which the same flow rate is used (250 ml/s) but different tidal volumes are applied.

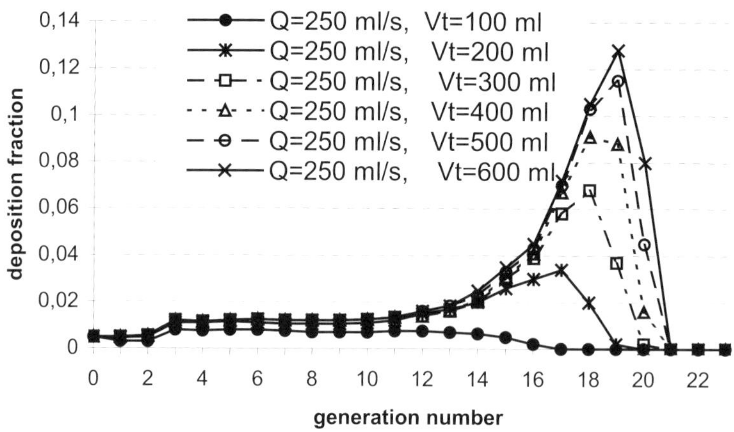

Figure 7 Aerosol deposition as function of generation number in a human adult lung with constant flow rate (Q=250 ml/s) but different tidal volumes

The tidal volume of 500 ml, which with a constant flow rate of 250 ml/s implies an inspiration time of 2 seconds, so assuming an equal expiration time a respiration frequency of 15 breaths/min, and therefore a minute volume of 7.5 l/min) represents almost normal quiet breathing. A small tidal volume implies high respiration rates while a high tidal volume implies a low rate. Results show that the maximal deposition is reached for the highest tidal volume, so the lowest respiration rate.

Again residence time for sedimentation dominates the deposition fraction. Most strikingly is that the place of deposition in the lung is rather independent of the tidal volume and lies around generation 18-19. For the smaller tidal volume a shift to the larger central airways, that is to lower generation numbers can be seen. This is to be expected since the smaller tidal volume will not reach the deeper airways. The disadvantage is that there is only a slight shift (to generation 17) but the deposition factor is significantly decreased compared with that reached with the higher tidal volumes.
To study whether the site of deposition can be influenced further the flow rate is varied (see Figure 8).

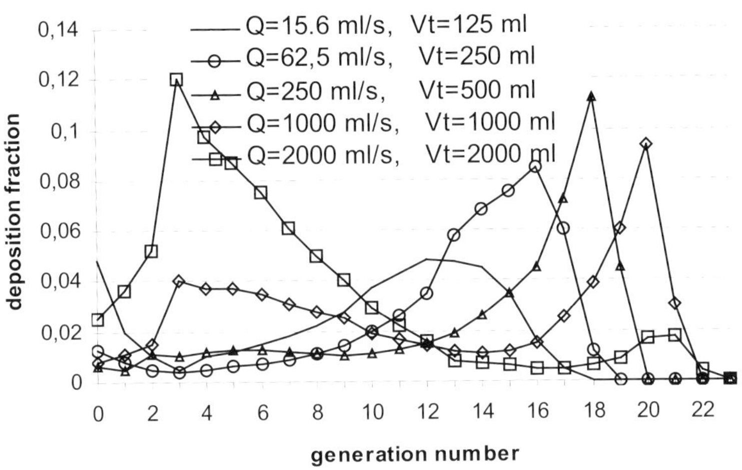

Figure 8. Aerosol deposition in the lung of a human adult as function generation number for different flow rates and different tidal volumes.

Again the line for normal respiration (Q=250 ml/s, Vt=500 ml, RR = 15/min) is shown. Doubling the tidal volume with a four times higher flow rate, doubles the respiration frequency. The deposition fraction becomes bimodal with a flat peak near generation 3-5 due to inertial impaction while the peak due to sedimentation shifts from generation 18 to 20. Again doubling tidal volume while quadrupling the flow rate shows that the sedimentation peak in the deep lung almost disappeared and only a clear peak due to inertial impaction around generation number 3 remains. On the other hand, slower breathing leads to a shift of the sedimentation peak to the higher generations, so deeper into the lungs. For example, with Q=15.6 ml/s and Vt=125 ml, so RR=3.75 breaths/min there is a peak around generation 12 with most of the deposition occurring between generation number 4 and 16. Upper airway and alveolar deposition is thereby avoided. However, taking a breath so slowly with such a small tidal volume is not physiological and certainly not possible if for instance a dry powder inhaler is used where higher airflows are necessary to generate particles with a MMAD of 3 μm. And the patient may be too ill from shortness of breath to breathe in this way. Not all possible combinations of flow rates, tidal volume and MMAD's have been investigated, so there might well be more favourable solutions.

5. DISCUSSION AND CONCLUSION

Most of the time model calculations yield different results to vivo situations because of a series of limitations. First, deposition calculations are performed using strictly monodiperse aerosols instead of choosing a distribution of diameters as found in real life aerosols produced by pressurised metered dose inhalers, dry powder inhalers or nebulizers. The difficulty is that these distributions may differ between medications. Secondly, breathing patterns in model calculations are often extremely simplified. Mostly a constant flow-rate for inspiration as well as expiration is used, although sometimes a sinusoidal breathing pattern is chosen. Even though the result of the model calculation in Figure 4 shows hardly any difference between the deposition fraction of a sinusoidal and a constant flow-rate pattern in real life, the breathing pattern of a person with a lung disease may differ drastically from the sinusoidal form. Everybody who measures lung function in (asthmatic) children knows that breathing patterns of children are irregular and especially the expiration phase is markedly different from the inspiration. Janssens showed in her vitro model that the amount of aerosols reaching the lungs with an irregular breathing pattern (awake) is less then with a regular breathing pattern (asleep) (Janssens thesis 2001). The third problem is the lung model itself. The most widely used is the Weibel lung model, which was based on the lung of an adult whose lung was inflated above its normal lung volume. Later more refinements and other models were made but standardised lung models for all age groups and diseases are still a Utopia. Inter- and intra-subject variability will always lead to average values. Finally, certain aspects are not modelled: these include electrostatic forces, hydrophobic or hydrophilic effects, interception, and oscillations by cardial activity. In addition most models are still one dimensional instead of three dimensional which makes a comparison of deposition results with radioactive scans cumbersome.

Children and neonates have a very low aerosol deposition fraction since most particles will be deposited in the upper airways. To avoid the problem of inertial impaction in the nose particles with a smaller MMAD (1 μm) might be used, but these particles will mainly deposit deep in the alveolar region and thereby possibly yield unwanted side effects. To optimise aerosol deposition at the place where the receptors for the medication are present, generation numbers must be known. Subsequently, combinations of breathing patterns, respiration rates and tidal volumes should be investigated for each specific lung model. Since standardised pathologic lung models and breathing patterns are hard to find, especially as function of age, the optimal combination of particle size, tidal volume and breathing pattern to maximize lung deposition and minimise side effects is still unknown. New techniques like bolus inhalation, hollow particles or particles with chemical properties that can influence the diameter after entering the lung might give better results for these young patients, in the future.

6. REFERENCES

Amirav I, Balanov I., Gorenberg M., Luder A.S., Newhouse M.T. Groshar D., β-Agonist aerosol distribution in respiratory syncytial virus bronchilitis in infants,*The Journal of nuclear medicine* 2002,43(4): 487-491

Anhoj J, Thorsson L, Bisgaard H, Lung deposition of inhaled drugs increases with age, *Am. J. Respir. Crit. Care Med.* 2000; 162(5):1819-22

Carlsen KH, Gerritsen J, Inhaled steroids in children: adrenal suppression and growth impairment, *Eur.Resp.J* 2002;19(6): 985-988

Chua HL, Colli GG, Newburry AM, Chan K, Bower GD, Sly PD, Le Souef PN, The influence of age on aerosol deposition in children with cystic fibrosis, *Eur Resp J* 1994;7:2185-2191

Diot P, Palmer L.B., Smaldone A., DeCelie-Germana J., Grimson R., Smaldone G., RhDNase I Aerosol deposition and related factors in cystic fibrosis., *Am J Respir Crit Care Med* 1997; 156:1662-1668

Gagliardi L., Rusconi F., Respiratory rate and body mass in the first three years of life. *Arch. Dis. Childh.* 1997;76:151-154

Hofmann W, Mathematical model for the postnatal growth of the human lung, *Resp. Physiol.* 1982; 49:115-129

Hofmann W, Martonen TB, Graham RC, Predicted deposition of nonhygroscopic aerosols in the human lungs as a function of subject age, *J. of Aerosol Medicine* 1989; 2(1):49-68

Hofmann W, Lung morphometry and particle transport and deposition: overview of existing model, in: *Aerosol Inhalation: Recent Research Frontiers*, Kluwer Academic Publishers,1996:91-102

Hughes JMB, Pride NB (editors) *Lung Function tests, physiological principles and clinical applications*, WB Saunders 1999, page 164

Janssens HM, *Aerosol therapy in young children*, PhD thesis, 2001, ISBN 9077017011

Janssens HM, de Jongste JC, Fokkens WJ, Robben SG, Wouters K, Tiddens HA, The Sophia anatomical infant nose-throat (SAINT) model: a valuable tool to study aerosol deposition in infants, *J Aerosol Med* 2001; 14(4):433-441

Martonen TB, Graham RC, Hofmann W, Human subject age and activity level: factors addressed in a biomathematical deposition program for extrapolation modeling, *Health Physics*, 1989; 57 (suppl1): 49-59

Martonen T, Yang Y, Deposition mechanics of pharmaceutical particles in human airways, from *Inhalation aerosols*, editors Hickey A.J. Vol 94 Series Lung Biology in Health and Disease, editor C. Lenfant, M. Dekker Inc. 1996

Phalen RF, Oldham MJ, Mautz WJ, Aerosol deposition in the nose as a function of the body size, *Health Physics*,1989; 57(suppl1), 299-305

Phalen RF, Oldham MJ, Methods form modelling particle deposition as function of age, *Resp Physiol.* 2001;128:119-130

Price J, Hindmarh P, Hughes S, Efthimiou J, Evaluating the effects of asthma therapy on childhood growth: what can be learnt from the published literature ?, Eur. Resp. J 2002;19(6):1179-1193

Schulz H, Brand P, Heyder J, Particle deposition in the respiratory tract, from *Particle Lung Interaction*, editors Gehr P, Heyder J, Vol 143 Series Lung Biology in Health and Disease, editor C. Lenfant, M. Dekker Inc. 2000

Shinwell ES, Karplus M, Reich D, Weintraub Z, Blazer S, Bader D, Yurman S, Dolfin T, Kogan A, Dolberg S, Arbel E, Goldberg M, Gur I, Naor N, Sirota L, Mogilner S, Zaritsky A, Barak M, Gottfried E, Early postnatal dexamethasone treatment and increased incidence of cerebral palsy, *Arch. Dis. Child, Fetal Neonatal Ed.* 2000, 83(3), F177-181

Stocks J, Sly PD, Tepper RS, Morgan WJ, *Infant Respiratory Function Testing,* John Whiley & Sons 1986

Taulbee DB, Yu CP, A theory of aerosol deposition in the human respiratory tract, *J.Appl Physiol,* 975;38(1) 77-85

Weibel ER, *Morphometry of human lungs*, Springer Verlag, Berlin 1963

7. AFFILIATIONS

F.H.C. de Jongh
1) Engineering Fluid Dynamics, Faculty of Mechanical Engineering, Twente University, Enschede
2) Neonatology, Academic Medical Centre, University of Amsterdam
3)Lungfunction lab, Medisch Spectrum Twente, Enschede, The Netherlands, email f.h.dejongh@amc.uva.nl / f.h.c.dejongh@utwente.nl

R.W.Griffioen
4) Pediatric Pulmonology, Academic Medical Centre, University of Amsterdam

YUNG SUNG CHENG

MODELING AEROSOL DRUG DELIVERY

1. INTRODUCTION

Delivery of a therapeutic agent by inhalation has seen increasing applications for many respiratory diseases, including asthma, COPD, and allergies. Aerosol delivery has advantages: it delivers medication directly to where it is needed, and it avoids the first-pass effect with minimum reduction of bioavailability. More recently, the inhalation route has been extensively researched as an alternative for systematic administration of proteins and peptides because of the large surface area in the pulmonary region and rapid absorption of the delivered drug from the alveolar region to the blood. Aerosol delivery is non-invasive and is effective in much lower doses than required for oral administration. Currently, there are several types of therapeutical aerosol delivery systems, including the pressurized metered dose inhaler (pMDI), the dry powder inhaler (DPI), the medical nebulizer, the solution mist inhaler, and nasal sprays.

Rigorous procedures are required to develop aerosol drugs and drug delivery devices. These include in vitro testing, toxicology testing using laboratory animals, determination of deposition patterns, pharmacokinetic (PK) studies, and clinical studies for safety and efficacy of the treatment. The in vitro test evaluating the particle size distribution, emitted dose, and dose uniformity is mostly for quality control purposes. Studies in human volunteers determine the deposition pattern of the aerosol delivered to the respiratory tract. The clinical PK studies quantifying the drug concentration in blood or urine provide information about systemic exposure. In the case of topical delivery of an aerosol drug for asthma and other respiratory diseases, PK data are more useful for safety and less relevant to clinical efficacy. However, for drugs administered by inhalation to achieve a systematic effect (for instance, insulin), PK data may be a key predicator of therapeutic response.

Following inhalation therapy, only a fraction of the dose reaches the expected targeted area. The initial deposition fraction defines the delivered dose, which is important in understanding the dose response of topical applications to the lung. The deposition pattern can be influenced by several factors: (1) particle characteristics, such as particle size, shape, density, charge, and hygroscopicity; (2) airway geometry as a function of gender, age, and diseases status; and (3) breathing pattern, including frequency, tidal volume, and breath-holding time. These factors influence not only the deposition efficiency but also the deposition pattern in the respiratory tract. For example, in the aerosol delivery of bronchodilators or corticosteroids to

L. Gradoń and J. Marijnissen (eds.), Optimization of Aerosol Drug Delivery, 165–188.

treat asthma patients, many aerosol devices provided only 5–25% of the drug to the target areas in the lung, whereas a majority of 40 to over 80% of the active compound was deposited in the oropharyngeal region. Oropharyngeal deposition provides minimum therapeutic benefit and may have local and systematic side effects. Newman et al. (2000) showed through case studies that the clinical effects of an asthma treatment can be correlated directly with the deposited dose in the lung. Lung deposition data are very useful to show the bioequivalence between two pMDIs and DPIs or between a new device and an existing device. Derom and Pauwels (1998) also obtained similar conclusions.

The delivery efficiency of an aerosol device is usually determined in human volunteer studies using 2D gamma scintigraphy, 3D single photoemission computed tomography, or positron-emission tomography. Scintigraphic imaging is either 2D (planar) or 3D (single photoemission computed tomography or photoemission computed tomography). Gamma scintigraphy was used in most human studies of aerosol drug delivery. The gamma scintigraphy techniques use a radiolabeled drug and provide quantitative information on the fraction of aerosol delivered to different regions of lung, including the naso- or oropharynx, lung, and expired air, as well as deposition in the device and/or spacer (Newman, 1998). The deposition pattern in the lung can be further divided into two or three regions of interest, generally including a central and a peripheral region, and often including a third "intermediate region" (Agnew, 1991). A so called "penetration index," such as a P/C ratio (deposition in the periphery and central region), is often calculated as an indication of the extent to which aerosol particles have penetrated through large and medium-sized airways to be deposited in the small airways and alveoli.

Lung deposition models, such as the one developed for the International Commission on Radiological Protection (ICRP, 1994) or National Council on Radiation Protection and Measurement (NCRP, 1997), model aerosol deposition in the human respiratory tract. These models have been used extensively to estimate radiation doses from occupational and ambient exposure to radioactive aerosols. There were also attempts to use these models or others to investigate factors that may affect aerosol drug delivery (Martonen, 1993; Pritchard et al., 1996; Price, 2000). It would be very useful to use the particle size information obtained in the in vitro studies to correlate the regional deposition pattern from clinical studies. This would provide additional means to facilitate drug development and evaluation, especially new inhalation devices and formulations. This report will review the clinical deposition data and compare the in vivo data with estimates of deposition patterns using the ICRP model.

2. IN VITRO SIZE MEASUREMENT

A pharmaceutical aerosol is traditionally measured in vitro using inertial devices, such as cascade impactors or impingers (Marple et al., 1998), which classify and collect particles into different aerodynamic size fractions. The material collected in individual stages can then be assayed to determine the mass fractions of active ingredient. Laser diffraction instruments, such as the Malvern 2600 or MasterSizer (Malvern Instruments, Malvern, UK), have also been used to measure particle size distribution of nebulizers and nasal sprays (Nerbrink et al., 1994; Clark, 1995; Cheng et al., 2001b). However, the optical technique can determine the physical size of the droplets, but does not provide the size distribution of the active component. When using the optical technique, one always assumes that the active component is uniformly distributed in droplets of different sizes. In the U.S., the Andersen cascade impactor (ACI) (Model Mark II, Andersen Instruments, Thermo Andersen, Smyrna, GA) and the Marple-Miller impactor (MSP Co., Minneapolis, MN) are commonly used. In Europe, a multi-stage liquid impinger (MLI, Copley Scientific Limited, Nottingham, UK) is frequently used. The ACI has eight stages and a backup filter, and is normally operated at 28.3 L min^{-1}, although it has been modified to operate at 60 and 90 L min^{-1} (Nichols et al., 1998). Particles are collected on the aluminum planchets or filters and must be extracted for chemical analysis. The Marple-Miller impactor has five stages and a backup filter, and can be operated at 30, 60, and 90 L min^{-1}. The collection stages were designed in the shape of a cup, which facilitates preparation for chemical analysis. The MLI has four stages containing the wash liquids. The particles are collected into the liquid directly. A new generation pharmaceutical impactor is available now (Model 170, MSP Co.). The main design features are ease of use and suitability for automatic operation. It has seven stages that collect particles inside the cups and has a backup filter.

A major feature of the in vitro size measurement is the inlet to the impinger or impactor. A glass or metal 90-degree bend is used as the inlet to simulate the oropharyngeal deposition. This is especially important for the pMDI and DPI because it is not uncommon for 25–80% of the aerosol to deposit in the oropharyngeal region. The commonly used artificial throat inlets include a glass throat for the impingers and a USP metal throat for cascade impactors. Dolovich and Rhem (1998) examined the effects of seven inlets in measuring the particle size of a salbutamol pMDI using an ACI. They found that the inlets and the impactor sampled significant differences in the amount of salbutamol. The fine particle fraction of the emitted doses varied almost three-fold, and the mass median aerodynamic diameters of the aerosols collected in the impactor ranged from 1.8–2.4 μm. Thus, in vitro sizing from different inlets may not be comparable and may introduce inconsistencies in correlations with in vivo data from different laboratories. A more standard methodology of using the sizing instruments, including the inlet, is needed.

3. FINE PARTICLE FRACTION

Attempts have been made to estimate lung deposition efficiency directly from in vitro measurement of particle size distribution. The most common index is the fine particle fraction (FPF) or respirable fraction, which is defined roughly as the mass fraction of particles with an aerodynamic diameter <5 μm diameter (Hickey et al., 1996; Newman, 1998). However, this is different from the respirable fraction or respirable mass as defined for industrial hygiene and environmental health (Phalen et al., 1986). The respirable fraction is defined as the portion of an aerosol mass available to the pulmonary region of the human respiratory tract during nose breathing (Phalen et al., 1986). The current international standard of respirable fraction can be expressed as a cumulative lognormal function for an aerodynamic diameter with a median diameter (d_{50}) of 4 μm and a geometric standard deviation (GSD) of 1.5 (Soderholm, 1989). Respirable samplers have been designed to have the sampling efficiency that simulates the curve. On the other hand, the FPF is more like a convention without a specified mathematical expression. Its value is often estimated by measuring in vitro particle size using inertia instruments (Marple et al., 1998). The most common inertial instruments used in the pharmaceutical industry are ACI and MLI. Depending on the instruments, the FPF is calculated differently. The FPF calculated from the ACI operating at 28.3 L min^{-1} is the portion of the particles <4.7 μm (fraction of mass collected from stages 3–7 and the back-up filter), whereas for the MLI operating at 60 L min^{-1}, it is <6.8 μm (fraction of mass collected at stages 3 and 4). Therefore, the FPF calculated from the MLI is greater than that of the Andersen impactor.

The FPF has been used to correlate the in vivo deposition obtained by gamma scintigraphy. Clark (1991) showed that changes in FPF produced by different spacer devices are mirrored by changes in lung deposition. In a more extensive study of pMDIs, DPIs, and soft mist inhalers (SMIs), Newman (1998) compared various measurements of FPFs from ACIs and MLIs with in vivo lung deposition data. Figure 1 shows the lung deposition fraction and the FPF (defined as % mass<5 μm), indicating correlation; however, the FPF always overestimates the true lung deposition fraction obtained by in vivo technique. However, the mass fraction of stage 4 on the MLI has a much better predictive value for in vivo deposition in the lung. The impinger data in Newman's analysis generally failed to predict correctly the relationship between the lung depositions from an MDI compared with another device. He concluded that FPF data alone are unreliable as a predictor of the relative deposition from two devices with markedly different spray characteristics (e.g., pMDI versus DPI). He also pointed out that the respirable fraction used in aerosol drug delivery is a misnomer.

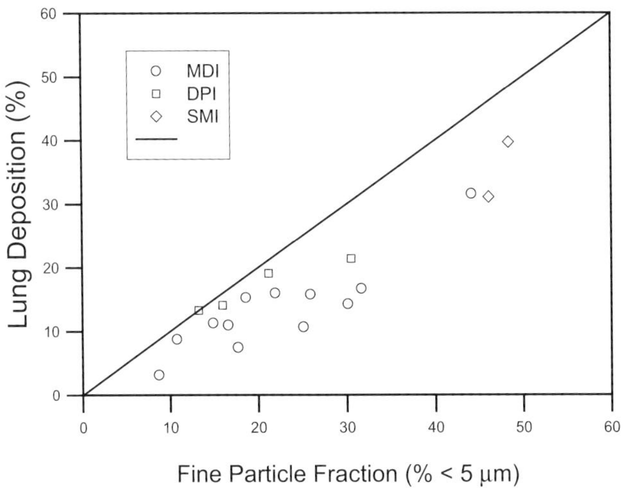

Figure 1. Correlation of lung deposition with the fine particle fraction (Newman, 1998). MDI = metered dose inhaler; DPI = dry powder inhaler; SMI = soft mist inhaler.

4. LUNG MODEL

Several lung models have been used to simulate deposition of pharmaceutical aerosols in the lung (Martonen, 1993; Pritchard et al., 1996; Price, 2000). In this paper, the ICRP 66 lung model (1994) is used to model the aerosol drug delivery. The ICRP model is a consensus model developed by a committee of international experts on aerosol deposition, lung morphology, and radiation dosimetry. The ICRP model has been used extensively in the health physics community for radiation dose calculation. The LUDEP software package (National Radiological Protection Board, Chilton, UK) was developed to implement the ICRP 66 respiratory tract model.

The morphometric model in the LUDEP software divides the respiratory tract into five anatomical regions: (1) ET1 (anterior nose), (2) ET2 (posterior nasal passage, mouth, larynx, and pharynx), (3) BB (trachea and bronchi, generations 1–8), (4) bb (bronchiolar region, generations 9–16), and (5) AL (alveolar region, generations 17–23). The deposition equations in each region of the respiratory tract were derived from empirical data of aerosol deposition in human volunteers and in some cases from human nasal and oral airway replicas (ICRP, 1994).

The input parameters used in the LUDEP model include the following: (1) particle size distribution; (2) inhalation flow rate, inhaled volume, and fraction breathed through the nose; and (3) gender and age of airway scaling. The current model included only the male adult subject because it did not have data for gender and age. In vitro measurement provided the particle size data in the form of mass median aerodynamic diameter and GSD of a log normal distribution or fraction of particles in a size increment of the impinger or impactor.

The procedure developed by Pritchard et al. (1996) and Price (2000) was used to estimate the deposition pattern using LUDEP software. For pulmonary delivery, mouth breathing was used. When a impactor or impinger was used for in vitro measurement in a human deposition study using gamma scintigraphy, the particle size measurement was also used to verify the radiolabeling procedure. Particle size distributions of the unlabeled and labeled drug were compared to ensure that they were similar and the size distribution did not shift because of the labeling procedure. Following the suggestion made by Pritchard et al. (1996), the fraction of particles collected in the impactor/impinger inlet (i.e., USP or glass throats) was assigned as a part of the deposition fraction in the ET2 region. The individual fraction of particles in the impactor/impinger stage was used to calculate deposition in the fraction in the ET2, BB, bb, and AL regions. The sum of the calculated deposition in the ET2 regions from the inlet and individual size increment was then the deposition fraction in the oropharyngeal region. The sum of the calculated deposition fraction in the BB, bb, and AL regions from each stage was then the lung deposition. The fraction of inhaled particles not depositing in the airway was assigned to the expired air. The only correction from the LUDEP calculation was the inhalability. The intake efficiency or inhalability was the percentage of air in the environment that would enter the nose or mouth. The LUDEP software assumed that particle inhalability decreased from 100% as the aerodynamic particle size increased. For the aerosol drug delivery, 100% inhalability was assumed in the calculation.

5. NEBULIZERS

Nebulizers were the first inhalers developed for inhalation administration of therapeutic compounds. The device has a reservoir containing several mL of solution or suspension, a mechanism to produce droplets, and air to disperse and carry the droplet for inhalation. Nebulizers include jet nebulizers and ultrasonic nebulizers. Air-blast or jet nebulizers use compressed air through small nozzles near the tip of a narrow feed tube to draw bulk liquid from the reservoir because of Bernoulli effects. The high-velocity air breaks up the liquid into droplets. The droplets impact on a baffle to make smaller droplets. A compressed air source is needed to provide the energy. In an ultrasonic nebulizer, the energy necessary to atomize a liquid comes from vibration induced by a piezoelectric transducer operating at the frequency of 1–3 MHz. A nebulizer continuously generates the aerosol from several minutes to about 15 minutes depending on the reservoir volume and the liquid generation rate.

Data from a gamma scintigraphic study of three Pari nebulizers (Newman et al., 1994) were used for comparison of values calculated by LUDEP. The droplet size distribution was measured using the Malvern 2600 laser diffraction technique (Malvern Instruments). The mass median diameters (MMDs) were between 4 and 5.5 μm. The GSD of 2.1 was assumed based on anther report (Barry and O'Callaghan, 1999). Ten healthy adult volunteers took part in the study. They were instructed to inhale from the nebulizer by relax tidal volumes for 15 minutes. Breathing information was not reported. A breathing rate or minute volume of 15 L and a frequency of 16 min^{-1} were assumed. The nebulizers were filled with 3 mL of ^{99m}Tc-labeled diethylenetriamine penta-acetic acid solution. Figure 2 shows the comparison of the deposition pattern for the Pari LC nebulizer operated in the intermittent mode. The MMD was 4.0 μm for the Pari LC nebulizer. The deposition pattern showed that 61.5% of the radioactivity remained in the nebulizer. The oropharyngeal deposition was 10.9%, and the lung deposition was 15.3%. The deposition calculated by the LUDEP software showed 8.2 and 16.7% deposited in the oropharyngeal region and lung with 13.5% in the expired air. The regional deposition patterns and fraction of expired air of the calculated values were within the range of experimental errors, indicating good agreement. Similar results were obtained for other two nebulizers having similar particle sizes (Table 1).

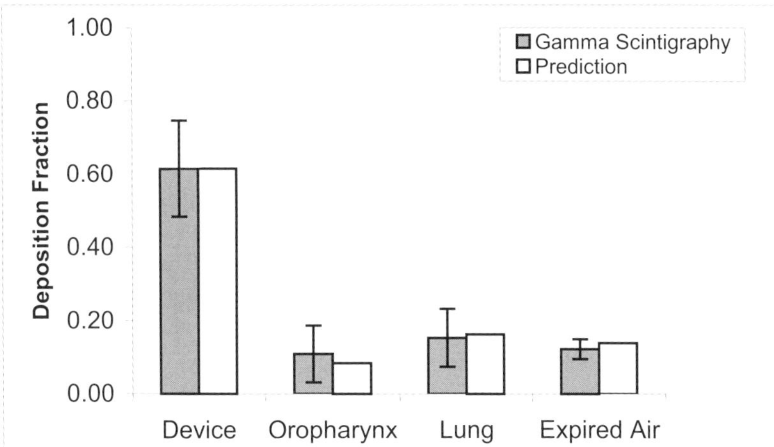

Figure 2. Deposition patterns of the Pari LC nebulizer from an in vivo gamma scintigraphic study (Newman et al., 1994) and calculation from the LUDEP software. The error bars are standard deviation of experimental measurement.

Table 1.

	Pari LL		*Pari LC (intermittent)*		*Pari LC (continuous)*	
MMD (μm)	*5.5*		*4.0*		*4.8*	
GSD	*2.1*		*2.1*		*2.1*	
	Measured	*Calculated*	*Measured*	*Calculated*	*Measured*	*Calculated*
Device	0.619	0.619	0.615	0.615	0.568	0.568
Oropharynx	0.138	0.117	0.109	0.084	0.142	0.115
Lung	0.111	0.157	0.153	0.162	0.128	0.181
Expired Air	0.132	0.107	0.122	0.139	0.162	0.136

Deposition data from another scintigraphic study for three air jet nebulizers and one ultrasonic nebulizer were also examined (Hardy et al., 1993). The droplet size distribution obtained in a Malvern 2600 laser diffraction instrument showed a wide range of MMDs from 4.9 μm (GSD = 2.1) for Pari Boy, 2.2 μm (GSD = 2.0) for Penta-Sonic (or Porta-Sonic), 1.2 μm (GSD = 1.7) for RespirGard II, and 0.8 μm for Pari IS-2 (GSD = 1.8). The GSDs were not given in the paper but were estimated from other reports (Sterk et al., 1984; Knoch et al., 1994; Vera et al., 1995). Deposition data in ten healthy adults showed quite variable deposition patterns mainly due to the particle size and residue that remained in the nebulizer as shown in Figure 3. For example, the tested ultrasonic nebulizer seemed to have a low emitted efficiency with 95% of the material left in the device, whereas in jet nebulizers 21–43% remained in the device. Comparison between the calculated and in vivo deposition patterns for all seven nebulizers (Figure 3) generally showed good agreement.

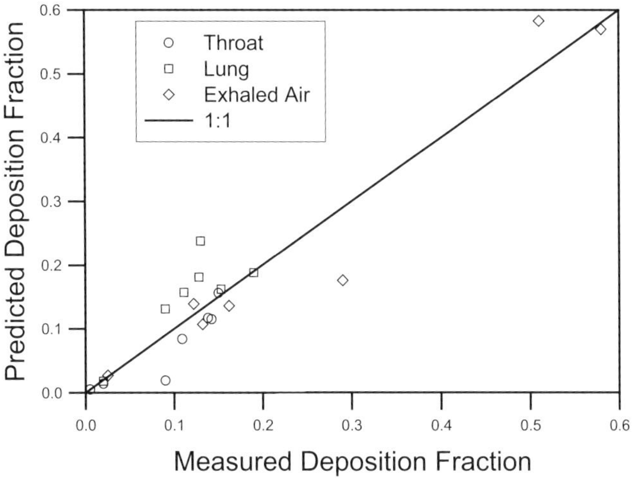

Figure 3. Comparison of in vivo and calculated deposition in the oropharynx, lung, and expired air for seven nebulizers (Hardy et al., 1993; Newman et al., 1994).

6. PRESSURIZED METER DOSE INHALERS

The pMDI is the most popular medical inhaler accounting for about 70% of worldwide sales. The advantages of the pMDI are its compactness, self-contained power source, reproducible multiple doses with each actuation, and its usability in the hospital or at home. A pMDI consists of three main components. A canister contains the therapeutic compound suspended or dissolved in a propellant. The metered valve is designed to deliver an accurately metered volume of the propellant/active drug, and the actuator includes the atomizing nozzle from which the metered dose is delivered and the mouth piece.

A pMDI usually contains 100–200 doses. When the canister is actuated, a metered volume, ranging from 25–100 μL, is released through the valve stem to the actuator. The propellant is vaporized rapidly because reducing pressure creates a cloud of expanding mixtures moving at high speed. The liquid ligament is torn apart to form droplets. The droplets slow down and evaporate as they move further away from the actuator. Typically, the initial velocity of the two-phase vapor liquid mixtures is approximately at 35 m sec^{-1} (Clark, 1996). The propellants traditionally used in the pMDIs are chlorofluorocarbons (CFCs), including Freon 11, 12, and 114. The CFC formulation will be phased out soon according to the Montreal Protocol. New formulae use hydrofluoroalkane (HFA), including HFA 134a and HFA227.

Gamma scintigraphic measurement of the deposition pattern of beclomethasone dipropionate (BDP) using CFC and HFA134a formulations was reported in healthy and asthmatic volunteers (Leach et al., 1998). The inspiratory flow rate was around 100 L min^{-1}, and the tidal volume was about 5.2 L. Figure 4 shows the size distribution obtained in an ACI with the USP throat. Size distributions of both unlabeled and labeled BDP were in agreement, indicating that the ^{99m}Tc label was uniformly distributed. The CFC-BDP formulation showed over 50% deposition in the USP inlet, whereas the HFA-BDP showed only 30% deposition. A high-speed plume from the pMDI contributed to the deposition in the USP throat. The particle size of the HFA-BDP was also smaller than the CFC-BDP. Figure 5 shows the in vivo deposition pattern. The CFC-BDP had 10, 81, 6.3, and 3.7% deposited in the actuator, oropharynx, lung, and expired air, respectively, whereas for the HFA-BDP, the deposition was 15, 25, 48, and 13%, in the actuator, oropharynx, lung, and expired air, respectively. For the HFA-BDP, deposition in the oropharynx was much lower than the CFC-BDP, resulting in higher lung deposition and expired air. A similar deposition pattern was obtained in asthmatic volunteers. The calculated deposition patterns shown in Figure 5 for lung and oropharyngeal deposition were outside the error bars of in vivo data, indicating that they were not in good agreement. Additional in vivo data were also used for comparison as shown in Figure 6, for deposition in the lung and oropharyngeal region (Newman et al., 1989a,b, 1991; Richards et al., 2001; Hirst et al., 2002). In general, the LUDEP method underestimated the oropharyngeal deposition and overestimated the lung deposition and expired drug.

From the deposition study, it is apparent that frequently, over 65% of the drug is deposited in the oropharyngeal region. The material deposited in the oropharynx may cause local and systematic side effects. In addition, the coordination between firing the device and the inhalation maneuver could be difficult for children and some adult patients. Holding chambers or spacers have been designed to improve the delivery of the pMDI (Dolovich, 1989). A spacer is placed between the pMDI actuator and the mouth. The aerosol plume is sprayed into the chambers and then subsequently inhaled by the patient. These additional features allow the spray to slow down and evaporate before reaching the respiratory tract, thus reducing oropharyngeal deposition. It is also reduced through impaction of larger particles on the walls or valves of the container. An example of the in vivo deposition of cromoglycate with and without a spacer (Aerotube) using 10 healthy volunteers for an inspiratory flow rate of about 30 L min^{-1} and 1.7 L tidal volume is shown in Figure 7 (Newman et al., 1991). Without the spacer, deposition was 11.4, 79.5, and 8.8% in the actuator, oropharynx, and lung, respectively. With the Aerotube, the deposition changed drastically with 59.1, 29.5, and 11.3% in the actuator/spacer, oropharynx, and the lung. As expected, the oropharynx deposition decreased and the deposition in the lung increased. Similar results were obtained in open spacers (Newman et al., 1989b). Again, the LUDEP calculation underestimated the oropharyngeal deposition and overestimated the lung deposition.

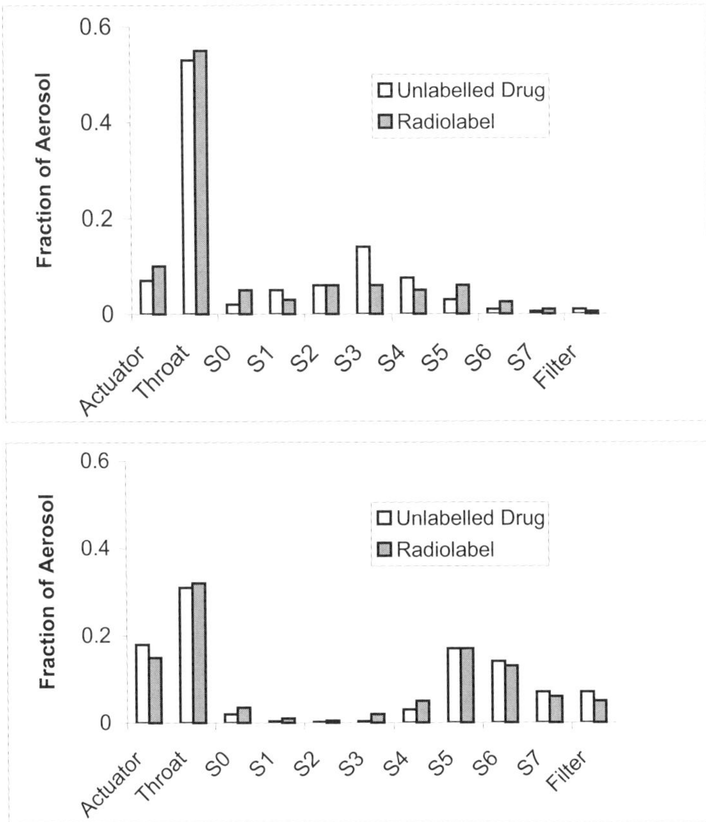

Figure 4. Size distributions of BDP pMDIs with CFC (top panel) and HFA (bottom panel) formulations obtained using an Andersen cascade impactor (Leach et al., 1998)

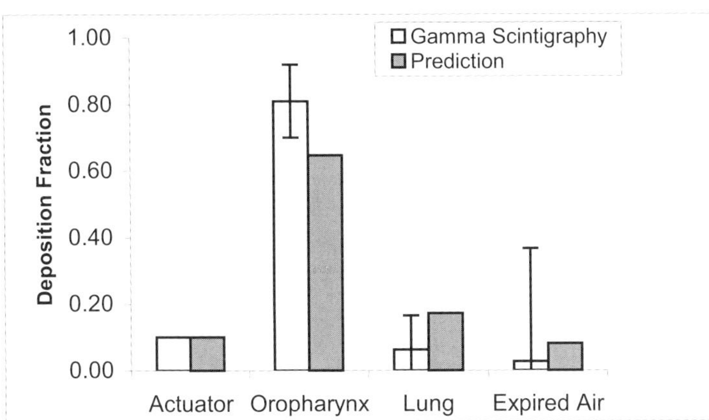

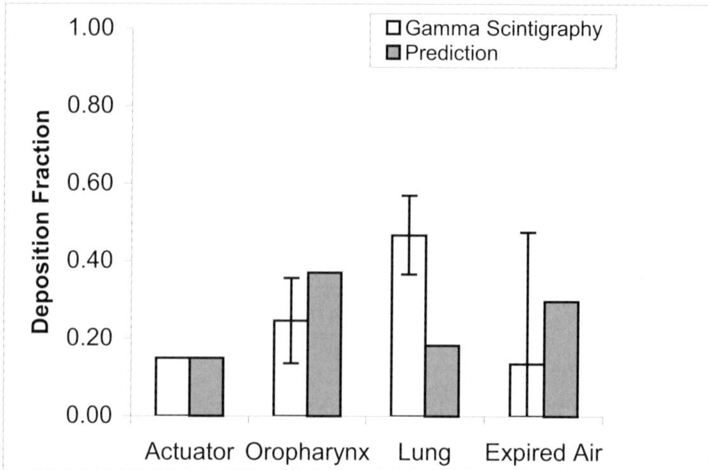

Figure 5. Comparison of in vivo and calculated deposition in the actuator, oropharynx, lung, and expired air for the pMDIs with CFC formulation (top panel) and HFA formulation (bottom panel) (Leach et al., 1998). The error bars are standard deviation of experimental measurement.

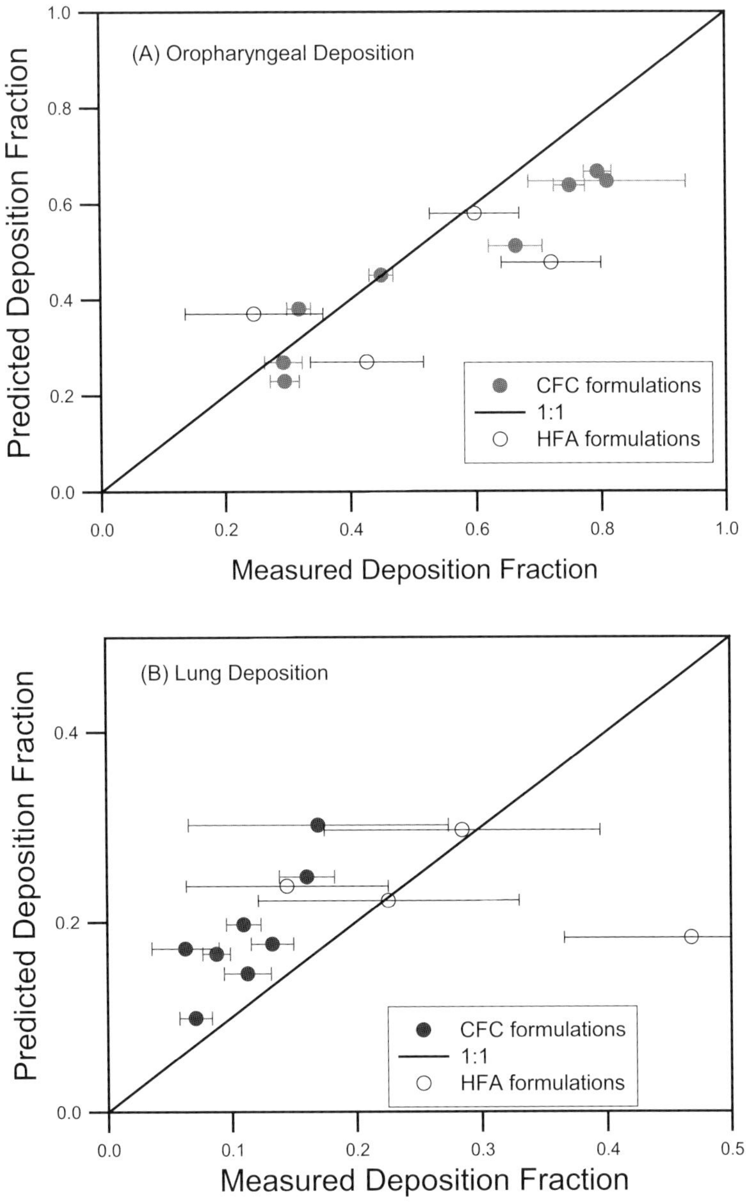

Figure 6. Comparison of in vivo and calculated deposition in the oropharynx and lung for several pMDIs (Newman et al., 1989b, 1991, Leach et al., 1998; Richards et al., 2001;Hirst et al., 2002). The error bars are standard deviation of experimental measurement.

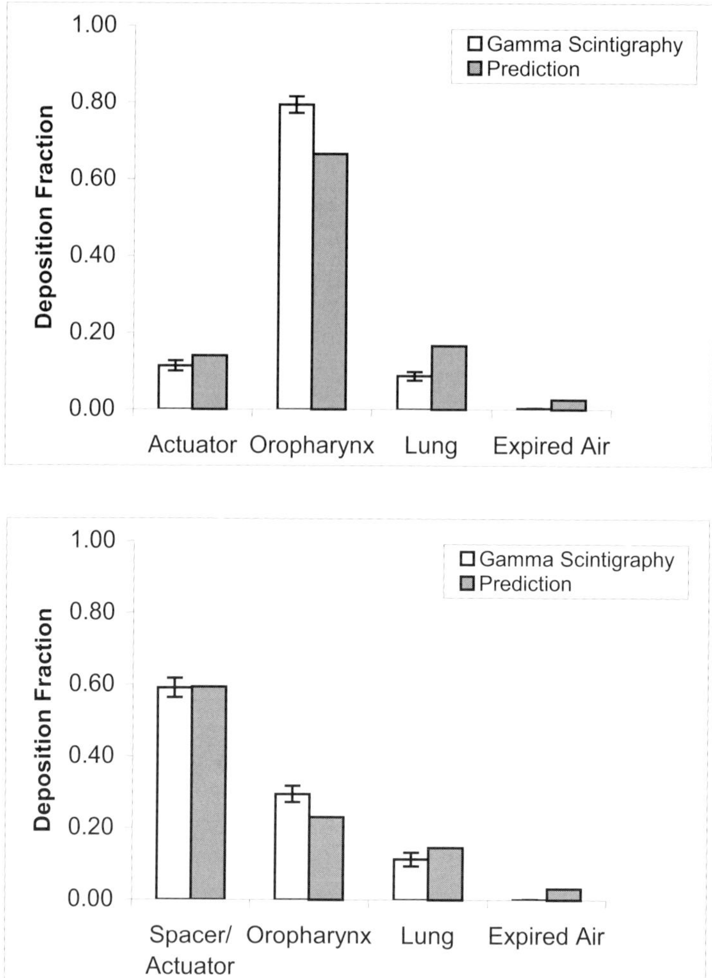

Figure 7. Comparison of in vivo and calculated deposition in the actuator, oropharynx, lung, and expired air for the Intal pMDIs (Newman et al., 1991). The top panel shows data without a spacer and the bottom panel shows data with a spacer. The error bars are standard deviation of experimental measurement.

7. DRY POWDER INHALER

Several DPIs have been developed as alternative devices to the pMDI. The early devices, including the Spinhaler and Rotahaler, were single dose devices. Newer devices, such as the Turbuhaler and Diskhaler, can deliver multi-doses. A DPI usually has a powder reservoir, a disk containing multiple blisters of therapeutical compounds, or simply a capsule for single-dose devices. The aerosolization mechanism can be passive or with power. Passive devices (Spinhaler, Rotahaler, Diskhaler, and Turbuhaler) use inspiratory flow to aerosolize the powder and the rotor, screen, or flow path to create turbulence in order to break up aggregates and reduce the particle size. A high peak inspiratory flow rate (>60 L min^{-1}) is recommended to increase air turbulence for enhancing powder deagglomeration. This requirement is difficult to achieve for children and some asthma patients with severe airway restriction. Some new devices under development use a powered component to aerosolize and disperse powder.

An example is the in vivo deposition of budesonide inhaled via the Turbuhaler in healthy subjects. Inspiratory flow rates of 58 (inhaled volume of 2.9 L) and 36 L min^{-1} (inhaled volume of 2.54 L) were used. Particle size was measured using the MLI with a glass throat inlet at a flow rate of 60 L min^{-1}. Figure 8 (bottom panel) shows the deposition patterns of 58, 28, and 0.1% at the oropharynx, lung, and expired air for a fast inhalation maneuver at 58 L min^{-1}. The oropharyngeal deposition was lower than CFC pMDI, and therefore the lung deposition was higher than CFC pMDI. The predicted deposition pattern was within the experimental error, indicating good agreement. Figure 8 (top panel) shows a deposition pattern of 67, 15, and 0.01% at the oropharynx, lung, and expired air for a slow inhalation maneuver at 36 L min^{-1}. The oropharyngeal deposition was higher, and the lung deposition decreased compared to those obtained at the fast inhalation maneuver. This could be attributed to an increased particle size at a low inspiratory flow (Cheng et al., in press). The predicted deposition shows lower oropharyngeal and higher lung deposition than the experimental data. A reason for this is that the particle size data used for the deposition calculation was the same set of data obtained by an MLI at 60 L min^{-1}, which did not reflect the possible particle size change at the lower flow rate of 36 L min^{-1}. Additional in vivo data were also used for comparison as shown in Figure 9, for the lung and oropharyngeal deposition (Newman et al., 1989a,b, 2001; Pitcairn et al., 1994, 1995). In general, the LUDEP method underestimated the oropharyngeal deposition.

8. NASAL SPRAYS

Several types of therapeutical agents, including corticosteroids, vasoconstrictors, and antihistamines, may be delivered to the nasal passages to produce local effects, such as treating rhinitis or allergies (Newman et al., 1987a; Thorsson et al., 1993). The nasal devices include the pMDI, DPI, and spray pumps. Both gamma scintigraphic measurements in human volunteers and nasal airway replicas have been used to study the deposition pattern (Newman et al., 1987a; Thorsson et al., 1993; Cheng et al., 2001b). A study of nasal deposition in a realistic nasal replica based on MRI scans of human volunteers was used for modeling purpose (Cheng et al., 2001b). Four types of nasal sprays were used. The droplet particle size distribution was measured using a Malvern MasterSizer S instrument (Malvern Instruments Ltd.). A constant inspiratory flow rate of 20 L min^{-1} was used. The droplet size was 48.3 μm (GSD of 1.67). The deposition pattern of the VP7 spray pump (anterior, middle, and posterior regions) and filter were 0.354, 0.644, 0.002, and 0.0, respectively. These results indicated that deposition was primarily in the anterior and middle parts of the nasal passage, and very little penetrated the nasal passage and was available for deposition in the lung or inspired air. Depositions in the middle and posterior regions of the nasal passage were classified as the ET2 region, and the anterior region was ET1, based on the ICRP 66 (1994) definition. Figure 10 shows a comparison of deposition patterns with calculations from LUDEP. Both the theoretical calculation and experimental data showed that of the 99% of the drug deposited in the nasal passage, very little was deposited in the lung or expired air. However, within the nasal passage, the LUDEP calculation indicates 50 and 50% deposition in the ET1 and ET2 regions, whereas the experimental data showed 35.4 and 66.4% deposition in the ET1 and ET2, as shown in the figure. Three other spray pumps also showed similar results; furthermore, the deposition pattern appeared to be a function of the spray angle (Cheng et al., 2001b). Similar results were obtained from an in vivo study of spray pumps (Newman et al., 1987a).

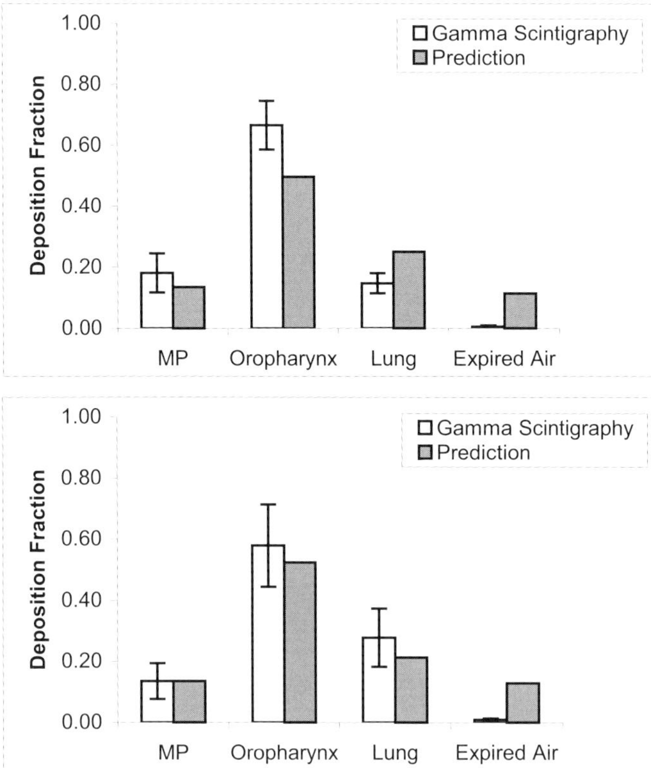

Figure 8. Comparison of in vivo and calculated deposition in the mouthpiece, oropharynx, lung, and expired air for budesonide in a Turbuhaler for inspiratory flow rate of 58 L min^{-1} (top panel) and 36 L min^{-1} (bottom panel) (Borgstrom et al., 1994). The error bars are standard deviation of experimental measurement.

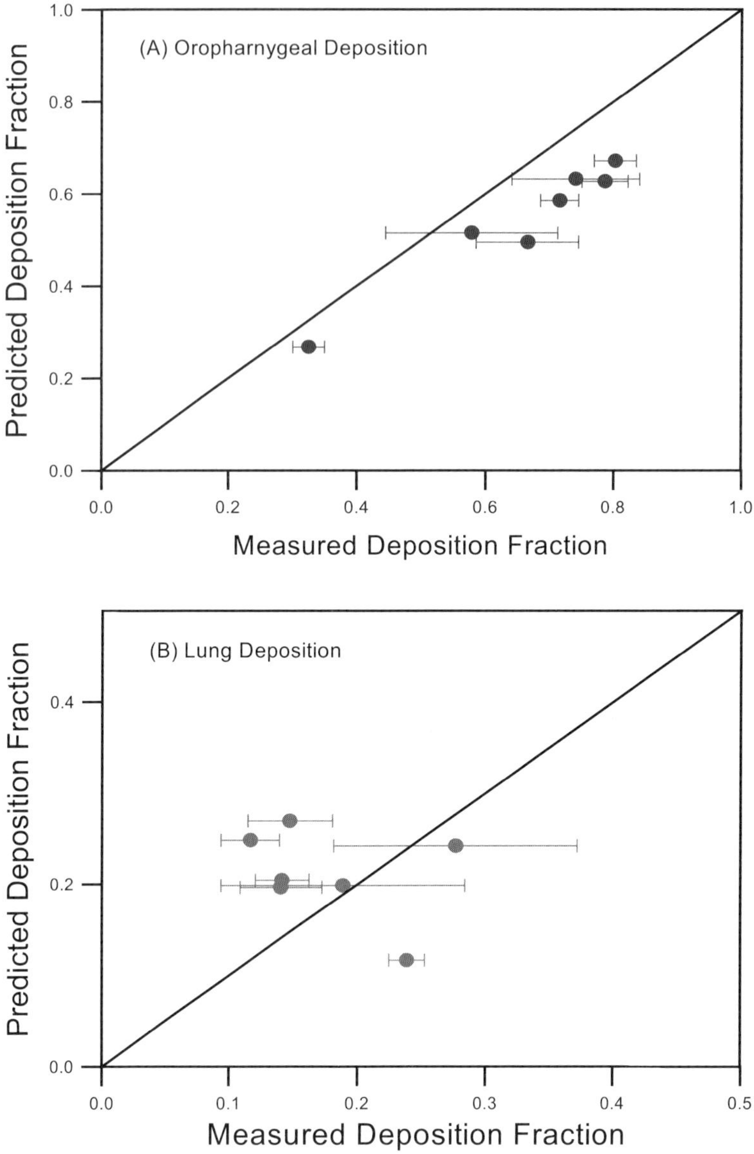

Figure 9. Comparison of in vivo and calculated deposition in the oropharynx and lung for several DPIs (Newman et al., 1989a, 2001; Borgstrom et al., 1994; Pitcairn et al., 1994, 1995). The error bars are standard deviation of experimental measurement.

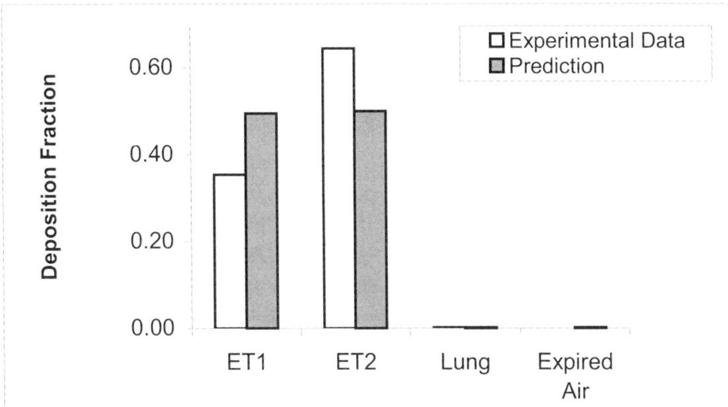

Figure 10. Comparison of deposition patterns of a nasal spray with a model prediction (Cheng et al., 2001b).

9. OROPHARYNGEAL DEPOSITION

Oropharyngeal deposition in pMDIs, DPIs, and even some small mist inhalers is a dominant fraction of the emitted dose as shown in the previous sections. The model calculation often did not agree with experimental data. Prediction of oropharyngeal deposition is based on in vitro measurement using either an impactor or impinger with a USP or glass throat inlet. It is therefore useful to compare the deposition in the glass or USP throat with the in vivo measurement. Figure 11 (top panel) shows the deposition efficiency (fraction of the emitted dose) in the oropharyngeal region as a function of flow rate from several HFA and CFC pMDIs without spacers. The oral deposition efficiency was higher for CFC-formulations (60–90%) than for HFA formulations (50–70%). As the flow rate increased from 30–90 L min^{-1}, the oral deposition decreased slightly. The deposition obtained with USP and glass throat inlets was substantially lower than experimental data. Figure 11 (bottom panel) shows the deposition efficiency in the oropharyngeal region as a function of flow rate from several DPIs. In the Turbuhaler and other tested DPIs, the oropharyngeal deposition efficiency is high, ranging from 56–91%. The oral deposition also decreased slightly as the flow rate increased, an indication that smaller particles were emitted at a higher inspiratory flow rate. In the same figure, depositions measured at the USP or glass throats were substantially slower, with the rate ranging from 10–40%.

In the model calculation, the oropharyngeal deposition included particle deposition in the USP or glass throat and the contribution of oral deposition from particles collected at the impactor or impinger stages. As shown in the previous sections, this prediction was generally lower than that of in vivo measurement. In addition to the artificial throat, the other contribution of the discrepancy may be due to the effect of the DPI or pMDI mouthpiece on deposition (Clark et al., 1998).

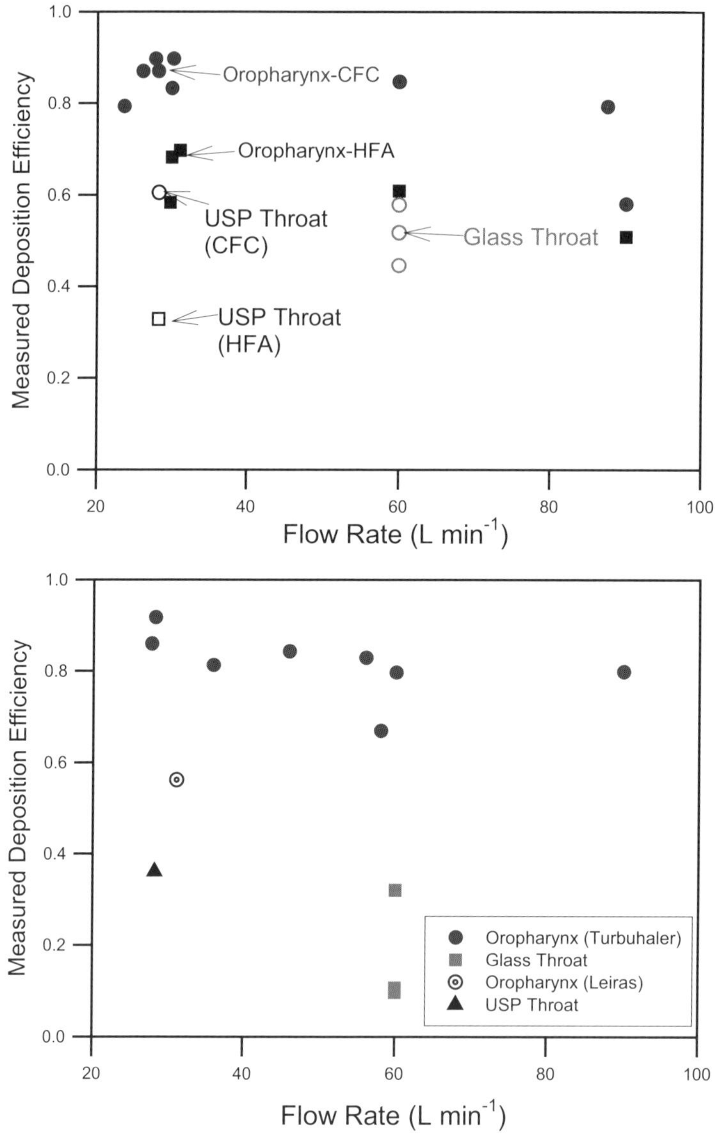

Figure 11. Deposition in the oropharynx and in glass or USP throat inlets. The top panel shows data obtained from pMDIs, and the lower panel shows data obtained from DPIs.

10. DISCUSSION

The FPF correlated somewhat to lung deposition, but could not accurately predict the lung deposition. This is no surprise because for the same FPF, one could still have quite different particle size distributions, which determine deposition pattern. It is still a useful tool for quality control as well as an index for designing a new formulation or device.

Modeling aerosol drug delivery using LUDEP software gave mixed results. For nebulizers, the model calculation usually agreed with in vivo deposition. This is not surprising because nebulizers were operated often continuously and subjects inhaled the tidal flow with breath holding. This is the same breathing maneuver used in the development of the LUDEP model. In these studies, the droplet size distribution was also determined using laser diffraction instruments. Based on the size distribution, deposition efficiency in the oropharyngeal region was <30%, not a dominant deposition site.

On the other hand, in studies with pMDIs and DPIs, the LUPDEP model usually underestimated the oropharyngeal deposition and overestimated the lung deposition. There could be several reasons for the discrepancy between the prediction and measurement. One reason is that in all pMDI and DPI deposition studies, the inertial device including the ACI and MLI with throat inlets were used for particle size measurement. We showed that the USP and glass throat inlets by itself underestimated oropharyngeal deposition because the flow profile in the simple geometry of artificial throats may not closely simulate the flow profile in the much more complex geometry of the oropharyngeal region. Use of a more realistic throat model provided much better agreement on the oropharyngeal deposition of pMDI (Cheng et al., 2001a). The oropharyngeal deposition in pMDIs and DPIs also decreased slightly with increased inspiratory flow rate. The ACI was operated at 28.3 L min^{-1}, and the MLI was operated at a 60 L min^{-1} fixed flow rate. These flow rates often were not the same as the breathing flow rate used in the in vivo deposition study. Therefore, the size measurement may not accurately reflect the particle size at the breathing flow rate.

Underestimation of the oropharyngeal deposition in the model calculation then led to more aerosols penetrating the lung for deposition. This was the main reason for overestimation of lung deposition. The breathing maneuver for both the pMDIs and DPIs also differed from the tidal breathing used in the ICRP model. In the pMDI and DPI, bolus inhalation for one or two puffs was used, and often after inhalation, a breath-holding period of up to 10 seconds was recommended and used in the deposition study. The LUDEP software could not account for deposition due to breath holding.

The model prediction of aerosol drug delivery can be improved if the oropharyngeal deposition for pMDIs and DPIs under different flow rates can be modeled more accurately. Additional improvement can also be obtained by considering breath holding, bolus inhalation, and perhaps the hygroscopicity of aerosol.

11. SUMMARY

Inhalation delivery of therapeutic compounds has become increasingly popular with expanded applications and new designs of medical inhalers. Both in vitro and in vivo testing are required in the development of an aerosol delivery system. The in vitro test includes measurements of the particle size and dose uniformity. The in vivo human clinical studies often include deposition pattern, pharmacokinetics, and clinical trials for safety and efficacy. The deposited dose of the inhaled aerosol can be related to the clinical efficacy, especially for topical application, such as bronchodilators to treat asthma attacks. However, predicting the deposition pattern based on in vitro measurement of particle size distribution would be very useful. Review of existing data indicated that a simple, fine particle fraction roughly defined as particles < 5 μm usually overestimates the lung deposition. For that reason, more elaborate modeling using an existing lung deposition model, such as the ICRP 66 model, was investigated. The results showed that the predicted deposition pattern in the respiratory tract using LUDEP software for the ICRP 66 model worked reasonably well for nebulizers. This is not surprising because the nebulizer was operated often continuously, and subjects inhaled tidal flow without breath holding. For pressurized metered dose inhalers (pMDIs) of both chlorofluorocarbon and hydrofluoroalkane formulations, the calculation underestimated deposition in the oropharyngeal region and overestimated the lung deposition. Similar results were obtained for dry powder inhalers (DPIs). Use of an artificial throat in the in vitro measurement could have contributed to the discrepancy in oropharyngeal deposition. Both the USP and glass throat substantially underestimated the oropharyngeal deposition. A second reason for the discrepancy could have been that the breathing maneuver for pMDIs and DPIs both require bolus inhalation with breath holding, but the LUDEP calculation did not account for the breath holding. For nasal spray pumps, LUDEP and the experimental data agreed that over 99% of deposition was in the nasal passage, but the pattern of deposition in the nasal airway was different.

12. REFERENCES

Agnew, J. E. (1991). Characterizing lung aerosol penetration. *J. Aerosol Med.* 4:237–249.

Barry, P. W., and C. O'Callaghan (1999). In vitro analysis of the output of salbutamol from different nebulizers. *Eur. Respir. J.* 1130:1164–1169.

Borgstrom, L., E. Bondesson, F. Moren, E. Trofast, and S. P. Newman (1994). Lung deposition of budesonide inhaled via Turbuhaler: A comparison with terbutaline sulphate in normal subjects. *Eur. Respir. J.* 7: 69–73.

Cheng, Y. S., C. Fu, D. Yazzie, and Y. Zhou (2001a). Respiratory deposition patterns of salbutamol PMDI with CFC and HFA-134a formulations in a human airway replica. *J. Aerosol Med.* 14:255–266.

Cheng, Y. S., T. D. Holmes, J. Gao, R. A. Guilmette, S. Li, Y. Surakitbanharn, and C. Rowlings (2001b). Characterization of nasal spray pumps and deposition pattern in a replica of the human nasal airway. *J. Aerosol Med.* 14:267–270.

Cheng, Y. S., D. Yazzie, J. Gao, D. Muggli, J. Etter, and D. E. Rosner (in press). Particle characterization and lung deposition patterns of a dry powder formulation of polylactic acid produced using supercritical fluid technology. *J. Aerosol Med.*

Clark, A. R. (1991). In vitro assessment of spacers and reservoir devices. In *Respiratory Drug Delivery II* (P. Byron, ed.), pp. 27–31, Boca Raton, FL: CRC Press.

Clark, A. R. (1995). The use of laser diffraction for the evaluation of the aerosol clouds generated by medical nebulizers. *Int. J. Pharm.* 115:69–78.

Clark, A. R. (1996). MDI: Physics of aerosol formation. *J. Aerosol Med.* 9:S19–S26.

Clark, A. R., S. P. Newman, and N. Dasovich (1998). Mouth and oropharyngeal deposition of pharmaceutical aerosols. *J. Aerosol Med.* 11:S116–S121.

Derom, E., and R. Pauwels (1998). Relationship between airway deposition and effects for inhaled bronchodilators. In *Respiratory Drug Delivery VI* (R. Dalby, P. R. Byron, and S. J. Farr, eds.), pp. 35–44, Buffalo Grove, IL: Interpharm Press.

Dolovich, M. (1989). Physical principles underlying aerosol therapy. *J. Aerosol Med.* 2:171–186.

Dolovich, M., and R. Rhem (1998). Impact of oropharyngeal deposition on inhaled dose. *J. Aerosol Med.* 11:S112–S115.

Hardy, J. G., S. P. Newman, and M. Knoch (1993). Lung deposition from four nebulizers. *Respir. Med.* 87:461–465.

Hickey, A. J., T. B. Martonen, and Y. Yang (1996). Theoretical relationship of lung deposition to the fine particle fraction of inhalation aerosols. *Pharm. Acta Helv.* 71:185–190.

Hirst, P. H., G. R. Pitcaim, J. G. Weers, T. E. Tarara, A. R. Clark, L. A. Dellamary, G. Hall, J. Shorr, and S. P. Newman (2002). In vivo lung deposition of hollow porous particles from a pressurized metered dose inhaler. *Pharmacol. Res. 19*: 258–264.

ICRP (1994). *Human Respiratory Tract Model for Radiological Protection, Publication 66, Annals of ICRP*. London, UK: Pergamon.

Knoch, M., E. Wunderlich, and S. Geldner (1994). A nebulizer system for highly reproducible aerosol delivery. *J. Aerosol Med.* 7:229–237.

Leach, C. L., P. J. Davidson, and A. Bouhuys (1998). Improved airway targeting with the CFC-free HFA-beclomethasone metered-dose inhaler compared with CFC-beclomethasone. *Eur. Respir. J.* 12:1346–1353.

Marple, V. A., B. A. Olson, and N. C. Miller (1998). The role of inertial particle collectors in evaluating pharmaceutical aerosol delivery systems. *J. Aerosol Med.* 11:S139–S153.

Martonen, T. B. (1993). Mathematical model for the selective deposition of inhaled pharmaceuticals. *J. Pharm. Sci.* 82:1191–1199.

NCRP (1997). *Deposition, Retention, and Dosimetry of Inhaled Radioactive Substances*. NCRP Report No. 125, Bethesda, MD: National Council on Radiation Protection and Measurements.

Nerbrink, O., M. Dahlback, and H. C. Hansson (1994). Why do medical nebulizers differ in their output and particle size characteristics? *J. Aerosol Med.* 7: 259–276.

Newman, S. P. (1998). How well do in vitro particle size measurements predict drug delivery in vivo? *J. Aerosol Med.* 11:S97–S104.

Newman, S. P., F. Moren, and S. W. Clarke (1987a). Deposition pattern of nasal sprays in man. *Rhinology* 26:111–120.

Newman, S. P., F. Moren, and S. W. Charles (1987b). The nasal distribution of metered dose inhalers. *J. Laryngol. Otol.* 101:127–132.

Newman, S. P., F. Moren, E. Trofast, N. Talaee, and S. W. Clarke (1989a). Deposition and clinical efficiency of terbutaline sulphate from Turbuhaler, a new multi-dose powder inhaler. *Eur. Respir. J.* 2:247–252.

Newman, S. P., A. R. Clark, N. Talaee, and S. W. Clarke (1989b). Pressurized aerosol deposition in the human lung with and without an "open" spacer device. *Thorax* 44:706–710.

Newman, S. P., A. R. Clark, N. Talaee, and S. W. Clarke (1991). Lung deposition of 5 mg Intal from a pressurized metered dose inhaler assessed by radiotracer technique. *Int. J. Pharm.* 74:203–208.

Newman, S. P., G. Pitcairn, G Hooper, and M. Knoch (1994). Efficient drug delivery to the lungs from a continuously operated open-vent nebulizer and low pressure compressor system. *Eur. Respir. J.* 7:1177–1181.

Newman, S. P., I. R. Wilding, and P. H. Hirst (2000). Human lung deposition data: The bridge between in vitro and clinical evaluation for inhaled drug products. *Int. J. Pharm.* 208:49–60.

Newman, S. P., G. Pitcairn, D. A. Adkin, M. T. Vidgren, and M. Silvasti (2001). Comparison of beclomethasone dipropionate delivery by Easyhaler dry powder inhaler and PMDI plus large volume spacer. *J. Aerosol Med.* 14:217–225.
Nichols, S. C., D. R. Brown, and M. Smurthwaite (1998). New concepts for the variable flow rate Andersen cascade impactor and calibration data. *J. Aerosol Med.* 11:S133–S138.
Phalen R. F., W. C. Hinds, W. John, P. J. Lioy, M. Lippmann, M. A. McCawley, and O. G. Raabe (1986). Rationale and recommendations for particle size-selective sampling in the workplace. *Appl. Ind. Hyg.* 1:3–14.
Pitcairn, G., G. Lunghetti, P. Ventura, and S. Newman (1994). A comparison of the lung deposition of salbutamol inhaled from a new dry powder inhaler at two inhaled flow rates. *Int. J. Pharm.* 102:11–18.
Pitcairn, G. R., T. Lankinen, E. Valkila, and S. P. Newman (1995). Lung deposition of salbutamol inhaled from the Leiras metered dose powder inhaler. *J. Aerosol Med.* 8:307–311.
Price, A. (2000). Validation of aerosol deposition models for pharmaceutical purposes: The way forward. In *Respiratory Drug Delivery VII*, pp. 197-208, Buffalo Grove, IL: Interpharm Press.
Pritchard, J. N., G. R. Layzell, and J. F. Miller (1996). Correlation of cascade impactor data with measurements of lung deposition for pharmaceutical aerosols. In *Drug Delivery to the Lungs VII*, pp. 101-104, London: The Aerosol Society.
Richards, J., P. Hirst, G. Pitcairn, S. Mahashbde, W. Abramowitz, A. Nolting, and S. P. Newman (2001). Deposition and pharmacokinetics of flunisolide delivered from pressurized inhalers containing non-CFC and CFC propellants. *J. Aerosol Med.* 14:197-208.
Soderholm, S. C. (1989). Proposed international standard for particle size-selective sampling. *Ann. Occup. Hyg. 33*: 301–320.
Sterk, P. J., A. Plomp, J. F. Van de Vate, and P. H. Quanjer (1984). Physical properties of aerosols produced by several jet and ultrasonic nebulizers. *Bull. Eur. Physiopathol. Respir.* 20: 65–72.
Thorsson, L., S. P. Newman, A. Weisz, E. Trofast, and F. Moren (1993). Nasal distribution of budesonide inhaled via a powder inhaler. *Rhinology* 31:7–10.
Vera, P., F. Blot, D. J. Gambini, M. H. Becquemin, F. Dumas, P. Beyne, I. Caubarrere, and L. Barritault (1995). Comparison of jet and ultrasonic nebulizers for alveolar targeting of methylprednisolone. *Nucl. Med. Commun.* 16:344–348.

13. AFFILIATIONS

Lovelace Respiratory Research Institute
Albuquerque, NM, USA

T. CIACH, L. DIAZ, E. VAN DEN IJSSEL AND
J. C. M. MARIJNISSEN

APPLICATION OF ELECTRO HYDRO DYNAMIC ATOMISATION IN THE PRODUCTION OF ENGINEERED DRUG PARTICLES

1. INTRODUCTION

The human body is sometimes in a need of medical assistance through either a medical procedure or exclusively through the administration of pharmaceuticals. This latter embraces the administration of medicine, receiving vaccines, and the need of particular pharmaceuticals in the case of permanent disorder, such as asthma and diabetes.

The administration process of a drug into the system of the body can vary in many ways. It can be delivered into the system orally or with the help of other measures such as an implantation, injection, inhalation or transdermal administration. Implantation and injection are not of great convenience. In both cases the body is affected by the repeatedly inserting of a foreign object and substance into the human body, which can be painful and which can result in scars and bruised limbs or other parts of the body.

Probably the oldest and also very convenient way of taking pharmaceuticals is by swallowing. However, due to the high concentration of digesting enzymes and low pH in our gastrointestinal (GI) tracks and also the nature of the GI membranes inhibiting permeation, only chemically resistant substances can be administered this way, Pauletti (1996). All the proteins and more complex substances are excluded or demand special treatment, Pauletti (1996). This way of drug administration is also quite slow which can be inconvenient in the case of acute pain relievers.

Transdermal drug administration became recently more and more popular. This technique is limited to compounds of molecular weight below 15 kDa and of rather lipophylic character, Banker (2002). This route is also considered as rather slow.

Another very promising way to deliver a medicine into the human body is inhalation. This technique is also very old, it has been used already thousands years ago to deliver substances formed by smoking tobacco or other herbs. With the development of knowledge and technology, this way of drug administration became very popular in lung disorder treatment like asthma. Because of the large internal area of the human lungs and its properties offering quick and effective absorption of chemical substances combined with low probability of infection, human lungs are now considered as a target for systemic drug delivery. Activity of digestive enzymes

L. Gradoń and J. Marijnissen (eds.), Optimization of Aerosol Drug Delivery, 189–204.

in the human lungs is low so vulnerable substances like vaccines or insulin can be easily absorbed and transported to the blood, Adjei (1997). In this technique, the size limit of the active molecule is much bigger than in the case of transdermal administration. Absorption from the lungs is very quick and effective. This is important, for example, in the case of morphine and its derivatives administered during acute pain caused by growing cancer.

The lung affords an excellent entry route for proteins and other macromolecular drugs. Many proteins absorb readily through deep lung tissue into the bloodstream by transcytosis, Banker (2002), Adjei (1997). This is a natural mechanism in which small invaginations grow into the surface of lung cells until they pinch off inside like tiny bubbles. They carry alveolar fluids as well as any drug present in those fluids. Studies show that most macromolecular drugs administered by inhalation achieve bioavailability levels of 15 % to 95 % in comparison to injection. Various scientists and companies report different values for different drugs and inhalers, Kaufman (2002), (web pages of companies introducing drugs for pulmonary delivery). That is why inhalation is a promising technique for modern systemic drug delivery as well as for insulin administration for diabetes.

Inhalation is a well-established form of treatment for lung disorder such as asthma or bronchitis. A large variety of different inhalers are present on the market. Probably the most popular type of inhaler has a medicine suspended in a liquid propellant under pressure. When the valve head is pressed a certain amount of suspension is atomised in the air, when the propellant evaporates, releasing the drug particles. Other types of inhalers are dry powder inhalers in which specially prepared powder is aerosolised with the energy of an air stream formed while a patient is inhaling. There are also other devices employing liquid solution of a drug and, as a source of energy, compressed air, ultrasonic vibrations or a high voltage. The efficiency of available devices varies a lot from one type to the other and is usually rather low. The design criteria for inhalers have to be based on the properties of the human respiratory system. This is not easy because only particles of a defined aerodynamic diameter can penetrate deeply into the lung as shown in fig. 1. As we see particles of approximately 3 μm aerodynamic diameter are probably the best for this purpose. Bigger particles are captured in the upper part of our respiratory system, and then they are removed with mucous and swallowed. Particles smaller then 3 μm are not captured in the lungs efficiently but it has to be mentioned that it is still not sure if nanoparticles, with a diameter below 100nm, can also serve as drug carriers for inhalation. This is mainly due to the problems with measurement of such small pieces of matter.

There are two main directions of research in the field of of pulmonary drug delivery systems. One is a development of more complex inhalers, equipped with microcomputers and employing sophisticated atomisation techniques. The second is the use of simple atomisation methods, like in dry powder inhalers, but the research is focused on the powder itself, to make it an easy to atomise and highly efficient drug carrier, for example pulmosphere (www.inhale.com).

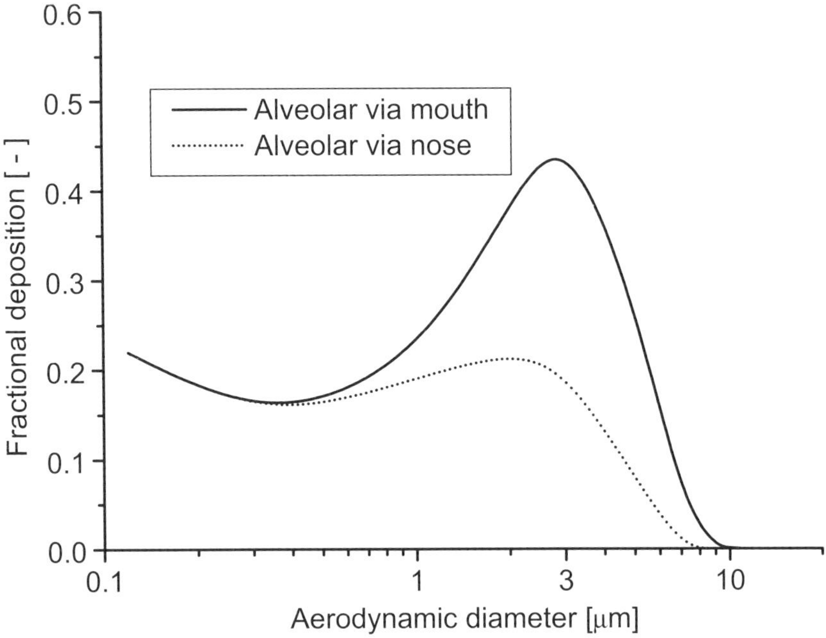

Fig.1. Fraction of particles deposited in the alveolar region of human lungs as a function of particle diameter for two ways of inhalation. Based on the model from the report of International Commission on Radiological Protection (ICRP,1994).

Sophisticated particles applied for drug delivery can offer many additional benefits like slow release of medicine or protection of vulnerable compounds Torchilin (2001). This is important because many of the new protein drugs are unstable and have a short lifetime *in vivo*. This means that they must be protected and delivered to the target tissue and then released to perform their tasks. Local delivery is also important in the case of toxic compounds such as modern anticancer drugs. Complex particles, because of the possibility of slow release of active compounds, can also improve the situation in case of frequently administered drugs. The time period between injections can be significantly increased. During that time slow release particles are deposited in muscles or circulate in vanes, due to slow decomposition of particle material, active compound is released. This allows keeping a constant concentration of medicine in the organism and avoiding overdosing or decreasing of medicine concentration below desired value. Also the concentration of a drug as a function of time can be easily modified to achieve better performance, which is important in the case of for example insulin.

Engineered particles also can be applied in pulmonary drug delivery. Vulnerable substances can be encapsulated and released with a predesigned rate. Modification of particles surface can make them easy to be atomised. This is very important because it allows application of simple and reliable inhalers and ensures accurate and stable dosing of a drug. Porous structures can decrease their density and, as a result, also the aerodynamic diameter to improve particle penetration into deep regions of the human respiratory tracks. Special surface properties can also significantly increase the residence time of particles in the human lungs. All these benefits can be accomplished by applying proper atomisation techniques and biodegradable polymers as building materials. In this paper, we describe the application of Electro Hydro Dynamic Atomisation as a tool in engineered particle production.

2. ELECTRO HYDRO DYNAMIC ATOMISATION

There are many techniques applied in particle production but we will focus on one method applying electrostatics as a driving force for droplet production. The phenomenon of the influence of the strong electric field on the liquid surface has been observed and reported in the sixteenth century by W. Gilbert in his book De Magnete (1600). Zeleny (1917) defined the phenomenon of electrospray as the case in which the electric field itself is the cause of the spraying of a liquid into fine droplets. Nowadays, we call this phenomenon Electro Hydro Dynamic Atomisation (EHDA), to avoid confusion with other spraying techniques where electric fields are applied.

EHDA refers to a process, where a liquid jet breaks up into droplets under the influence of electrical forces. Depending on the strength of the electric stresses in the liquid surface relative to the surface tension and the kinetic energy of the liquid jet leaving the nozzle, different spraying modes can be obtained. For micro-particle production, the cone-jet mode seems to be the most promising. This form of EHDA operation is also the most investigated. In this mode the process is very stable and small, mono disperse particles can be formed (fig.2).

In the cone jet mode, a liquid is pumped through a nozzle at a low flow rate. An electric field is applied over the nozzle and some counter electrode. This electric field induces a surface charge in the growing droplet at the nozzle. Due to this surface charge and the electric field, an electric stress is created in the liquid surface. If the electric field and the liquid flow rate are in the appropriate range, then this electric stress will overcome the surface tension stress. In that case, the electric stresses transform the droplet into a conical shape. The tangential electric field accelerates the charge carriers at the liquid surface toward the cone apex. In a liquid, the charge carriers are mainly ions. These ions collide with the surrounding liquid molecules. This results in an acceleration of the surrounding liquid. As a result, a thin liquid jet emerges at the cone apex. This jet can break up into a number of main droplets with a narrow size distribution, and a number of smaller secondary droplets and satellites. The number of secondary droplets and satellites can be of the same order of magnitude as the number of main droplets. However, the total volume of these droplets is much smaller than the volume of the main droplets. Due to the excess of surface charge in the liquid cone and jet, the droplets are highly charged. If solid particles must be formed by solvent evaporation, the droplets should be discharged beforehand to avoid disintegration (Rayleigh explosion due to mutual repulsion of electrical charges).

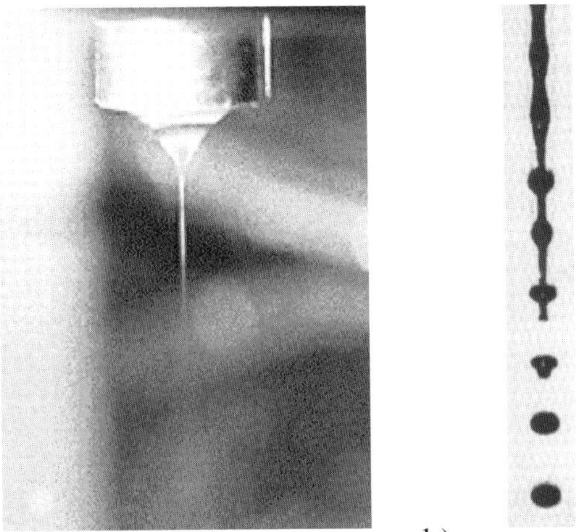

a) b)

Fig. 2. a) Electrospraying of the liquid in the cone-jet mode. Nozzle diameter is 5mm. b) Magnification of the droplet formation point on the jet.

Several attempts have been made to model the cone-jet mode, including analytical and numerical models. Taylor (1964) was the first to calculate analytically a conical shape, the 'Taylor cone'. For this cone shape the surface tension stress is balanced with the electrical normal stress. Taylor assumed a cone without a liquid jet. Consequently, there was no electrical current or liquid flow through this liquid cone. In Taylor's case the cone had to be a perfect cone. Only the angle at the cone apex could be used as a parameter for modelling. For this simplified case, Taylor found a cone angle of 98.6°.

Combining dimensional analysis with experimental results Gañán-Calvo (1997) and De la Mora (1994), derived relations for the droplet size and the current as a function of liquid flow rate and liquid properties (conductivity, surface tension, density, viscosity and relative permittivity), the so-called scaling laws. Hartman *et all* (1999) developed a physical numerical model of the process and refined the scaling laws.

As it has been mentioned before droplets produced by EHDA carry a high electrical charge. To avoid Rayleigh disintegration of the droplets, which happens when the mutual repulsion of electrical charges exceeds the confining force of surface tension and to make the droplets manageable, they have to be completely or partially neutralized. This can be accomplished in the set-up shown in figure 3.

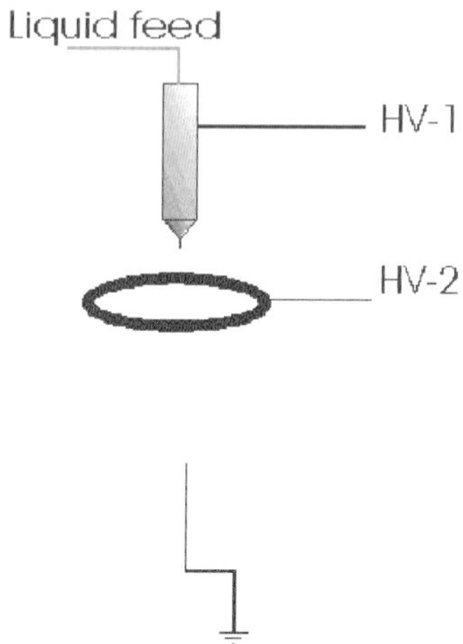

Fig. 3. Typical electrospray set-up with a ring electrode and discharging counter electrode.

The EHDA spraying nozzle is positioned opposite a counter ring electrode. Both the nozzle and the ring are connected to a high voltage source of the same polarity, with the nozzle the highest voltage. This creates the electric field that causes the formation of the cone and the jet, which breaks up forming small droplets. Opposite the nozzle and the ring a grounded needle electrode is placed. The electric field around the needle tip induces a corona discharge. Oppositely charged ions resulting from this neutralize the charge on the droplets.

3. POWDER PRODUCTION SET-UP AND APPLIED MATERIALS

Modern polymer chemistry offers a choice of materials that can slowly decompose inside a human body without leaving any harm or unusual remains. These biopolymers can be an excellent material for drug encapsulation or controlled release. In our experiments we selected poly-DL-lactide-co-glycolide (PLGA) (50:50, Aldrich) and polycaprolactone (PCL, Aldrich). Polyethylene glycol (PEG, Aldrich) was used as an additive to modify the decomposition rate. These polymers are biodegradable and have been approved by the American Food and Drug Administration authorities as drug components. PLGA is a rigid transparent material of good mechanical properties, which decomposes very slowly in the human body. PCL has worse mechanical properties and decomposes much quicker. PEG is a waxy solid dissolvable in water. In some experiments this polymer was added to increase the hydrolysis rate of the PLGA or PCL particles, which controls the release rate of the drug. As an example of a drug paclitaxel (taxol) was used, a medicine against certain types of cancer, won from yew trees. As a solvent in the described experiments a dichloromethane acetone mixture (4:1 weight) was employed. Also release experiments with dicarboxyfluoresacein as a tracer to check decomposition and active compound release rate were done. This substance was chosen because of its fluorescence properties making analysis much easier. The controlled-release medicine particles were made via EHDA in a set up as shown in figure 4.

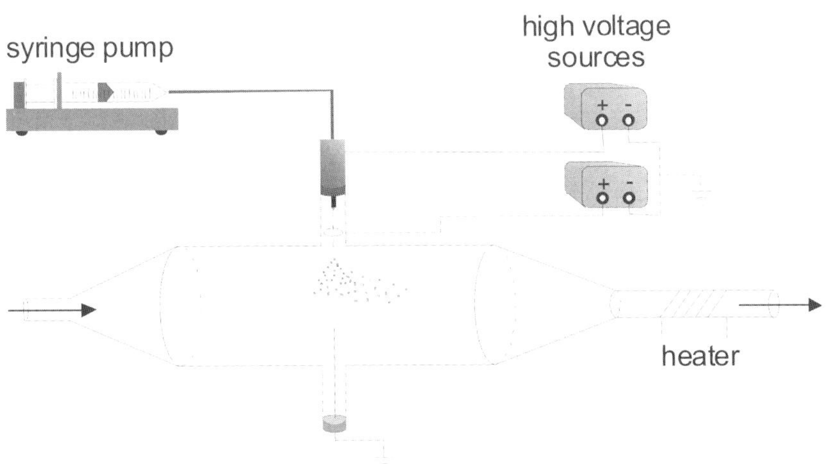

Fig.4. Particle production set-up.

The set-up consists of a cylindrical glass tube (ID =10 cm) with tapered ends. One end acts as an inlet for filtered air and the other end ducts the produced particles via a heater to the collection side. The heating, after which the produced particles are captured on a filter, is necessary to evaporate the solvent. Normally dichloromethane and acetone evaporate very quickly especially in the form of small droplets but if the polymer is present in the solution a jelly outer layer is formed which slows down the evaporation drastically. The EHDA spraying nozzle is positioned in a glass side tube (ID = 2.5 cm) in which also the counter electrode ring is placed close to the main glass cylinder. The distance between the nozzle-tip and the ring is 15 mm and the voltage difference between the liquid nozzle and the ring is about 7 kV.

The liquid is supplied via a syringe pump with a flow rate of 1 mL/h. The spray solution consists of 1% (wt) polymer dissolved in a mixture of dichloromethane and acetone 4:1 weight ratio. All the parameters are fixed such that the system operates in the cone-jet mode. The needle is inserted in the glass cylinder opposite to the spraying nozzle. As can bee seen in figure 4 here is also made use of a glass side tube. The voltage difference between the ring and the needle creates a corona discharge at the needle tip supplying negative ions for droplet neutralization. The airflow through the system equals 60 L/min. Particles are collected on a membrane filter (not shown on the picture) placed downstream of the set-up. Air is drawn through the filter by a vacuum pump maintaining a small under-pressure in the spraying chamber. The filters with collected particles were left overnight for thorough drying in a dryer at 40°C.

To investigate the drug release characteristics, particles together with the collection filter were immersed in a 200 mL buffer solution of pH7 at room temperature with a small addition of sodium azide to prevent bacteria growth. To measure the paclitaxel release into the liquid as a function of time, samples of 1ml of the solution were taken at certain time intervals and after passing them through a membrane filter analysed on the paclitaxel content with liquid chromatography. The results are described in the following chapter. Because the analyses for the paclitaxel content were very difficult and time consuming we decided to change the analytic technique in future experiments. We applied a fluorescent dye as a tracer for the release investigation. This allows using fluorometry, which is a very sensitive, convenient and quick analytical method. In this second set of experiments we made particles from a 2 %(wt) solution of PCL in dichloromethane. The solution contained also 2 %(wt) of the fluorescent dye - carboxyfluorescein, related to the weight of the polymer. To estimate the release rate of our tracer, filter with collected particles was immersed in the buffer solution, like in the previous case. Samples of this solution were taken by a syringe through a membrane filter, after gentle agitation. These samples were analysed on the content of the fluorescent dye by fluorometry. Results of the release investigation are shown in the following chapter.

Low-density particles were made in the same set-up as presented in fig. 4. Only different materials were used. High porosity shell-like particles were obtained from a dilute Budesonide solution in a water – ethanol mixture. This material is a drug used in asthma treatment and was supplied by the manufacturer Astra-Draco, Sweden. Polymer balloon-like particles were made from PLGA dissolved in a dichloromethane ethylacetate (3:1, wt) mixture. Spongy particles were produced from a PEG water solution with an addition of blowing agent and surfactant. A detailed description of low density particles production will be given in chapter 5.

4. SLOW RELEASE DRUGS

Figure 5 shows PLGA particles containing paclitaxel as produced with the set up of figure 4.

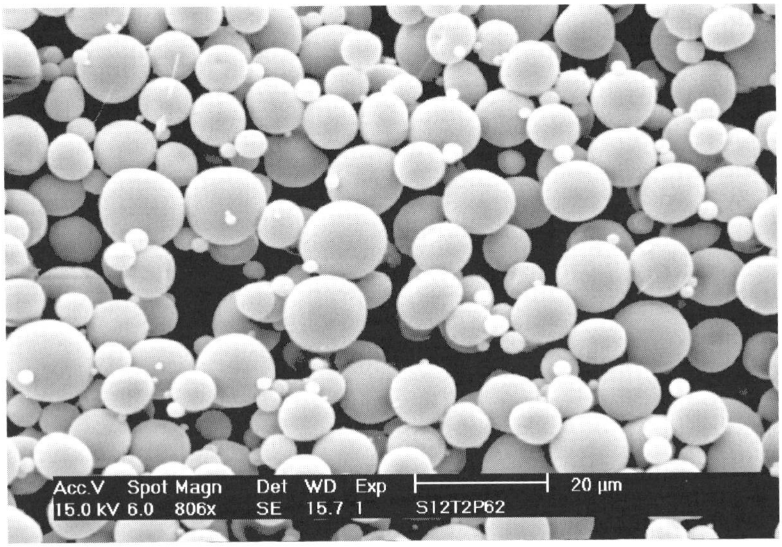

Figure 5. PLGA particles containing paclitaxel produced by EHDA.

As can be seen the size distribution is narrow but additional small particles are present. The contribution of these particles in the total mass of the system is negligible but some effort will be made in the future to avoid formation of these small particles.

In the buffer solution slow release of the medicine takes place. The involved mechanisms are supposed to be hydrolytic decomposition of the polymer matrix followed by dissolution of medicine entrapped in the polymer. In addition, diffusion of active compound to the surface and dissolution probably also takes place. The results of the paclitaxel release as a cumulative release with time is presented in figure 6.

It is clear that the cumulative release of the medicine is rather linear with time, with some faster release in the first few days and a slowing down after about 35 days. This initial burst of the active substance could originate from decomposition of small particles and/or from the drug available on the particle surface. For a higher time span (some 30 days) the release rate is more or less constant. Such a long release time probably cannot be applied in pulmonary drug delivery because of the much quicker clearance mechanisms existing in the human lungs. It means that particles will be removed from the lungs before they will accomplish medicine release; then they will end up in the stomach and medicine will be digested. That is why PCL, a faster decomposing polymer, was used for the next experiments.

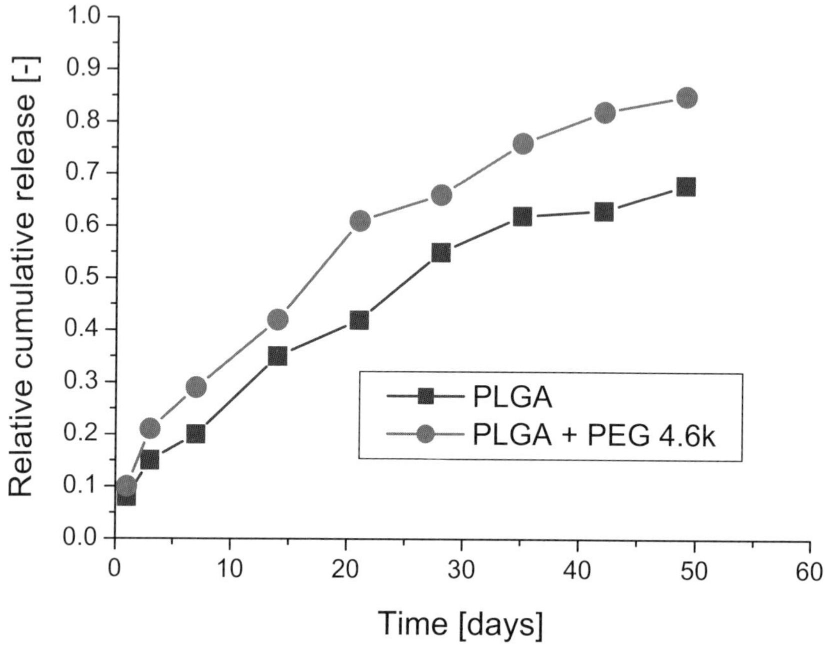

Figure 6. Relative cumulative release of paclitaxel from polymer microparticles made of PLGA (polylactide-co-glycolide) and PEG (polyethylene glycol). Conditions: room temperature, pH=7.

For these experiments we use carboxyfluorescein as a tracer, to determine active compound release. Results of fluorescent dye release from 5 μm PCL particles are shown in figure 7.

In the case of PCL particles, the release rate of active compound is much faster than with PLGA. The release is practically finished after 4-5 days. That is also the time when we see that particles disappear from the filter and suspension. The time scale of the process is in the range of the residence time of the particles in human lungs, Gradoń (1996).

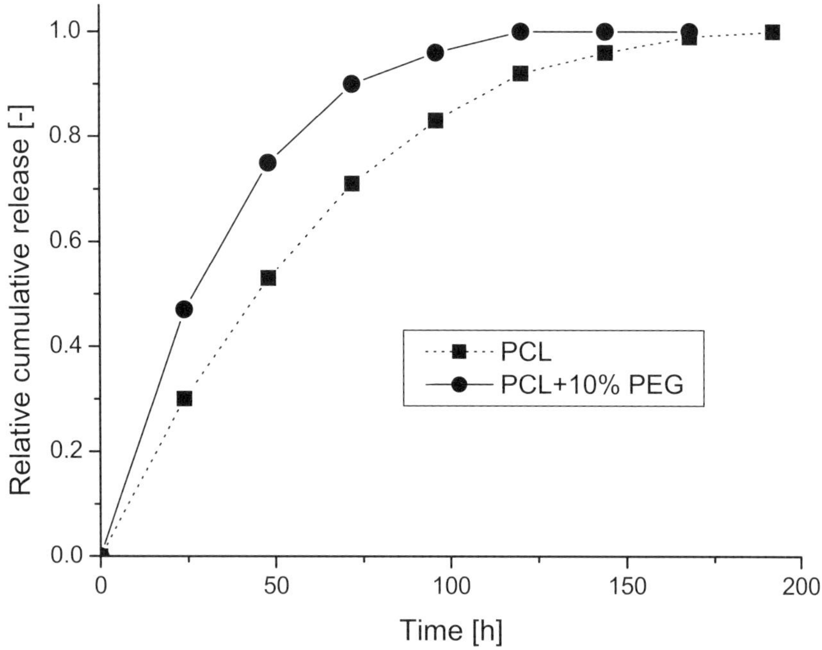

Fig 7. Carboxyfluorescein release from PCL (poly caprolactone) particles with and without addition of PEG (poly ethylene glycol).

As can be seen from figures 6 and 7, choosing a proper polymer or polymer mixture can serve s a tuning method for particle decomposition time.

5. LOW DENSITY PARTICLES

The aerodynamic size of particle is the most important parameter for inhalation, because it determines the deposition place. If we have a solid spherical drug particle and we blow the particle up, without changing the mass, to a bigger sphere, then the sphere diameter increases. However, as the density of the particle decreases, the aerodynamic diameter of the particle will also decrease. This phenomenon offers enormous possibilities in inhalation drug delivery. By manipulating the chemical composition and the drying conditions, EHDA is a method to produce hollow and blown up particles.

Hollow or balloon like particles can be obtained with the right evaporation conditions. In reality also other factors play an important role, such as mechanical properties and porosity of the formed solid shell as well as the surface tension of the solution and the presence of surface-active compounds. If we do not choose the composition of the droplets or the conditions of solvent evaporation properly, we can get the wrong particle structure such as small solid particle or remains of collapsed shells. There is still no scientific way to accurately predict if the solution we have will produce nice hollow spheres. The only way is still trial and error. An example of particles obtained 1 %(wt) solution of Budesonide® in a water-ethanol (1:10, wt) mixture, is shown in figure 8.

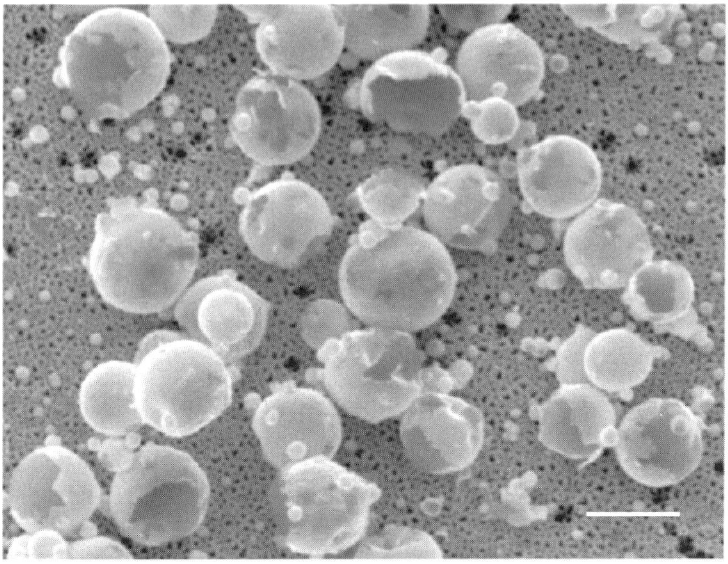

Fig. 8. Budesonide® particles produced by EHDA.
The white bar length in the picture is 5 μm.

On the picture we can see nice shell-like particles. The calculated volumetric fraction of the walls in relation to the whole particle is about 1 %. Some of the particles have broken walls. Among big shells we see small particles, which may have been formed from satellite droplets.

Not only brittle materials can form shell like particles, also polymers such as the biodegradable polymers (PCL, PLGA) can be produced in the form of hollow spheres. In this case they are rather balloon like and because of the mechanical properties of polymers they don't break. An example of such a balloon polymer particle can be seen in figure 9. As we can see most of the particles have dents or concave walls. Probably the solvent didn't evaporate completely from the polymer before the particles were collected so that they still were soft. The fact that these particles are of low density was checked by the measurement of their aerodynamic diameter with an Aerosizer (Amherst). Also measurements were done with an optical particle measuring instrument (Topas). From the combined measurements it was tried to calculate the porosity. However, we were not able to get an accurate value of the porosity. It is estimated to be between 0.7 and 0.9.

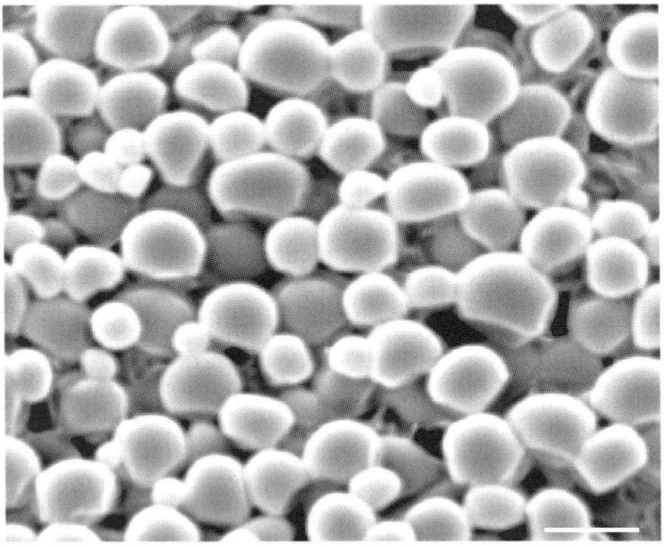

Fig. 9. PLGA hollow particles made by EHDA.
The white bar length in the picture is 5 μm.

Another way to obtain low-density particles is to blow them up by releasing a gas inside the polymer structure after particle formation. As a substance that can release gas we use $NaHCO_3$ or $(NH_4)_2CO_3$. These blowing agents decompose at elevated temperature (about 60 °C) releasing carbon dioxide. At this temperature the polymer is already soft. The gases are formed inside the polymeric structure of the particle and the whole process can be compared to baking a cake where the biodegradable polymer is the dough.

To verify this idea we used a solution of PEG (10 kDa MW.) containing 0.5 %(wt) of $NaHCO_3$ and 0.1 %(wt) of surfactant (related to the weight of the polymer). At a first test we created droplets on a 50 μm wire by immersing the wire in the solution. This resulted in tiny droplets hanging at the end of the wire. After evaporation of the solvent we put the wire with the particles for five minutes in an oven at 60 °C. At this temperature polymer became soft and the inflating agent decomposed, releasing CO_2. The gas expanded the particle. An example particle is shown in figures 10 and 11.

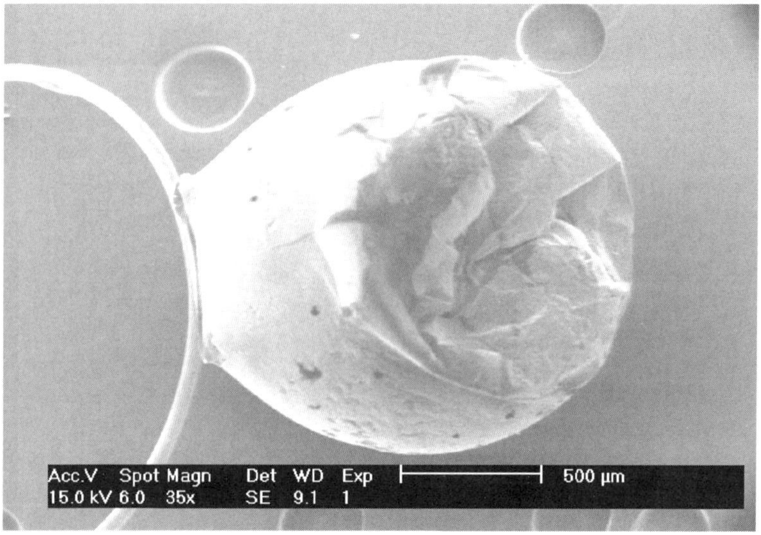

Fig. 10. Inflated PEG particle attached to the wire.

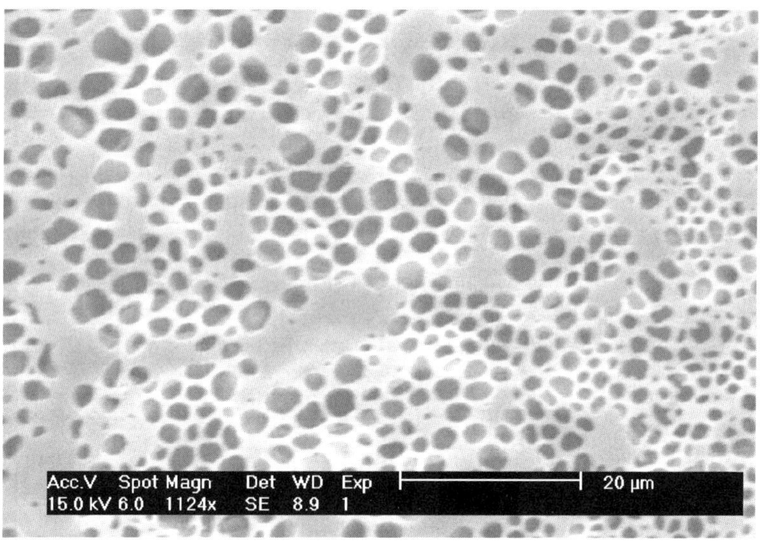

Fig.11. Surface of the inflated PEG particle.

We can see in figures 10 and 11 that the particle has a spongy structure with pores. The measured porosity of this particle is about 80 %. The particle is slightly collapsed.

Currently we try to accomplish the same process in the aerosol state by heating airborne particles. Particles are produced by EHDA and after solvent evaporation they pass a heated chamber where, we expect that gas is released inside the polymeric particles.

5. CONCLUSIONS

Electro Hydro Dynamic Atomisation as a technique for particle production for drug delivery is presented here. This method is very gentle and versatile and allows us to produce particles of a narrow size distribution from a variety of different compounds. Particles for controlled release of medicine can be obtained in this way; the particles could be delivered either by injection or by inhalation. The release rate of the active compounds can be set in the range from few hours up to weeks. The chemical composition of the particles is a tool to control the release rate of medicine. It is also shown that EHDA can serve as a method to produce porous particles for deep-lung delivery of medicines. Two ways of obtaining low-density particles are presented. They can be made either hollow or porous. Hollow or shell like particles can be made by controlling the precursors and the evaporation conditions. Blown-up particles can be made by releasing CO_2 from carbonates inside polymeric particles.

6. REFERENCES

Adjei L., Gupta P., (1997) Inhalation Delivery of Therapeutic Peptides and Proteins, Dekker.

Banker G., Rodhes C., (2002) Modern Pharmaceutics, Dekker, New York.

Chase H.P. (2002), Understanding insuline dependent diabetes, www.uchsc.edu.

Gilbert W., (1600), Gvilielmi Gilberti, De Magnete, Londini, Anno MDC.

Ganán-Calvo A.M.,Dávila J. Barrero A. (1997), Current and droplet size in the electrospraying of liquid scaling laws. J. Aerosol Sci., 28, No. 2, 249-275.

Gradoń L., Podgorski A., (1996), The retention of inhaled particles in the human respiratory system, Aerosol Inhalation, Kluwer Academic Press.

Hartman R., Brunner, D. Camelot, J. C. M. Marijnissen, B. Scarlett, (1999), Electro Hydro Dynamic Atomization in the Cone-Jet Mode Physical Modeling of the Liquid Cone and Jet, Journal of Aerosol Science, Vol. 30/7, 823-849.

ICRP, International Commission on Radiological Protection, www.icrp.org, reports, number of reports giving ingestion and inhalation dose coefficients, originally made for radiological protection.

Kaufman F., (2002), www.medforum.nl/reviews/background_on_inhaled_insulin_therapy.htm.

De la Mora F. J., Loscertales I.G., The current emitted by highly conducting Taylor cones, J. of Fluid Mechanics, 260, 155-184, 1994.

Pauletti G.M. at all, (1996), Structural requirements for intestinal absorption of peptide drugs, Journal of Controlled Release, 41. 3-17.

Taylor G., (1964) Disintegration of water drops in an electric field, Proc. R. Soc. A280, 383-397.

Torchilin V.P., (2001),Structure and design of polymeric surfactant based drug delivery systems, J. of controlled release, V73, S137-172.
Zeleny, J. (1917), Surface Phenomena Induced by The Electric Field, Phys. Rev. 10, 1.

7. AFFILIATIONS

Dr Tomasz Ciach worsk at the Faculty of Chemical and Process Engineering, Warsaw University of Technology, Waryńskiego 1, 00-645 Warsaw, Poland, and part time at The University of Delft, Delft, Particle Technology Group, Faculty of Chemical Technology and Material Science, Julianalaan 136, 2628 BL Delft, The Netherlands, ciach@ichip.pw.edu.pl.
Lidia Diaz and Esther van den IJssel obtained their master degree at the University of Delft, Particle Technology Group, Faculty of Chemical Technology and Material Science, Julianalaan 136, 2628 BL Delft, The Netherlands.
Dr Jan Marijnissen is an associate professor at the Particle Technology Group, Faculty of Chemical Technology and Material Science, Julianalaan 136, 2628 BL Delft, The Netherlands,

T.R. SOSNOWSKI, L. GRADOŃ,
F. ISKANDAR and K. OKUYAMA

INTERACTION OF DEPOSITED AEROSOL PARTICLES WITH THE ALVEOLAR LIQUID LAYER

1. INTRODUCTION

The kinetics of retention of inhaled particles in the human respiratory system is determined by the course of two complex processes – particle deposition and clearance rates.

The clearance mechanisms, especially for the pulmonary zone of the lungs, have not been recognized in the whole and some of them seem to be hypothetical. Obersdörster (1988) indicated intristic mechanisms of removal of deposits from the pulmonary regions of the respiratory system, which are: a) chemotaxis, ingestion and phagocitosis of foreign insoluble particles by alveolar macrophages (AM), b) endocytosis of deposits by tissue cells and subsequent exocytosis to the interstitium, and c) direct hydrodynamic transport of deposits due to flow induced by activity of the pulmonary surfactant (PS).

Pulmonary surfactant, consisting primarily of phospholipids and apoproteins, is produced in the alveolar region by type II epithelial cells. The direct measurement of surface tension in alveoli demonstrates periodic variations between 0 and 0.025 N/m during breathing (Bachofen et al., 1987). By comparison, surface tension in the trachea remains relatively constant near a value of 0.03 N/m. Consequently, a time-mean surface tension gradient exists with the potential to transport liquid out of the lung periphery via the airway liquid layer. Although the origin of this gradient is not fully understood, it is likely to be due to a variety of factors including the localized production of surfactant in the alveolar zone, degradation, reabsorption or fouling of surfactant along the airway tree, transepithelial liquid flux and ventilation (Fariday, 1976). According to the hypothesis by Podgórski and Gradoń (1990, 1993), liquid and deposits, both free and ingested by AMs, can be removed from alveoli because of the specific properties of the surfactant. The effective flow results from the surface tension gradient dynamically generated in the liquid lining of the distal airways during the breathing cycle. It was confirmed in studies by Espinosa and Kamm (1997).

Since the high activity of the surfactant is essential for this hydrodynamic transport (Sosnowski et al., 1995, Gradoń et al., 1996), we hypothetize that there may exist a relationship between the quality of PS and the rate of the alveolar clearance. The appropriate activity of the surfactant is also important for the clearance for other reasons: PS is believed to stimulate the activity of AMs

L. Gradoń and J. Marijnissen (eds.), Optimization of Aerosol Drug Delivery, 205–216.

(Baugham et al., 1987; Gradoń and Podgórski, 1995), and to facilitate phagocitosis due to opsonization of deposited particles.

Interfacial activity of the surfactant, considered to be a major indicator of its quality, can be assessed using physicochemical methods. These measure the ability of PS to vary the surface tension adequately during experimental simulations of the breathing cycle. Two devices - Langmuir Film Balance (LFB) and Pulsating Bubble Surfactometer (PBS) – have been used in our previous studies for experimental analysis of the surfactant activity.

The most important conclusion to emerge from our earlier investigation (Gradoń et al., 1996), is the existence of a very strong influence of surfactant monolayer on hydrodynamic conditions in the alveolar hypophase. Due to appearance of a surface tension gradient, the active monolayer "organizes" the interface and allows for a fast transfer of a surface disturbance to regions that are relatively far from the place where the local disturbance occurred.

From the physiological viewpoint, the ordering effect of interface due to presence of the surfactant monolayer is important for displacement of deposited particles and for the rate of alveolar macrophages attraction. We proposed (Gradoń and Podgórski, 1995) that a surface tension gradient induced by PS film may act as a stimulus for AM's receptors. On the other hand, the purely physical phenomenon known as Marangoni effect, which may arise due to surfactant adsorption on a surface of deposited particle, can facilitate the migration of the AM towards the spot of deposition (Sosnowski, 2001). In view of the results presented in our previous works, it can be concluded that the local event of deposition of an aerosol particle on the alveolar lining layer interface promote the signal, which is transferred over a whole alveolar area. It arrives to an AM immediately and is independent of other properties of deposits. Concentration gradient for chemoattraction may come later, but this first stimulus may be very important for a specific immunological response.

There is also the very important influence of the deposited particle structure on the behaviour of the surfactant molecules approaching it and then, on the transmission of this effect along the surfactant monolayer. We have stated that the distribution of the surface tension gradient strongly depends on the initial surface concentration and can be affected by the structural and surface properties of the deposited particles. The aim of this paper is to measure the activity of surfactant monolayer following its interaction with particles of defined morphology.

2. EXPERIEMENTAL TECHNIQUES

2.1. Particle Formation

Silica particles of controlled size, shape and morphology were produced for further investigation. The experimental setup used to produce the nanostructured particles from colloidal suspensions is shown in Figure 1 (Chemical Engineering Department, Hiroshima University).

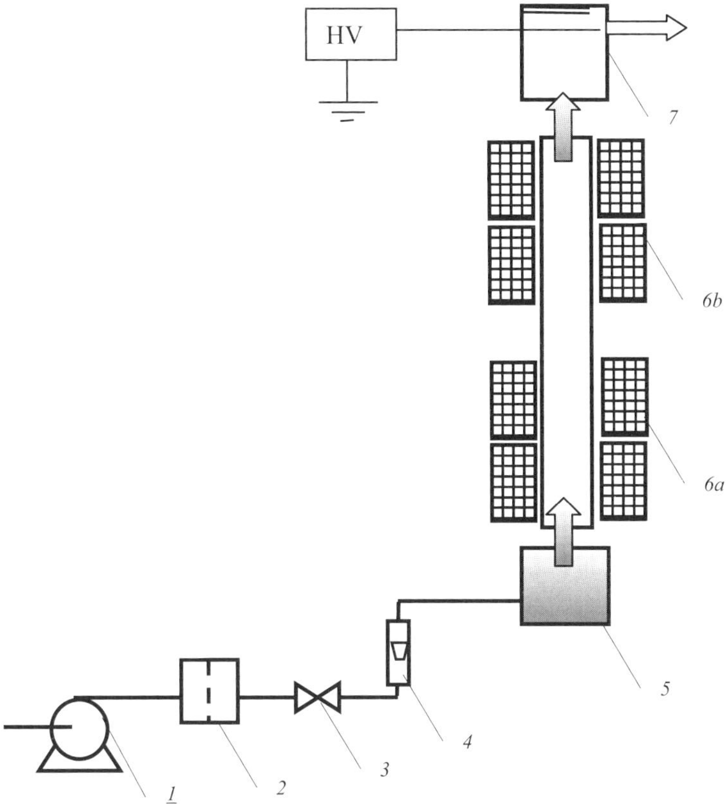

Figure 1. System for production of the submicron silica particles.
1 – air pump, 2 – air filter, 3 – control valve, 4 – flowmeter, 5 – ultrasonic nebulizer of colloidal suspension, 6a , 6b – two sections of tubular reactor (furnace), 7 – electrostatic precipitator for collection of particles.

The system consists of three sections:

1. spray generator (elements 1-5 in Figure 1),
2. vertical tubular furnace (6),
3. collector of aerosol particles (7).

Nanosized silica colloidal suspensions were used as a precursor for producing submicron size silica particles. The spray generator used was an ultrasonic spray nebulizer with 1.75 MHz resonator (Omron Corp.). The atomised droplet size (from water as precursor) measured by light-scattering particle size analyser (Malvern Instruments Corp.) was in the range 1 to 11 μm, and the mean droplet size was around 4.5 μm. The droplets were carried by a stream of air from the pump into the tubular reactor. The flow rates were varied in the range 1 to 4 l/min.

The reactor size was 13 mm in inner diameter and 1000 mm in length. The temperature of the reactor was varied from 200 to 1000°C. The prepared particles were collected using a filter maintained at 150°C to prevent water condensation.

The silica colloidal suspensions (silica sol) were obtained from Nissan Chemical Industry Ltd. The nominal particle sizes, as measured by TEM were 5 ± 1 nm, 15 ± 5 nm, 25 ± 5 nm, 50 ± 10 nm and 85 ± 15 nm. The colloidal silica particles were diluted with water to give the expected concentration. In some experiments to control the final shape of particles, 0.1% of surfactant was mixed with the precursor before spraying by ultrasonic nebulizer. The morphology of prepared particles were observed using Hitachi S-5000 Field Emission Scanning Electron Microscope. The detailed description of operational condition for obtaining required particles is presented in another paper by Iskandar et al. (2002). The examples of particles used in further experiments are shown in Figure 2.

Figure 2. SEM photographs of compact silica particles of different shapes obtained from 25 nm primary particles: compact spherical (CS) particles on the left and compact toroidal (CT) particles on the right.

Silica particles obtained in experiments described above had different size and shapes and had a surface porosity related to the diameter of colloidal particles used for the production of larger submicron particles.

For a detailed analysis of the influence of particle structured porosity on the interaction with surfactant monolayer, the silica mesoporous particles were also used. Iskandar et al. (2001) obtained such particles using their original technology. This method is a modified technique used for the formation of compact silica particles, as shown in Figure 1. According to it colloidal mixtures of silica nanoparticles and polystyrene latex (PL) nanoparticles were sprayed into a vertical reactor that contained two temperature zones (6a and 6b in Figure 1). The droplets were carried from the low to the high temperature zone. The solvent in droplets was evaporated in the first zone (6a) to produce a powder composite consisting of silica and PL nanoparticles. The PL nanoparticles in the powder were evaporated in the second zone (6b) to produce a silica powder consisting of mesopores of controlled size. SEM pictures of mesoporous silica particles are shown in Figure 3.

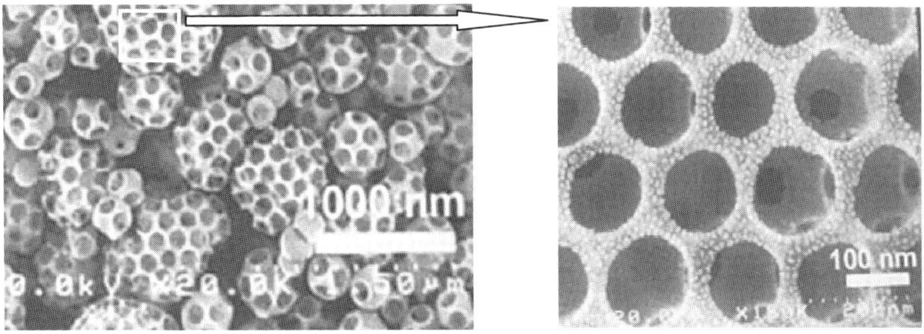

Figure 3. SEM photographs of silica mesoporous particles.

It can be seen that spherical particles in the submicron range have many open spaces what significantly increases their total surface area and reduces their density.

These three types of particles - compact spherical (CS), compact toroidal (CT), and mesophorous spherical (MS) - were used in further experiments on their behavior on contact with the pulmonary surfactant.

2.2. Particle-surfactant interactions

Investigations of interactions of solid particles with the pulmonary surfactant in vitro were done using the technique described previously by Sosnowski et al. (2000). Dynamic surface activity of the surfactant material was studied with oscillating bubble technique using PBS apparatus (Electronetics, USA). The operation principle is depicted in Figure 4.

Small air bubble (d = 0.8 mm) attached to the capillary tube is oscillated in the surfactant solution by means of a piston, which moves at the rate and amplitude corresponding to physiological variations of interfacial area of an alveolus during breathing (typically: 20 min^{-1} and interfacial area changes by 53%). The surfactant-rich liquid is kept in physiological temperature (37 ± 1 °C). The pressure inside the sample chamber is continuously recorded and recalculated into instantaneous values of the surface tension based on the Young-Laplace equation:

$$\sigma = \frac{r\Delta P}{2} \quad (1),$$

where r denotes the radius of the bubble at the given instant, and ΔP – the difference between the atmospheric pressure and the pressure inside the bubble at the same time.

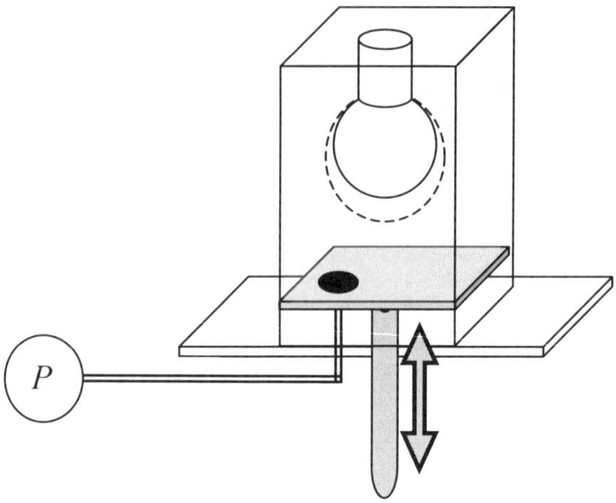

Fig.4. The principle of dynamic surface tension measurement in the Pulsating Bubble Surfactometer (PBS).

By measuring dynamic changes of the surface tension during breathing-like oscillations of a bubble it is possible to evaluate the physicochemical properties of surfactant samples, which are important for the physiological functions discussed earlier, including rate of pulmonary clearance. In our experiments we studied the properties of Curosurf (Chiesi Pharmaceuticals, Italy) - the natural substitute of the pulmonary surfactant, widely used in the clinical practice (e.g., Marraro et al., 1999). The original drug was diluted with a sterile physiological salt solution to obtain the final concentration of phospholipids equal 8 mg/ml. Such material was used for the control study in the PBS and for exposure studies with three different types of silica particles (CS, CT and MS as denoted before). Dose of particles used in the exposure studies was 1 mg per ml of surfactant solution, similarly to the one used by us previously in the investigations of a potential toxicity of inhaled carbon black in occupational exposures (Sosnowski et al., 2000). Each sample was incubated for 2 hours at 37 °C, then the dynamic surface tension of the sample was measured with the PBS device.

Based on the results of dynamic tensiometry, the quality of the surfactant can be assessed in several ways (Sosnowski and Podgórski, 1999). One of the most important parameters is the minimum value of the surface tension in the oscillation cycle, σ_{min}. A well-recognized property of the physiologically efficient surfactant is a very low value of σ_{min} , in the order of a few mN/m. Another important feature is the existence of a large hysteresis of the surface tension during compression and expansion of the interfacial area of a pulmonary bubbles. This hysteresis reflects the specific balance of viscous and elastic components in the complex structure of alveolar liquid covered by the surfactant. It plays an essential role in breathing mechanics, allowing for even air distribution in the expanding lungs. The hysteresis is also important for clearance processes as a factor that ensures a time-dependence of the surface tension gradient responsible for flows of the alveolar fluid. From a physical viewpoint, the area of the σ-A hysteresis loop can be considered as the energy converted in the system to produce the mechanical effect – superficial flow, acting as the intrinsic clearance process.

A quantity used for characterization of the hysteresis size is the normalized hysteresis area, HA_N introduced originally by Notter et al. (1982), and defined as:

$$HA_N = \frac{\left[\int_A \sigma dA\right]_{\text{expansion}} - \left[\int_A \sigma dA\right]_{\text{compression}}}{A_{\max} - A_{\min}} \quad (2).$$

The HA_N value is high for the surfactant of the adequate properties, while it is reduced for physiologically inefficient surfactants e.g., in RDS or other surfactant-dependent lung dysfunctions. In order to compare changes of hysteresis loop after exposure to potentially toxic agent, it is convenient to use the reduced value of normalized hysteresis area, i.e. ratio of actual HA_N to the value for the control sample HA_{N0} (Podgórski et al., 2001):

$$HA_{Nr} = \frac{HA_N}{HA_{N0}} \quad (3).$$

As long as such reduced value does not decline below 1, the surfactant material is believed to preserve its physicochemical properties and good physiological functionality.

3. EXPERIMENTAL RESULTS

Examples of surface tension changes recorded during oscillations of a bubble in the control Curosurf sample and the sample after a two hour exposure to silica particlesare presented in Figures 5 and 6.

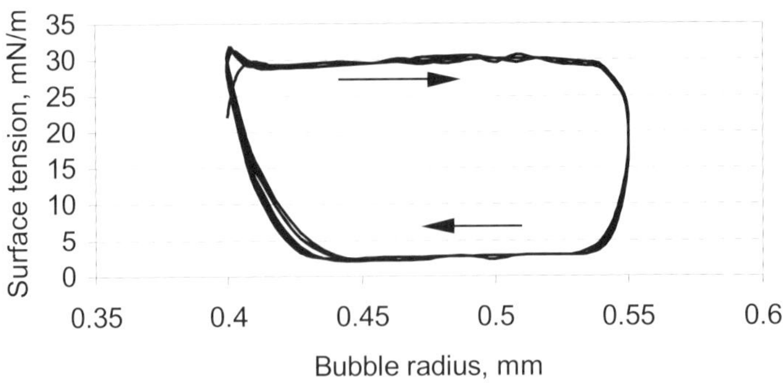

Fig.5. Surface tension measured during bubble pulsations in pure Curofurf (8 mg/ml).

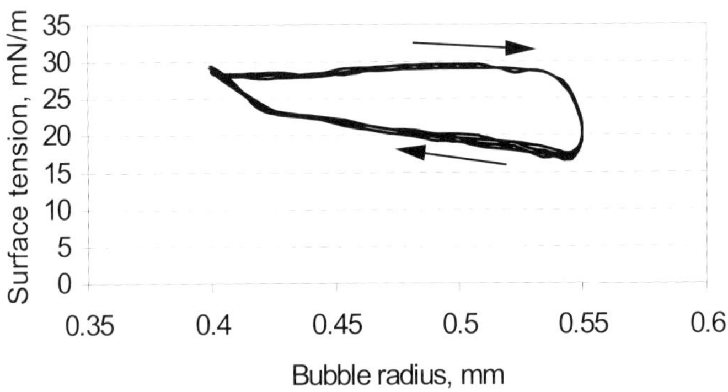

Fig.6. Surface tension measured during bubble pulsations in Curofurf (8 mg/ml) after 2-hour exposure to MS silica particles.

The changes in minimum surface tension measured in several independent experiments are illustrated in Figure 7. One can observe that particle shape and morphology influence the surface-active properties of the surfactant. Practically no changes in σ_{min} were observed for the compact spherical particles (CS) – the minimum surface tension in the compression-expansion cycle remained below 5 mN/m, similarly to what was measured for the control (pure) surfactant sample. For doughnut-shaped (toroidal) compact particles – CT, the value of minimum surface tension increased to approximately 10 mN/m, while for mesoporous spherical (MS) particles - to approximately 16 mN/m.

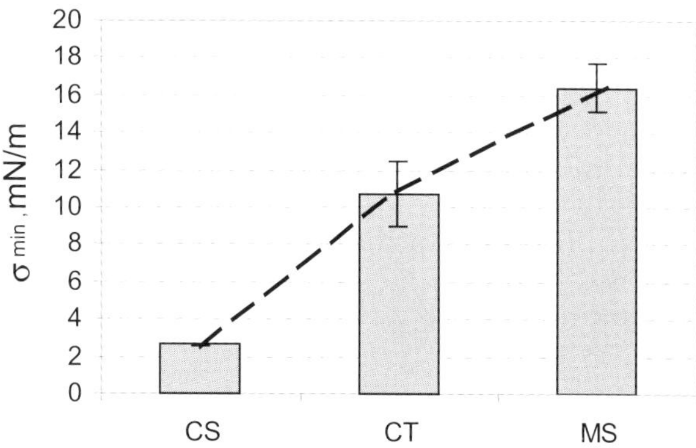

Fig.7. Changes of minimum surface tension of Curofurf (8 mg/ml) after contact with particles of different morphology.

These changes suggest that the amount of the dissolved most active components of the surfactant is reduced, most probably due to their adsorption on the silica particles mixed with the solution. Interestingly, when we recall the curves depicted in Figures 5 and 6, we will find out that the maximum value of the surface tension during the oscillation cycle is unchanged.

Measured changes of the surface tension hysteresis area are presented in Figure 8 as the value of the reduced normalized hysteresis area, HA_{Nr}, calculated according to Equation 3.

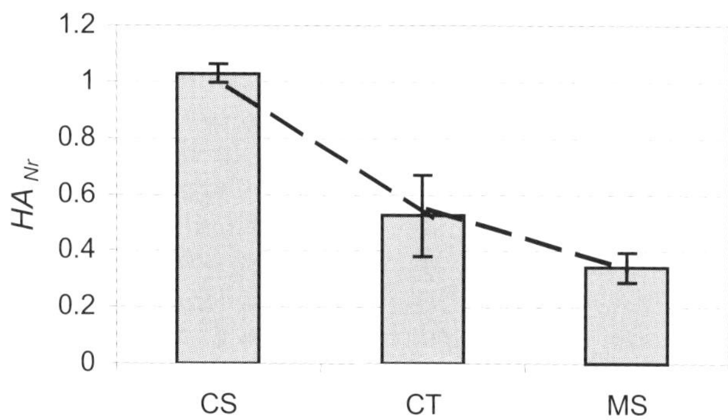

Fig. 8. Changes of reduced hysteresis area for Curosurf (8mg/ml) after contact with particles of different morphology.

Again, it is evident that the particle shape and morphology are related to the character of dynamic interfacial activity of the pulmonary surfactant. Compact spherical particles at the dose studied here do not influence the surface tension hysteresis. The increase of the total surface area for toroidal silica particles results in decrease of HA_{Nr} value by 50%, what again may be related to a higher adsorption of surfactant molecules on these particles. Such a concept seems to be supported by the largest reduction of hysteresis area (by 70%) for mesoporous particles, which possess the highest interfacial area allowing for the extensive adsorption of the surfactant from the solution.

The results presented here suggest that the morphology of particles deposited in the pulmonary region may be the crucial factor determining the character of their interaction with the surfactant, and - as a consequence - their residence time in that region of the respiratory system.

4. DISCUSSION

Experimental data confirm that there is a relationship between morphology of particles that remain in contact with the pulmonary surfactant and the physicochemical activity of the surfactant under dynamic conditions of breathing. Highly porous particles of a large interfacial area inactivate the surfactant in respect of its capability of adequate regulation of the surface tension value during different phases of breathing cycle. The minimum surface tension and size of the surface tension hysteresis loop are altered in a way that prevents the surfactant fulfilling its physiological role efficiently, especially – its role in promoting the pulmonary clearance. As suggested recently (Sosnowski et al., 2000; Sosnowski, 2001), there may be a direct effect of inhaled and deposited particles on the rate of the surfactant-mediated pulmonary clearance. Some aerosol deposits, due to the local depletion of surfactant concentration by adsorption are capable of inducing hydrodynamic signal (micro-flow) that will help a macrophage to localize the foreign body in the system. However, when too many deposits arrive to the alveolus, this signal will be ambiguous, and, in addition, the total concentration of the surfactant can be depleted. This will alter the specific surfactant activity, as demonstrated in this study, and this can take effect in the overall slow-down of clearance mechanisms within alveoli, so the particles will remain longer in the region of their deposition.

Such a scenario allows us to consider the possibility of designing a novel type of inhalation drugs, which would remain in the respiratory zone for the extended time period. If the drug formulation can be carried in the form of particles of a high sorption capacity, then such particles should stay longer in the region of their deposition because of a lower clearance rate induced by surfactant adsorption on them. Such technique could also be successful for drugs that require slow release and need to remain for an extended time in the respiratory zone. From our experiments it is clear that submicron particles with mesoporous structure would be the best for such purpose, but also other structures characterized by high interfacial area suitable for surfactant adsorption can be useful. The technical problem to be overcome can be found in the technical aspects of controlled synthesis of such particles from biocompatible materials that might be used as carriers for inhalation drugs.

5. CONCLUSIONS

The study was focused on investigation of interactions of submicron particles of different morphology and surface properties with the pulmonary surfactant under controlled, in vitro conditions. The application of the oscillating bubble technique allowed the comparison of the influence of three different types of silica particles: compact spherical, compact toroidal and mesophorous spherical, on physicochemical activity of the surfactant samples under breathing-like conditions. It was demonstrated that particles characterised by a large interfacial area (mesoporous particles) induced the highest reduction of the dynamic surface activity of the surfactant material. This was explained as the result of extensive adsorption of surfactant components on the surface of those particles. The observed effects of the reduction of surfactant activity, for example the reduction of surface tension hysteresis loop, may be responsible for the impairment of pulmonary clearance that would extend the time of the contact of a particle with the alveolar liquid. If environmental particles are inhaled such an effect is undesirable and it may be responsible for the high toxic potential of inhaled submicron aerosol particles at the alveolar level. On the other hand, if we take advantage of such mechanism to extend the residence time of inhaled drug particles in the pulmonary zone, we will be able to achieve a better outcome of inhalation therapy. The results of our investigations indicate a possibility of using new types of particles as carriers for inhalation drugs with a prolonged residence time in the respiratory tract.

6. REFERENCES

Bachofen, H.S., M. Schűrch,, M. Urbinelli, M., and E.R. Weibel. 1987. Relation among alveolar surface tension, surface area, volume and recoil pressure. *J. Appl. Physiol.* 62, 1878-1887.

Baugham, R. P., D.J. Mangels, S. Strohofer, and B.C. Corser. 1987. Enhancement of macrophage and monocyte toxicity by the surface active material of lung lining fluid. *J. Lab. Clin. Med.* 109, 692-697.

Espinosa, F., and R. Kamm. 1997. Thin layers flow due to surface tension gradients over a membrane undergoing uniform periodic strain. *Annals of Biomedical Engineering* 25, 913-925.

Fariday, E.E. 1976. Effect of ventilation on movement of surfactant in airways. *Respir. Physiol.* 27, 323-334.

Gradoń, L., and A. Podgórski. 1995. Displacement of alveolar macrophages in the airspace of human lung. *Med. Biol. Engng. Computing* 33, 575-581.

Gradoń, L., A. Podgórski, and T.R. Sosnowski. 1996. Experimental and theoretical investigation of transport properties of DPPC monolayer. *J. Aerosol Med.* 9, 357-367.

Iskandar, F., L. Gradoń and K. Okuyama. 2002. Control of the morphology of nanostructured particles prepared by the spray drying of nano-particle sol. *J. Colloid Interf. Sci.* - in press

Iskandar, F., N. Mikrajuddin, and K. Okuyama. 2001. Controllability of pore sized and porosity of self-organized mesoporous silica particles. *Nano Letter* 1, 231-237.

Marraro, G.A., M. Luchetti, E.M. Galassini, and G. Abbiatti. 1999. Natural surfactant supplementation in ARDS in paediatric age. *Minerva Anestesiol.* 65 Suppl. 1, 92-97.

Notter, R.H., R. Taubold, and R.D. Mavis. 1982. Hysteresis in saturated phospholipid and its potential relevance in vivo. *Exp. Lung Res.* 3, 109-127.
Obersdörster, G. 1988. Lung clearance of inhaled insoluble and soluble particles. *J. Aerosol Med.* 1, 289-330.
Podgórski, A., and L. Gradoń. 1990. Dynamics of pulmonary surfactant system and its role in alveolar cleansing. *Ann. Occup. Hyg.* 34, 137-147.
Podgórski, A., and L. Gradoń. 1993. An improved mathematical model of hydrodynamical self-cleansing of pulmonary alveoli. *Ann. Occup. Hyg.* 37, 347-365.
Podgórski, A., T.R. Sosnowski, and L. Gradoń. 2001. Deactivation of the pulmonary surfactant dynamics by toxic aerosols and gases. *J. Aerosol Med.* 14, 455-466.
Sosnowski, T.R. 2001. Sorption-induced Marangoni microflows in the pulmonary surfactant system. *Inż. Chem. Proces.* 22, 251-267.
Sosnowski, T.R., L. Gradoń, and A. Podgórski. 1995. Transport of aerosol deposits at the active surface of liquid layer. *J. Aerosol Sci.* 26S, 541-542.
Sosnowski, T.R., L. Gradoń, and A. Podgórski. 2000. Influence of insoluble aerosol deposits on the surface activity of the pulmonary surfactant: a possible mechanism of alveolar clearance retardation ? *Aerosol Sci. Techn.* 32, 52-60.
Sosnowski, T.R., and A. Podgórski. 1999. Assessment of the pulmonary toxicity of inhaled gases and particles with physicochemical methods. *Int. J. Occup. Safety Ergon.* 5, 433-449.

7. AFFILIATIONS

Dr Tomasz R. Sosnowski and Prof. Leon Gradoń work at the Department of Chemical and Process Engineering, Warsaw University of Technology, Waryńskiego 1, 00-645 Warsaw, Poland.
Dr Ferry Iskandar and Prof. Kikuo Okuyama work at the Department of Chemical Engineering, Graduate School of Engineering, Hiroshima University, Kagamiyama, Higashi-Hiroshima 739-8527, Japan.

8. ACKNOWLEDGMENT

Part of this work was supported by the Polish State Committee for Scientific Research (KBN) under the project No. 7 T09C 009 20.

E. DEROM

PULMONARY DEPOSITION AND EFFECTS OF AEROSOLIZED DRUGS IN PULMONARY PATIENTS

1. INTRODUCTION

Over the two last decades, clinicians have conducted numerous studies to document both the beneficial and the adverse effects of aerosolized drugs in pulmonary patients and to determine their exact place in the treatment of pulmonary disorders. This body of literature has been recently summarized in a number of consensus documents on the management of asthma and chronic obstructive pulmonary disease (COPD), the two most prominent pulmonary disorders. In these statements, the important role of inhaled drugs in the treatment of these disorders has been unequivocally reported.

In two recent articles, published in the prestigious British Medical Journal, it was claimed that the choice of the inhaler used to administer these inhaled drugs was of no clinical relevance, and that new inhaler devices are not more effective than established ones (1;2). Conversely, a steadily growing body of literature has focused on a variety of issues: the behavior of aerosolized particles within the airways, the prediction of particle deposition within the tracheobronchial tree using mathematical models, the likelihood to improve selective targeting of drugs to particular areas within tracheobronchial tree, and the relevance of these issues in clinical practice.

Although the clinical effects of aerosolized drugs have been recognized since early in the twentieth century, investigations designed to understand how drugs delivered by nebulisers, pressurized-metered dose inhalers (pMDI) or dry powder inhalers (DPI) eventually reach the airways to exert their beneficial effects are scarce. This chapter aims to bridge between basic and clinical science. It represents an attempt to provide more insight into the following question: is the design of an inhaler of relevance for the clinical and systemic effects of the drug it contains?

2. CLINICAL EFFECTS OF INHALED DRUGS

2.1 The inhalation process

It is common knowledge that a drug can only reach the airways, if it is aerosolized, and subsequently inhaled. Detailed analysis of the inhalation process in a human subject shows that it consists of three steps: (i) dose generation and delivery from the inhaler, (ii) pulmonary deposition of part of the delivered dose, and (iii)

L. Gradoń and J. Marijnissen (eds.), Optimization of Aerosol Drug Delivery, 217–242.

functional response to the deposited dose. The first step includes the generation of a sufficient amount of fine particles. The second step is affected by the inhalation flow used during inhalation, - and for the pMDI the co-ordination of actuation with inhalation - and by physiological factors, such as airway narrowing, airway closure, and mucous plugging. Design and functioning of the device affects both the first and the second step. The functional response of the dose deposited in the lungs - the third step - is dependent on different factors, such as potency of the drug, potential for inhaler-induced bronchoconstriction, and site of aerosol deposition within the tracheobronchial tree.

2.2. Pulmonary deposition

Aerosols used for medical purposes are invariably poly- (or hetero-) disperse. They generally contain particles with a MMAD between 2 and 10 μm. Drugs must be targeted to their receptor sites, wherever they may be, in order to be effective. Aerosol therapy is thus in essence a problem of aerosol deposition, since all particles that touch a surface are likely to deposit the site of deposition being the site of initial contact. The ICPR model of Pritchard, which proposes an estimation of regional mass deposition for oral inhalation of polydisperse aerosols, suggests that particles with a MMAD of 9 μm or more are mainly deposited in extrathoracic regions (oropharynx) and the central airways. Particles ranging between 5 and 7 μm deposit on the upper and lower bronchial airways, from the trachea down to the larger bronchi. Particles with an aerodynamic diameter of about 3 μm or less may reach the alveoli, whereas a substantial part of the very fine particles (MMAD of 1 μm or less) are too small to settle down and will leave the lung during expiration (3). Fine particle mass is generally defined as the amount of drug contained in particles of less than 5 μm. As deposition within the tracheobronchial tree is highly dependent on particle size, drug targeting may be improved by using monodisperse aerosols. Pieter Zanen will discuss the implication of the use of monodisperse aerosols elsewhere in these proceedings (4).

Particle characteristics are not the only element that determines pulmonary deposition. Indeed, factors such as inhalation technique (inhalation flow, breath holding time, initial acceleration of inhalation flow), airway diameter and alteration in particle size (evaporation, hygroscopic growth and particle agglomeration) are also of some relevance. For example, a more proximal deposition has been reported in patients with airways obstruction (5), compared to healthy subjects.

2.3. Lung residence time

The functional effect of an inhaled drug results from its interaction with cellular receptors. Slow absorption from the lung prolongs exposure to airway receptors and may enhance the functional effect. For example, β_2-agonists have been characterized as those that directly activate the receptor and appear rapidly in the systemic circulation (salbutamol), those that directly activate the receptor and are taken up into a membrane depot (formoterol), and those that are taken up in a membrane depot before interacting with the receptor (salmeterol). These differences in residence time are reflected in the kinetics of airway smooth muscle relaxation in vitro (6) and in the time course of the bronchodilatation in asthmatic patients (7). For salmeterol, the enhanced retention in lung tissue appears to be at the expense of the onset of action in patients with asthma, which is slower than salbutamol (8) and formoterol (7;9).

Retention in lung tissue of inhaled glucocorticosteroids has not been studied extensively. Slow absorption from the lung and prolonged exposure of inhaled glucocorticosteroids to the airways may enhance the anti-inflammatory effect without a concomitant increase in systemic glucocorticosteroid activity. The prolonged pharmacodynamic response of budesonide (10;11) has been attributed to its association with intracellular formation of long-chain fatty acid esters (12). Preliminary data indicate that liposome encapsulation of drugs also offers prolonged retention and drug action in the lower airways (13).

2.4. Pulmonary clearance

Particles deposited between the larynx and the terminal bronchioles may be cleared via two different mechanisms: mucociliary transport and pulmonary absorption. Insoluble substances or large particles deposited on the mucociliary escalator are cleared within less than 24 h. Mucociliary transport may, however, be somewhat slower in patients with pulmonary disease (14). Insoluble particles deposited in the alveoli are cleared via alveolar macrophages, the interstitial space, and lymphatics. Pulmonary absorption is the most prominent and most effective mechanism by which soluble substances are cleared (15;16). All currently prescribed drugs deposited in the airways follow the pathway of soluble substances. They bind to receptors located either on the cell membrane or within the cell, or are retained in the interstitial tissue for some time before absorbed by the pulmonary circulation. The pulmonary bioavailability of an inhaled drug is defined as the amount of drug originating from the lung that will eventually reach the systemic circulation. It should be noted that maximum serum concentrations of an inhaled drug generally occur within the hour of inhalation, even when gastro-intestinal absorption is blocked. Soluble particles deposited in the alveoli are eliminated by alveolar absorption, which is even more rapid than bronchial absorption (17). Any particle remaining in the lungs after 24 h is deemed to be alveolar in site with a half-life which may range between weeks and months.

2.5. Systemic elimination

A drug that is absorbed via the lung membrane or gastro-intestinal mucosa (and escapes the first-pass metabolism (see Fig. 1 and section 2.6 for further details)) becomes systemically available. Once in the systemic circulation, this drug runs the fate of an intravenous dose and is eliminated by hepatic and renal clearance.

Clearance and volume of distribution determine elimination half-life. Systemic accumulation of an inhaled substance will occur, if the dosing interval is shorter than its elimination (18). Accumulation of inhaled bronchodilators does not occur, as they are cleared rapidly. Differences in elimination half-life between glucocorticosteroids have, however, been reported. For example, elimination half time of fluticasone exceeds seven hours (18;19), whereas that of triamcinolone, flunisolide, and budesonide is less than three hours (18;20). Suppression of cortisol secretion in patients treated with inhaled fluticasone and budesonide in both healthy volunteers and in asthmatic patients has been attributed to this difference in half-life (21;22).

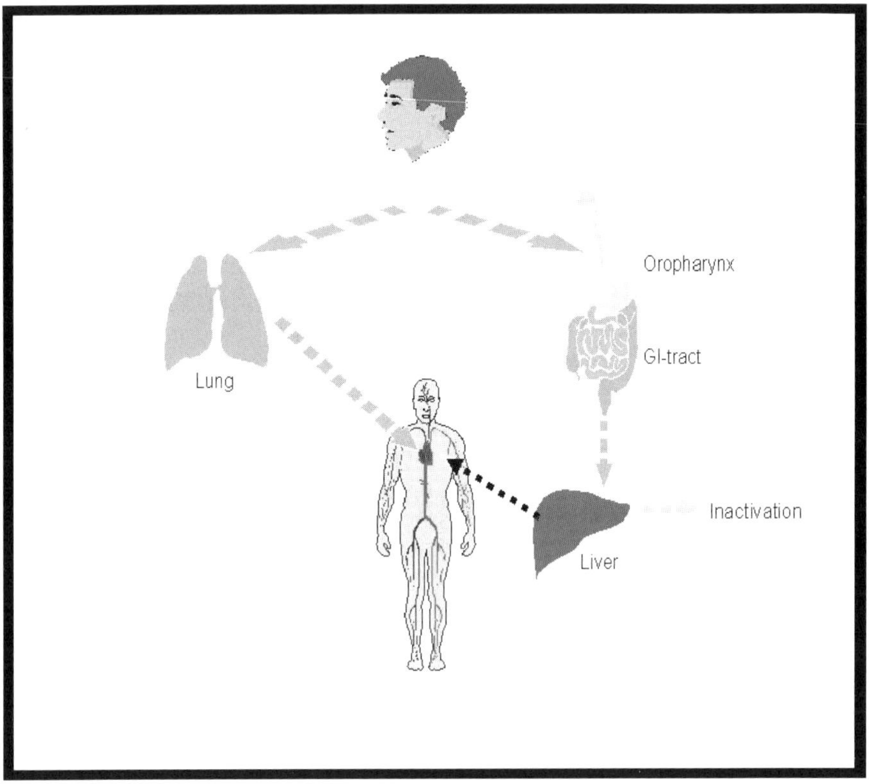

Figure 1. Fate of an inhaled aerosol. One portion of the aerosol generated by the device (not shown) is deposited in the oropharynx, whereas another part is deposited in the airways. Part of the aerosol absorbed through the gastro-intestinal mucosa is inactivated. The systemic

bioavailability of the inhaled drug consists of the portion absorbed from the lung and the portion absorbed via the gastro-intestinal tract, not inactivated by the first-pass metabolism.

2.6 Extrapulmonary deposition

Only a limited amount of the dose emitted by a pMDI or a DPI will reach the airways, the major fraction being deposited in the oropharynx (Fig. 1). This extrathoracic fraction may contribute to local side effects and, if swallowed, to systemic effects. The latter is, however, only true for inhaled drugs that are absorbed in the gastro-intestinal tract and not inactivated by the hepatic first-pass metabolism. In general, only a limited amount of the drug, deposited in the oropharynx and subsequently swallowed reaches the systemic circulation. However, important differences in terms of first-pass metabolism appear to exist between drugs belonging to a same category. For example, more than 30% of orally administered beclomethasone, 11% of budesonide and 1% of fluticasone will eventually reach the systemic bioavailability (18;23;24). This fraction, called the oral bioavailability of the inhaled drug, will, together with the pulmonary bioavailability, contribute to the systemic effects of the drug. Consequently, systemic effects of inhaled drugs, which are almost completely inactivated by the first-pass metabolism, will completely result from the fraction of the drug that is deposited in, and subsequently absorbed from the airways.

3. CLINICAL EFFECTS OF INHALED DRUGS IN OBSTRUCTIVE PULMONARY DISORDERS

3.1. The role of inhaled therapy in the treatment of obstructive pulmonary disorders

Inhalation therapy with bronchodilating and anti-inflammatory drugs is generally considered to be the cornerstone in the treatment of obstructive pulmonary disorders. This has been stated in the recent WHO guidelines on asthma, known as the Global Initiative for Asthma or GINA Guidelines (25) and the recent WHO guidelines on COPD, known as the Global Initiative for Chronic Obstructive Lung Disease or GOLD Guidelines (26). Inhaled therapy is preferred for several reasons: (1) their favorable therapeutic index, which means that an excellent therapeutic effect can be obtained with minimal side effects; (2) the instantaneous onset of action (for bronchodilating drugs only); (3) the maximal bronchodilating effect (greater for bronchodilators than what can be reached by their systemic administration); (4) the use of drug which are not absorbed in the gastro-intestinal tract (e.g. cromones).

Inhaled medications for obstructive lung diseases are used to reverse and prevent symptoms of airflow limitation. They consist of controllers and relievers. Relievers are inhaled short-acting bronchodilators, such as short-acting β_2-agonists and anticholinergics, which act relatively quickly to relieve bronchoconstriction and its accompanying symptoms, such as shortness of breath, wheezing, chest tightness, and exercise intolerance. Inhaled β_2-agonists represent the treatment of choice to relieve asthma patients from an acute attack. Controller medications, which should be taken daily on a long-term basis, are useful in getting and keeping the disease under control. They are indicated in patients with persistent asthma and COPD. The two most important controller medications are inhaled glucocorticosteroids and inhaled long-acting β_2-agonists. Inhaled glucocorticosteroids reduce local airway inflammation, whereas inhaled long-acting β_2-agonists exert a prolonged bronchodilating effect. Both drugs have been shown to reduce chronic symptoms, to improve pulmonary function, and to enhance quality of life in patients with asthma and COPD patients.

According to GINA, the main goals of treatment of asthma are to achieve and maintain control of symptoms, to prevent exacerbations and hospitalizations, to maintain normal activity levels, including exercise, to normalize lung function, and to prevent asthma mortality (25). In practice, control of asthma symptoms means to reduce the occurrence of asthma symptoms to less than once a week, and nighttime symptoms to less than twice a month. Patients with *intermittent* asthma may get perfect symptom control and protection of their airways against bronchoconstriction caused by a variety of stimuli, such as allergens, fog, or exercise, if they inhale a short-acting β_2-agonist when needed. Conversely, patients with *persistent* asthma need daily inhalations of a moderate dose of glucocorticosteroids for adequate symptom control. If low doses of inhaled steroids are insufficient, a combination of inhaled long-acting β_2-agonists and inhaled glucocorticosteroids or higher doses of inhaled steroids are prescribed. Such a treatment not only results in excellent control of symptoms, but also prevents against asthma exacerbations (27). Clinical experience indicates that the aims of treatment can be reached with inhaled drugs in more than 95% of all patients with asthma.

The effective management of COPD includes relief of symptoms, improvement of exercise tolerance, prevention of exacerbations, and enhancement of quality of life (26). Short-acting bronchodilators, such as short-acting β_2-agonists and short-acting anticholinergics are considered to be first choice to improve symptoms such as breathlessness or exercise intolerance. There is increasing evidence that inhaled steroids, inhaled long-acting β_2 agonists, and inhaled long-acting anticholinergics may be beneficial to patients with moderate to severe COPD, particularly since their use may lead to a substantial decrease in exacerbations and improvement of quality of life. As in asthma, inhaled therapy appears to be first choice in the majority of patients with COPD.

Table 1. Different formulations containing the same active component, available in Belgium in 2002

Drug Category	*Active Component*	*Formulation*
Short-acting β_2-agonists	Salbutamol	Ventolin®, Airomir®, Docsalbuta® (all pMDIs)
	Terbutaline	Bricanyl® (pMDI, Turbuhaler)
Long-acting β_2-agonists	Salmeterol	Serevent® (pMDI, Diskus)
	Formoterol	Foradil® (Aerolizer), Oxis® (Turbuhaler)
Glucocorticosteroids	Fluticasone	Flixotide® (pMDI, Diskus)
	Budesonide	Pulmicort® (pMDI, Turbuhaler)
	Beclomethasone	Becotide®, Aldecin® Beclophar®, QvarTH (all pMDIs)

3.2. Formulation of inhaled drugs

In each country, several inhaled anticholinergics, β_2-agonists and glucocortico steroids are marketed. In some instances, the same substances are formulated in different inhalers. For example, in Belgium, salbutamol is available in three different pMDIs and terbutaline in a pMDI and a DPI, the Turbuhaler. Formoterol is formulated in two different dry powder devices, Turbuhaler and Aerolizer. For inhaled steroids, two to four different formulations are available for each substance (Table 1). Some of these drugs are available in even greater number of different formulations in countries, where generic formulations have been popularized.

The existence of different formulations on the market represents a challenge for the clinician and the pharmaceutical industry. The clinician has to choose and prescribe both the drug (and the device in which it is formulated) which is likely to be most successful in his patient. He is supposed to make choices, taking various factors, such as age, easy use, ability to learn the correct use, number of inhalations a day, and portability of the inhaler into account (28). He expects to receive good information concerning the bio-equivalence of different formulations containing a same drug from the pharmaceutical industry, which is essential to prescribe the correct dose to his patient.

3.3. Bio-equivalence of formulations of inhaled drugs: methodological problems

3.3.1. Reliever medications

Reliever medications, used by the patients to treat symptoms caused by acute bronchoconstriction, are expected to relax airway smooth muscle, thereby reducing airways obstruction. Reversibility of airways obstruction in COPD patients is generally poor and incomplete. Conversely, airways obstruction is almost completely reversible after inhalation of a β_2-agonist in asthma patients. It is generally accepted that spirometric measurements, such as the FEV_1 or peak expiratory flow (PEF), can be used as a marker of activity of an inhaled bronchodilator in patients with asthma. Since it allows estimating the severity of airways obstruction as well as the degree of reversibility to a bronchodilating drug, spirometric measurements make it possible to determine the equivalence of two formulations in patients with asthma.

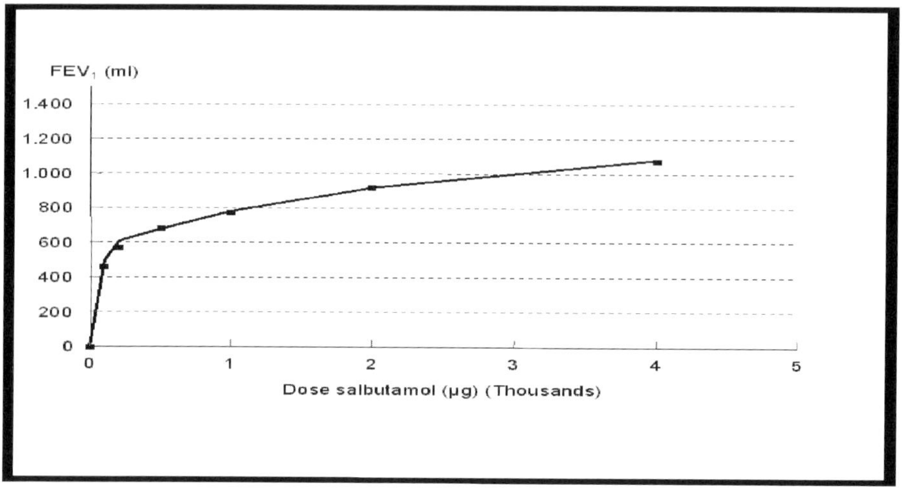

Figure 2. Bronchodilating effects of cumulative doses of a β_2-agonist in a patient with asthma.

However, a word of caution is necessary here. Indeed, the relationship between inhaled dose (and hence pulmonary deposition) and the airway effects of inhaled β_2-agonists in patients with asthma is not linear, but consists of an initial slope, followed by a plateau (Fig. 2) (29). The existence of a sigmoid dose-response relationship has two important consequences. First, a higher dose of drug inhaled by the patient and deposited in the lung will only induce a higher response, if the lower dose of the drug yielded a response that is still on the slope of the dose response curve. Important differences between inhalation techniques or delivery systems could be missed, if the comparison is made with doses that are on the plateau and not on the slope of the sigmoid dose-response curve. In bio-equivalence studies, more than one dose level should be included, in order to verify that the response is measured on the slope of the dose-response curve. Valid conclusions cannot be drawn from studies performed at only one dose level for each of the formulations under investigation (30).

The use of the FEV_1 or PEF as a marker of clinical efficacy is far more difficult in patients with COPD, since the reversibility of their airways obstruction is generally very poor.

3.3.2. Controller medications

If a rapid change in FEV_1 or PEF may be considered as a reliable marker of the effect of reliever medications, no such rapid effect is expected to occur with inhaled glucocorticosteroids. The usual approach to compare two inhalers containing such drugs is to conduct studies lasting for at least 4 weeks, in – preferably – steroid-naïve asthma patients allocated into parallel groups, who are treated at the lowest possible dose. The high variability of the response of patients to inhaled steroids, their relatively flat dose-response curve, and the relatively imprecise indirect endpoints available to estimate their efficacy (symptom scores, daily measurements of peak expiratory flow, extra use of bronchodilator medication) makes it necessary to recruit large groups of patients. Moreover, our-fold increase in dose is necessary to obtain a statistically significant difference in effect (31). Possibly, bronchoprotection against AMP could represent a more sensitive marker for the anti-inflammatory and clinical effects of inhaled steroids in asthma patients (32). Studies in which relevant endpoints, such as health care contacts, number of exacerbations necessitating a change in treatment are measured are even more difficult to conduct, and should last over at least 6 to 12 months to be conclusive. In fact, over the last decade, only one such study has been published, in which the primary aim was to document the impact of the inhaler on control of asthma symptoms over a lengthy period of one year (33).

The immediate bronchodilating effect of long-acting bronchodilators have sometimes been used to assess the bio-equivalence of different formulations in asthma patients. Although a well-designed study may provide valuable information on the equivalence of different formulations containing long-acting β_2-agonists, it remains unclear to which extent the immediate bronchodilating effect is good marker for their efficacy as controller medication.

For reasons outlined in the previous section, studies on the bio-equivalence between different formulations of inhaled glucocorticosteroids and long-acting β_2-agonists in patients with COPD are even more difficult to conduct than in patients with asthma, since the clinical benefit of these drugs are not translated into spectacular improvement in PEF or FEV_1.

Thus, it appears to be not easy at all to assess the clinical effects of inhaled controller and reliever medications in patients with asthma and COPD. For that reason, clinical pharmacologists have been looking for alternative tools, such as pulmonary deposition, to predict the effects of new formulations of inhaled drugs or compare novel agents with established ones.

4. IN VIVO MEASUREMENTS OF PULMONARY DEPOSITION OF INHALED DRUGS

There has been an increasing interest in the in vivo assessment of drug delivery to the lungs over recent years. This interest originates from a desire to understand the strengths and limitations of the current systems and to explore the basic mechanisms, which determine deposition of medical aerosols in patients, in order to develop improved systems in the future. Indeed, some of the currently used delivery systems have been developed years before techniques were available to assess pulmonary deposition.

Two basic approaches allow obtaining absolute measurements of lung dose: imaging techniques and pharmacokinetic techniques. Two-dimensional planar gamma scintigraphy is the most frequently used imaging technique. Three-dimensional single photon emission computed tomography (SPECT), and position emission tomography (PET) are in development, but have still to be validated (34). Pharmacological techniques are based on serial measurements of plasma concentrations over time or on excretion of drug components in urine.

4.1. Imaging techniques

Imaging techniques permit to localize deposition of an inhaled drug within the different compartments of the respiratory tract, to make an estimate of extrapulmonary deposition, to establish differences between healthy subjects and pulmonary patients (5) and to detect non homogenous distribution of pulmonary deposition of an inhaled drug (35;36). Disadvantages include health risk to the subjects, the limited number of measurements that can be made in one subject, the complex corrections for attenuation by body tissues and background activity, and, last but not least, the validity of the radio-active label (37). Indeed, the addition of the label may affect neither the total amount of drug delivered by the inhaler nor the particle size distribution. To be valid, the distribution of drug and label should be the same, and the properties of radiolabeled formulation should always be verified using in vitro techniques before starting the measurements. Since the image can be changed significantly by altering the gain on the camera, imaging techniques may give rise to misleading results.

4.2. Pharmacokinetic techniques

Pharmacokinetic techniques have gained recent attention because of the developing of analytical techniques allowing measurements of small quantities of drugs in plasma and urine. The biggest challenge of the pharmacokinetic technique is to distinguish the portion of the drug that reaches the systemic circulation via the lungs from the portion of the drug reaching the systemic circulation via the gastro-intestinal tract. If none of the swallowed and gastro-intestinally absorbed drug reaches the systemic circulation, it is reasonable to assume that the amount of drug which is present in the systemic circulation represents the lung dose, provided that no metabolization occurs in the airways (38). For other drugs, the gastro-intestinal absorption can be blocked using activated charcoal (39).

The technique has been refined by Borgström et al., who administered a low dose of deuterium-labeled drug intravenously, after which urine was collected over 48 to 72 hours. The ratio of the concentrations of the unlabeled and the labeled terbutaline reflected the ratio between pulmonary bioavailability of the inhaled drug and the amount of drug administered intravenously. This technique eventually allows a precise estimation of the pulmonary deposition of the drug (39).

Pharmacokinetic techniques have yielded slightly lower deposition values than the radiolabelled technique (16). This has been attributed to mucociliary clearance of drug deposited in the central airways. Indeed, a small portion of the drug, measured by imaging technique, may be eliminated via mucociliary clearance, and hence, be unavailable for pulmonary absorption. Pharmacokinetic techniques have a number of advantages, such as the use of an unaltered, original formulation and the possibility to perform serial measurements. However, pharmacokinetic techniques do not provide any information about the exact site of the tracheobronchial tree at which deposition occurs. Finally, incomplete urine collections may sometimes render the interpretation of the data more complex, although this issue appears to be an issue of lesser relevance in well-monitored studies.

5. PULMONARY DEPOSITION OF PRESSURIZED METERED–DOSE INHALERS AND DRY POWDER INHALERS IN HEALTHY VOLUNTEERS AND PATIENTS WITH ASTHMA

In order to inhale an aerosolized drug, the patient should activate the inhaler, from which a dose is subsequently emitted. Some of this metered dose is retained in the inhaler, adapter, spacer or mouthpiece (retained dose). Only a fraction of the delivered dose will reach the airways, since some of it may be lost into the air or on the face, another fraction will be deposited in the oropharynx, and a small part of drug will be exhaled.

In clinical studies, pulmonary deposition of a drug delivered from a given formulation is generally expressed as a fraction of a reference dose, which may be the metered or delivered dose. The nominal dose can be based on either the metered dose (the amount contained in the metered unit) or the delivered dose (the amount drug leaving the device). This issue is important, as, for example, budesonide delivered via a pMDI attached to a Nebuhaler, with a high retention in the spacer has a pulmonary deposition of 34% with reference to the normal metered dose compared to 76% with reference to the dose delivered from the spacer (40;41;41). In this section, nominal dose will be used as dose reference, unless otherwise mentioned.

The amount of topically available drug, relative to the nominal dose is determined by a number of factors: characteristics of the aerosol (particle size distribution, speed and shape of the aerosol plume) delivered by the inhaler, handling of the device, inhalation technique of the patient, and patient-related physiological factors, such as airway disease (narrowing or closure of airway closure, mucous plugging). Data presented here are obtained in healthy volunteers or stable asthma patients, who used the optimal inhalation technique for each of the devices.

Differences in pulmonary deposition between devices have been reported and are summarized in elsewhere (30). For pMDIs, pulmonary deposition varies between 10-11% and 20-25% of the nominal dose, but can be raised to 38% for pMDI with large volume spacer (42). A new generation of pMDIs, containing beclomethasone dipropionate, exhibits an even higher pulmonary deposition of about 40% (43) of the nominal dose or 55-60% of the emitted dose (44;45), because this pMDI produces particles with a lower MMAD and the speed of the plume leaving the canister is lower. For dry powder inhalers, pulmonary deposition has been estimated to reach 5% for Spinhaler 16% for Diskus (46;47), and 20 to 30% for Turbuhaler (30) or Respimat (48;49).

6. CAN IN VIVO DEPOSITION MEASUREMENTS BE USED AS A SURROGATE FOR THE CLINICAL RESPONSE TO INHALED ASTHMA DRUGS?

6.1. Effects of inhalation technique

In studies with pMDIs containing radiolabelled Teflon particles, Newman demonstrated that pulmonary deposition was markedly improved by the use of a longer breathholding time (10 vs. 4 s), a low inhalation flow (30 vs. 80 L/min), and by actuating at the beginning of inspiration (15;15;50). Others have subsequently confirmed these findings (51). Flow dependent deposition of pMDIs was further demonstrated in studies showing that the use of a breath-actuated pMDI increased pulmonary deposition in "poor" co-ordinators (52).

In patients with reversible airway obstruction (53), the greatest bronchodilatation was obtained when the pMDI containing terbutaline was actuated during a slow inhalation (25 to 30 L/min) and followed by a breathholding lasting 10 s. A significant correlation between mean bronchodilator response for the different inhalation modes and pulmonary deposition of labeled Teflon particles, obtained in the aforementioned studies, was reported (15). Studies with other β_2-agonists inhaled via pMDI have given similar results (54) and confirmed the clinical relevance of flow dependency of pulmonary deposition for pMDIs.

The inhalation flow affects pulmonary deposition, but not the properties of the aerosol delivered by a pMDI. Conversely, the intrinsic properties of the aerosol emitted from a DPI – a breath actuated device – are highly dependent on the inhalation effort of the patient (55), and the inhalation flow has always to be sufficiently high to produce an aerosol containing the optimal amount of fine particles (56;57;57).

The impact of flow-dependency on pulmonary deposition has been assessed for Rotahaler in healthy volunteers using the charcoal block method: the use of a low and high inhalation flow resulted in a pulmonary deposition of 3.6% and 7.0% of the nominal dose, respectively (58). Studies with Turbuhaler have yielded depositions ranging between 16.8% in a first and between 20 and 27% in later studies, when an inhalation flow of approximately 60 L/min was applied (16;57;57;59;59). Pulmonary deposition was reduced by half, once the inhalation flow fell below 30 L/min (57).

Although in one study with children, a significantly greater increase in FEV_1 was obtained when salbutamol was inhaled via Rotahaler at a higher flow of 90 to 120 L/min than at an inhalation flow of 30 to 50 L/min (60), most studies in which the effects of terbutaline inhaled via Turbuhaler were assessed, showed that at a low inhalation flow of 30 L/min, a less pronounced improvement of expiratory flow was obtained than at a higher inhalation flow (60 L/min) (61). In another study, it could be shown in children, who inhaled terbutaline via Turbuhaler at flows of 13, 22, 31, and 60 L/min, that its bronchodilating effect was flow-dependent, once the inhalation flow fell below the critical value of 30 L/min (62). It should, however, be underlined that even an extremely low flow, 13 L/min, resulted in a bronchodilating effect.

6.2. Differences in pulmonary deposition between devices

Numerous studies have been conducted, in which the efficacy of inhalers containing the same drug has been compared and their relative effectiveness established. Both cumulative and non cumulative dose designs have been used, and clinical response tends to be higher after the same total dose has been given as a cumulative dose compared with a single dose. Non-cumulative dose designs seem to be more sensitive (30).

The best way to establish the relationship between pulmonary deposition and clinical effect is to measure both variables simultaneously. This has been done in only a small number of studies. In two studies, terbutaline was administered via Turbuhaler and pMDI in mild to moderate asthmatics. For Turbuhaler, a peak inspiratory flow of 60 L/min was aimed for, whereas patients were asked to generate a flow of 90 L/min for the pMDI. In these studies, inspiratory volume and breathholding time were also standardized. In the first study, the effect of 0.5 mg of terbutaline inhaled via Turbuhaler or via pMDI on the increase in FEV_1 was similar (63). Pulmonary deposition was 22.6% of the nominal dose after Turbuhaler and 8.8% after pMDI. As the discrepancy between the difference in deposition and the similarity of the bronchodilating effect might have been caused by the administration of too high doses, a second study was conducted (64), which also included a 0.25 mg terbutaline dose. In that study, pulmonary deposition was similar to that obtained in the first study. The 0.25 mg terbutaline dose given by Turbuhaler resulted in a significantly higher FEV_1 response than 0.25 mg inhaled via pMDI. The increase in FEV_1 after the 0.5 mg dose given via pMDI and the 0.25 mg dose inhaled via Turbuhaler were similar, indicating that the amount of drug reaching the effector site, the lungs, determined the clinical effect. In addition, the difference in clinical effect between 0.25 and 0.5 mg given by Turbuhaler did not reach statistical significance, indicating that already 0.25 mg terbutaline administered via Turbuhaler resulted in a close to maximum effect.

In a third study, it was demonstrated that the protective effect of Terbutaline against histamine and methacholine was dependent on pulmonary deposition of the drug rather than on the prescribed dose. Indeed, the protective effect of 0.25 mg of terbutaline inhaled via Turbuhaler significantly exceeded that of 0.25 mg terbutaline inhaled via pMDI, but equaled that of 0.5 mg terbutaline inhaled via pMDI. This observation was again attributed to the higher pulmonary deposition of terbutaline with Turbuhaler (65).

Several clinical studies have focused on the bronchodilating effects of salbutamol inhaled via pMDI and via Diskhaler, Rotahaler, and Turbuhaler, without measuring pulmonary deposition. Conflicting results have appeared for multi-dose comparisons between Rotahaler and pMDI, as in some studies identical dose-response curves were obtained (66;67;67), whereas the difference was in favor of the pMDI in other studies (68). For Diskhaler and Autohaler, only inconclusive single dose studies have been published, whereas at equal nominal doses, salbutamol delivered via Turbuhaler generally gave responses that were superior to that given via Diskhaler (69) and pMDI (70).

A good knowledge of the two most important factors affecting pulmonary deposition of inhaled drugs is of clinical relevance, because the eventual pulmonary deposition rather than the delivered dose determines the clinical effect of inhaled bronchodilators. For years, improper inhalation technique has been recognized as a frequent cause of treatment failure (71). Lack of insight of the deposition characteristics of the prescribed formulation is, however, at least as important. Inhalers characterized by a high pulmonary deposition may represent an advantage, as a switchover from "device A" with low pulmonary deposition to "device B" with high pulmonary deposition might lead to an improvement in symptoms. Conversely, attempts should always be made to reduce the prescribed dose of the drug in patients whose disease is well under control, if a device with a higher pulmonary deposition is used, in order to avoid unnecessary exposure of the patient to intrapulmonary, and hence, systemic concentrations of the drug.

Most of the work on the relationship between pulmonary deposition and clinical effect has been done with β_2-agonists, since the immediate effect these drugs is easy to measure. Although it is more difficult to conduct comparative studies with inhaled glucocorticosteroids and anticholinergics, it is very likely that the same concepts hold also true for these drugs, as some studies suggest (72;73).

7. ARE SYSTEMIC EFFECTS AND SIDE EFFECTS OF INHALED DRUGS DEPENDENT ON THE CHOICE OF THE DEVICE?

7.1. Local side effects of inhaled drugs

As outlined in section 2.6, the major fraction of the drug emitted from a pMDI or a DPI will be deposited in the oropharynx, where it may exert local side effects. Since β_2-agonists do not exert local side effects of any clinical significance and the local side effects of anticholinergics are mainly bad taste and mouth dryness, extra-pulmonary deposition does not represent a clinically relevant issue that should affect the choice of an inhalation system. Conversely, it has been shown that oral candidiasis, one of the most frequent local side effects of inhaled glucocorticosteroids occurs in 5% of the patients using a pMDI and is somewhat related to the dose deposited in the oropharynx. Indeed, reduction of the deposition by use of large volume spacers (74) or a Turbuhaler combined with mouth rinsing (75) resulted in a decreased incidence of candidiasis. Hence, systems characterized by a high pulmonary and a low pharyngeal deposition of inhaled glucocorticosteroids, such as the combination of a pMDI with a large volume spacer, the reformulated HFA-containing pMDI or the recently developed Respimat (48;49) could represent an advantage in this regard.

7.2. Systemic effects and therapeutic index

The pulmonary bioavailability, the amount of aerosolized drug reaching the effector site, the lung, determines the elicited effect. The sum of the pulmonary bioavailability and the oral bioavailability (the fraction deposited in the oropharynx that will be absorbed in the GI tract, and is not inactivated by the first-pass metabolism) will be responsible for the systemic effects. Each combination of active drug and device is unique in terms of division of delivered drug between lungs and oropharynx, and hence, of balance between the desired local activity and undesired systemic activity.

It is possible to evaluate the balance between pulmonary bioavailability and systemic bioavailability if the following data are known: retention of the drug in the inhaler, the percentage pulmonary and oropharyngeal deposition of the drug, the degree of gastro-intestinal absorption, and the inactivation by the first-pass metabolism. From these data the therapeutic index or L/T ratio (L = local bioavailability, T = systemic bioavailability) can be derived (Figs. 3 and 4). For terbutaline the L/T ratio averages 0.55 for a pMDI and 0.81 for Turbuhaler (76). For salbutamol the L/T ratio averages 0.35 for pMDI, 0.15 for Rotahaler 0.24 for Diskhaler and 0.45 for Turbuhaler (76). The differences in value between salbutamol and terbutaline are a reflection of inherent differences in gastrointestinal absorption and first-pass metabolism between these drugs. For budesonide similar calculations have yielded 0.59 for pMDI, 0.84 for Turbuhaler and 0.95 for pMDI with Nebuhaler (77).

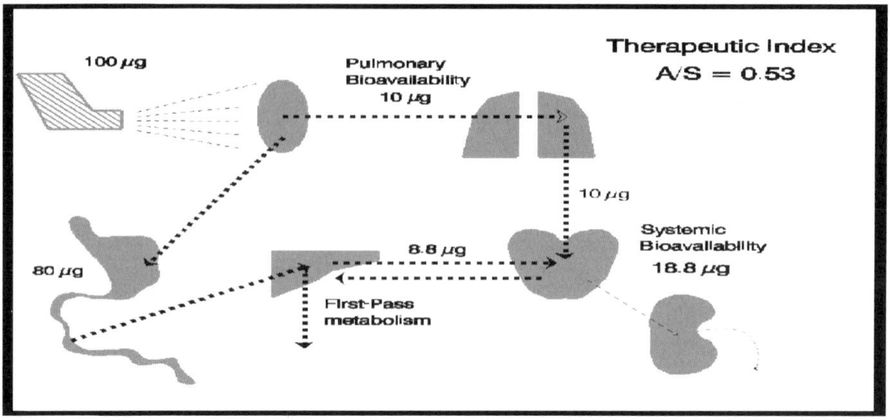

Figure 3. Therapeutic index of a drug inhaled via pMDI, derived from the pulmonary and systemic bioavailability.

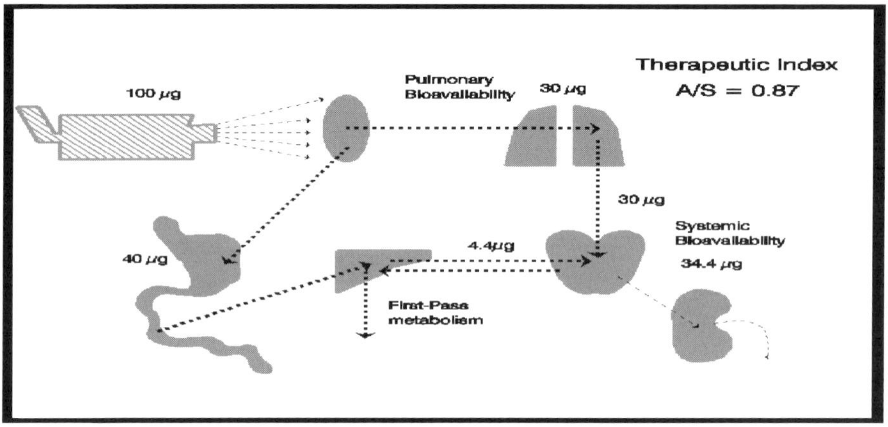

Figure 4. Therapeutic index of the same drug as in Fig. 3, inhaled via a pMDI connected to a large volume spacer. The change in device profoundly affects the therapeutic index.

A higher L/T ratio will always indicate a more favorable ratio with respect to the balance between desired and undesired effect. This may result either from a good targeting ability of the formulation, or from a low oral bioavailability. A high L/T ratio also implies that it is the pulmonary bioavailability that determines the systemic activity.

The therapeutic index will remain relatively elevated and will be minimally affected by the oral bioavailability if formulated in a device characterized by a high pulmonary deposition (Fig. 5). Conversely, the therapeutic index is highly dependent on oral bioavailability, if formulated in inhaler with low pulmonary deposition. In such cases, the impact of the extrathoracic fraction on the systemic bioavailability is only prominent for drugs with an oral bioavailability exceeding 10%.

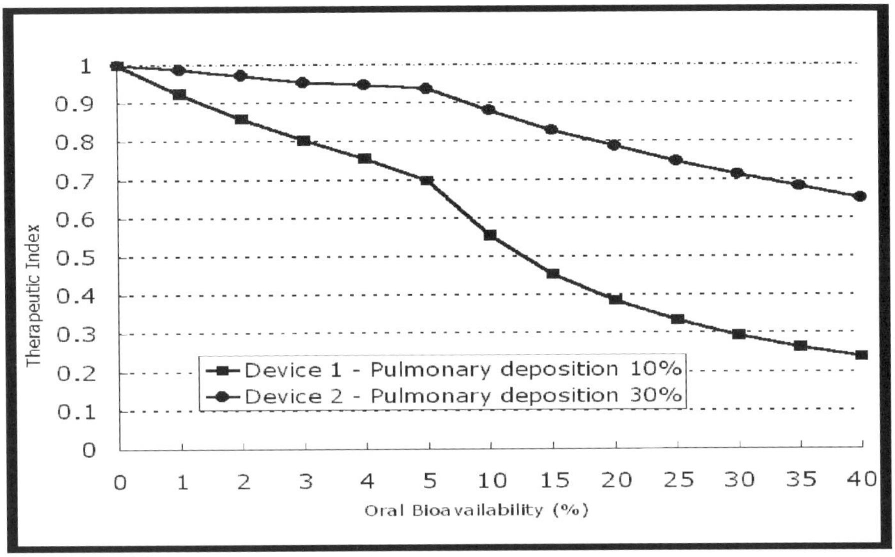

Figure 5. Relationship between therapeutic index and oral bioavailability, for devices with different pulmonary deposition.

7.3. Systemic effects of inhaled drugs

As outlined in the previous sections, a certain amount of drug emitted by an inhaler and deposited in the oropharynx will be absorbed in the gastro-intestinal tract. Part of this will eventually reach the systemic circulation, depending on the degree of gastro-intestinal absorption and inactivation by the hepatic first pass-metabolism. For example, 7% of a nominal dose of terbutaline delivered via a pMDI (39) and 35% of a dose of salbutamol (78) reach the systemic circulation by gastro-intestinal absorption. Differences in gastro-intestinal absorption explain the differences in systemic concentrations between β_2-agonists. These differences in systemic bioavailability between β_2-agonists is, however, not of clinical relevance, since the systemic effects of these drugs are generally mild (tremor and palpitations), as long as recommended doses are used. Moreover, the clinical relevance of systemic bioavailability of β_2-agonists remains unclear, since there is no clear relationship between systemic concentration of β_2-agonists and their side effects.

Measurable systemic effects, such as suppression of endogenous cortisol secretion by the adrenal glands, have been reported with inhaled glucocorticosteroids. These drugs have been claimed to cause growth retardation, osteoporosis, diabetes, and open angle glaucoma. However, only easy bruising and skin atrophy have generally been recognized as a clinically relevant side effect of these drugs (79). The extrathoracic fraction of fluticasone which is inactivated for 99% by first-pass metabolism, does not contribute to the development of systemic effects. When high doses of inhaled beclomethasone are prescribed, its extrathoracic fraction of beclomethasone may, however, contribute to the development of systemic effects, as 20 to 35% (23;24) of the swallowed portion of the drugs will eventually reach the systemic circulation. This explains why less suppression of endogenous cortisol secretion occurred in patients treated with high doses of beclomethasone, inhaled via pMDI attached to a large volume spacer (80) and why rinsing the mouth or blocking the GI uptake with charcoal reduced plasma concentrations of budesonide inhaled via Turbuhaler in children by approximately 20% (81).

Differences in therapeutic indices also explained why pulmonary deposition of budesonide via Turbuhaler was approximately twice that of a pMDI, whereas the systemic bioavailability increased by only 50% (40).

The L/T ratio is only clinically useful to compare different formulations containing drugs, which have a potential to exert systemic effects and exhibit a high oral bioavailability. Comparisons between drugs should not be made, as different substances may differ in terms of relative activities in lungs and systemic circulation. For example, inhaled glucocorticosteroids, such as fluticasone, with minimal gastro-intestinal bioavailability has been shown to exert systemic effects (22), despite a very high L/T. These systemic effects has been attributed to the long terminal half-life of this drug (19). The L/T ratio should thus always be verified clinically.

8. PREDICTION OF PULMONARY DEPOSITION OF INHALERS USING IN VITRO MEASUREMENTS

In recent years, attempts have been made to predict the in vivo behavior and clinical effect of aerosolized drugs by comparing in vitro measurements of emitted dose and size properties, delivered via different devices (82). Some regulatory authorities, such as the FDA, are very keen on this issue, in particular on reproducibility of dose and composition of the aerosol leaving the device. Multistage impactors are used to assess particle mass and mass distribution of an aerosol, and methods are available to estimate the mass median aerodynamic diameter (MMAD) of the aerosol as well as the dose of delivered from, and retained within an inhalation system.

Measurements of the total amount of drug emitted as well as its size properties were originally developed to estimate the quality of a manufactured aerosol. In vitro measurements have also been used to investigate whether a generic formulation is equivalent to the originator in terms of dose and size properties. In addition, such measurements make it possible to assess whether the addition of a radiolabel to the drug alters its composition or particle size distribution, and to examine the potential effects of handling or mishandling of a device on aerosol output and quality (see section. 4.1).

Fine particle dose is, however, not the only variable influencing pulmonary deposition. Inertia of the particles, the applied negative pressure and inhalation, flow breathing pattern, and alterations in flows within the respiratory tract caused by disease-related changes in local geometry of the airways, will inevitably modify aerodynamic behavior and pulmonary deposition of aerosolized drugs. Moreover, for pMDIs, in vitro studies do not take hygroscopic growth in the airways or the differences in plume geometry into account.

The ability of the cascade impactor to predict the behavior of aerosolized drugs in vivo in a reliable way has been recently questioned. Indeed, it has been demonstrated that several generic pMDIs with fine particle fractions comparable to the original formulation (as measured in vitro) appeared to behave differently in vitro once attached to a large volume spacer (83;84;84;85;85). In another study, the inter- and intra-individual variability of pulmonary deposition of terbutaline inhaled via pMDIs and Turbuhaler was assessed in healthy volunteers and in patients with asthma. The results showed that the intra-individual variability in pulmonary deposition of terbutaline inhaled via Turbuhaler equaled that of the pMDI, both in healthy volunteers as in patients (86), despite the fact that the in vitro dose intra-device variability was greater with Turbuhaler than with the pMDI formulation (87). Furthermore, the inter-subject variability in lung deposition was lower with Turbuhaler than with pMDI, indicating that the use of Turbuhaler was more subject independent than the pMDI. Thus, lung deposition is more predictable with Turbuhaler than with pMDI, despite a somewhat lower in vitro reproducibility.

Although a good description of the particle size properties of an aerosolized drug may give some predictive information on its gross behavior within the tracheobronchial tree, pulmonary deposition studies will, however, always be necessary to bridge between in vitro measurements and the clinical effect. Therefore, determinations of therapeutic equivalence of different formulations should not rely on in vitro measurements alone. Pharmacokinetic and pharmacodynamic measurements should always be performed to evaluate whether two formulations are equivalent.

It remains that in vitro studies have provided useful information in the assessment of the impact of inappropriate handling of a device on aerosol delivery. For instance, in vitro studies have demonstrated the importance of storing devices at different temperatures and humidity (88), the relevance of priming or shaking the pMDI canister before actuation (89), and the effects of delay, multiple actuations and spacer electrostatic charge (90;91) on drug output from a spacer device(91).

9. CONCLUSIONS

Whole lung deposition data are predictive for the clinical response to asthma drugs for the currently prescribed inhalers and drugs used to treat asthma. A good understanding of these issues is of relevance for the prescribing physician who may consider dose-reduction or expect superior control of symptoms when switching over to a formulation with higher lung deposition. Aerosolized glucocorticosteroids are the only inhaled anti-asthma drugs that have the potential for exerting systemic effects. Strategies to minimize systemic effects of inhaled glucocorticosteroids include minimizing oropharyngeal deposition and the use of inhalation devices, characterized by a high pulmonary deposition and a low oropharyngeal deposition. These strategies will, however, not be effective in patients treated with glucocorticosteroids, which exhibit a minimal oral bioavailability, since the systemic effects of these glucocorticosteroids are mainly determined by their pulmonary bioavailability and pharmacokinetic properties.

10. REFERENCES

(1) Brocklebank D, Wright J, Cates C. Systematic review of clinical effectiveness of pressurised metered dose inhalers versus other hand held inhaler devices for delivering corticosteroids in asthma. BMJ 2001; %20;323(7318):896-900.

(2) Ram FS, Wright J, Brocklebank D, White JE. Systematic review of clinical effectiveness of pressurised metered dose inhalers versus other hand held inhaler devices for delivering beta (2)agonists bronchodilators in asthma. BMJ 2001; %20;323(7318):901-905.

(3) International commission on radiological protection (ICRP). Human respiratory tract model for radiological protection. Ann ICRP 1994; 24:36-52.

(4) Zanen P, Go LT, Lammers JW. The efficacy of a low-dose, monodisperse parasympathicolytic aerosol compared with a standard aerosol from a metered-dose inhaler. Eur J Clin Pharmacol 1998; 54(1):27-30.

(5) Newman SP, Hirst PH, Pitcairn GR. Understanding regional lung deposition data in gamma scintigraphy. In: Dalby RN, Byron PR, Farr SJ, editors. Respiratory Drug Delivery VI. Interpharm Press, Inc: Buffalo Grove, IL, 1998: 9-16.

(6) Linden A, Rabe KF, Lofdahl CG. Pharmacological basis for duration of effect: formoterol and salmeterol versus short-acting beta 2-adrenoceptor agonists. Lung 1996; 174(1):1-22.

(7) Derom EY, Pauwels RA. Time course of bronchodilating effect of inhaled formoterol, a potent and long acting sympathomimetic. Thorax 1992; 47(1):30-33.

(8) Ullman A, Svedmyr N. Salmeterol, a new long acting inhaled $beta_2$ adrenoceptor agonist: comparison with salbutamol in adult asthmatic patients. Thorax 1988; 43(9):674-678.

(9) Palmqvist M, Persson G, Lazer L, Rosenborg J, Larsson P, Lötvall J. Inhaled dry-powder formoterol and salmeterol in asthmatic patients: onset of action, duration of effect and potency. Eur Respir J 1997; 10(11):2484-2489.

(10) McFadden ER, Casale TB, Edwards TB, Kemp JP, Metzger WJ, Nelson HS et al. Administration of budesonide once daily by means of turbuhaler to subjects with stable asthma. J Allergy Clin Immunol 1999; 104(1):46-52.

(11) Campbell LM. Once-daily inhaled corticosteroids in mild to moderate asthma: improving acceptance of treatment. Drugs 1999; 58 Suppl 4:25-33.

(12) Miller-Larsson A, Mattsson H, Hjertberg E, Dahlback M, Tunek A, Brattsand R. Reversible fatty acid conjugation of budesonide. Novel mechanism for prolonged retention of topically applied steroid in airway tissue. Drug Metab Dispos 1998; 26(7):623-630.

(13) Saari M, Vidgren MT, Koskinen MO, Turjanmaa VM, Nieminen MM. Pulmonary distribution and clearance of two beclomethasone liposome formulations in healthy volunteers. Int J Pharm 1999; 181(1):1-9.

(14) Currie DC, Pavia D, Agnew JE, Lopez-Vidriero MT, Diamond PD, Cole PJ et al. Impaired tracheobronchial clearance in bronchiectasis. Thorax 1987; 42(2):126-130.

(15) Newman S, Steed K, Hooper G, Källén A, Borgström L. Comparison of gamma scintigraphy and a pharmacokinetic technique for assessing pulmonary deposition of terbutaline sulphate delivered by pressurized metered dose inhaler. Pharm Res 1995; 12(2):231-236.

(16) Borgström L, Newman S, Weisz A, Morén F. Pulmonary deposition of inhaled terbutaline: comparison of scanning gamma camera and urinary excretion methods. J Pharm Sci 1992; 81(8):753-755.

(17) Bennett WD, Ilowite JS. Dual pathway clearance of 99mTc-DTPA from the bronchial mucosa. Am Rev Respir Dis 1989; 139(5):1132-1138.

(18) Hochhaus G, Mollmann H, Derendorf H, Gonzalez-Rothi RJ. Pharmacokinetic/pharmacodynamic aspects of aerosol therapy using glucocorticoids as a model. J Clin Pharmacol 1997; 37(10):881-892.

(19) Thorsson L, Dahlstrom K, Edsbäcker S, Källén A, Paulson J, Wiren JE. Pharmacokinetics and systemic effects of inhaled fluticasone propionate in healthy subjects. Br J Clin Pharmacol 1997; 43(2):155-161.

(20) Ryrfeldt Å, Andersson P, Edsbäcker S, Tonnesson M, Davies D, Pauwels R. Pharmacokinetics and metabolism of budesonide, a selective glucocorticoid. Eur J Respir Dis Suppl 119 1982; 122:86-95.

(21) Lönnebo A, Grahnén A, Jansson B, Brundin RM, Ling-Andersson A, Eckernas SA. An assessment of the systemic effects of single and repeated doses of inhaled fluticasone propionate and inhaled budesonide in healthy volunteers. Eur J Clin Pharmacol 1996; 49(6):459-463.

(22) Derom E, Van Schoor J, Verhaeghe W, Vincken W, Pauwels R. Systemic effects of inhaled fluticasone propionate and budesonide in adult patients with asthma. Am J Respir Crit Care Med 1999; 160(1):157-161.

(23) Kelly HW. Establishing a therapeutic index for the inhaled corticosteroids: part I. Pharmacokinetic/pharmacodynamic comparison of the inhaled corticosteroids. J Allergy Clin Immunol 1998; 1998(102):S36-S51.

(24) Daley-Yates PT, Price AC, Sisson JR, Pereira A, Dallow N. Beclomethasone dipropionate: absolute bioavailability, pharmacokinetics and metabolism following intravenous, oral, intranasal and inhaled administration in man. Br J Clin Pharmacol 2001; 51(5):400-409.

(25) WHO. Global Initiative for Asthma;Pocket guide for asthma management and prevention. National Heart LaBI, editor. 1998. Washington.

(26) WHO. Global Initiative for Chronic Obstructive Lung Disease; NHLBI/WHO workshop report. National Heart LaBI, editor. 2001. Washington.

(27) Pauwels RA, Lofdahl CG, Postma DS, Tattersfield AE, O'Byrne P, Barnes PJ et al. Effect of inhaled formoterol and budesonide on exacerbations of asthma. Formoterol and Corticosteroids Establishing Therapy (FACET) International Study Group. N Engl J Med 1997; 337(20):1405-1411.

(28) Pedersen S. Inhalers and nebulizers: which to choose and why. Respir Med 1996; 90(2):69-77.

(29) Lipworth BJ, Clark RA, Dhillon DP, Brown RA, McDevitt DG. Beta-adrenoceptor responses to high doses of inhaled salbutamol in patients with bronchial asthma. Br J Clin Pharmacol 1988; 26(5):527-533.

(30) Pauwels R, Newman S, Borgström L. Airway deposition and airway effects of antiasthma drugs delivered from metered-dose inhalers. Eur Respir J 1997; 10(9):2127-2138.

(31) Barnes PJ, Pedersen S, Busse WW. Efficacy and safety of inhaled corticosteroids. New developments. Am J Respir Crit Care Med 1998; 157(3 Pt 2):S1-S53.

(32) Taylor DA, Jensen MW, Kanabar V, Engelstätter R, Steinijans VW, Barnes PJ et al. A dose-dependent effect of the novel inhaled corticosteroid ciclesonide on airway responsiveness to adenosine-5'-monophosphate in asthmatic patients. Am J Respir Crit Care Med 1999; 160(1):237-243.

(33) Pauwels RA, Hargreave FE, Camus P, Bukoski M, Ståhl E. A 1-year comparison of turbuhaler vs pressurized metered-dose inhaler in asthmatic patients. Chest 1996; 110(1):53-57.

(34) Snell NJ, Ganderton D. Assessing lung deposition of inhaled medications. Consensus statement from a workshop of the British Association for Lung Research, held at the Institute of Biology, London UK on 17 april 1998. Respir Med 1999; 93:123-133.

(35) Dolovich MB, Sanchis J, Rossman C, Newhouse MT. Aerosol penetrance: a sensitive index of peripheral airways obstruction. J Appl Physiol 1976; 40(3):468-471.

(36) Itoh H, Ishii Y, Maeda H, Todo G, Torizuka K, Smaldone GC. Clinical observations of aerosol deposition in patients with airways obstruction. Chest 1981; 80(6 Suppl):837-840.

(37) Dolovich M. In vitro measurements of delivery of medications from MDIs and spacer devices. J Aerosol Med 1996; 9 Suppl 1:S49-58.:S49-S58.

(38) Richards R, Dickson CR, Renwick AG, Lewis RA, Holgate ST. Absorption and disposition kinetics of cromolyn sodium and the influence of inhalation technique. J Pharmacol Exp Ther 1987; 241(3):1028-1032.

(39) Borgström L, Nilsson M. A method for determination of the absolute pulmonary bioavailability of inhaled drugs: terbutaline. Pharm Res 1990; 7:1068-1070.

(40) Thorsson L, Edsbäcker S. Lung deposition of budesonide from a pressurized metered-dose inhaler attached to a spacer. Eur Respir J 1998; 12(6):1340-1345.

(41) Thorsson L, Edsbäcker S, Conradson TB. Lung deposition of budesonide from Turbuhaler is twice that from a pressurized metered-dose inhaler P-MDI. Eur Respir J 1994; 7(10):1839-1844.

(42) Thorsson L, Kenyon C, Newman S, Borgström L. Lung deposition of budesonide in asthmatics: a comparison of different formulations. Int J Pharm 1998; 168:119-127.

(43) Borgström L. The pharmacokinetics of inhaled hydrofluoroalkane formulations. J Allergy Clin Immunol 1999; 104(6):S246-S249.

(44) Leach C. Effect of formulation parameters on hydrofluoroalkane-beclomethasone dipropionate drug deposition in humans. J Allergy Clin Immunol 1999; 104(6):S250-S252.

(45) Leach CL, Davidson PJ, Boudreau RJ. Improved airway targeting with the CFC-free HFA-beclomethasone metered-dose inhaler compared with CFC-beclomethasone. Eur Respir J 1998; 12(6):1346-1353.

(46) Mackie AE, McDowall JE, Falcoz C, Ventresca P, Bye A, Daley-Yates PT. Pharmacokinetics of fluticasone propionate inhaled via the Diskhaler and Diskus powder devices in healthy volunteers. Clin Pharmacokinet 2000; 39 Suppl 1:23-30.:23-30.

(47) Brindley C, Falcoz C, Mackie AE, Bye A. Absorption kinetics after inhalation of fluticasone propionate via the Diskhaler, Diskus and metered-dose inhaler in healthy volunteers. Clin Pharmacokinet 2000; 39 Suppl 1:1-8.:1-8.

(48) Newman SP, Steed KP, Reader SJ, Hooper G, Zierenberg B. Efficient delivery to the lungs of flunisolide aerosol from a new portable hand-held multidose nebulizer. J Pharm Sci 1996; 85(9):960-964.

(49) Newman SP, Brown J, Steed KP, Reader SJ, Kladders H. Lung deposition of fenoterol and flunisolide delivered using a novel device for inhaled medicines: comparison of RESPIMAT with conventional metered-dose inhalers with and without spacer devices. Chest 1998; 113(4):957-963.

(50) Newman S, Pavia D, Garland N, Clarke SW. Effects of various inhalation modes on the deposition of radioactive pressurized aerosols. Eur J Respir Dis 1982; 63 Suppl. 119:57-65.

(51) Farr SJ, Rowe AM, Rubsamen R, Taylor G. Aerosol deposition in the human lung following administration from a microprocessor controlled pressurised metered dose inhaler. Thorax 1995; 50(6):639-644.

(52) Newman SP, Weisz AW, Talaee N, Clarke SW. Improvement of drug delivery with a breath actuated pressurised aerosol for patients with poor inhaler technique. Thorax 1991; 46(10):712-716.

(53) Newman S, Pavia D, Clarke S. Simple instructions for using pressurized aerosol bronchodilators. J Roy Soc Med 1980; 73:776-779.

(54) Lawford P, McKenzie D. Pressurized aerosol inhaler technique: how important are inhalation from residual volume, inspiratory flow rate and the time interval between puffs? Br J Dis Chest 1983; 77:276-281.

(55) Dolovich M. Inhalation technique and inhalation devices. In: Pauwels R, O'Byrne PO, editors. Beta2-agonists in asthma treatment. New York, Basel, Hong Kong: M.Dekker, Inc, 1997: 229-256.

(56) Pitcairn G, Lunghetti G, Ventura P, Newman S. A comparison of the lung deposition of salbutamol inhaled from a new dry powder inhaler, at two inhaled flow rates. Int J Pharm 1994; 102:11-18.

(57) Newman S, Morén F, Trofast E, Talaee N, Clarke S. Terbutaline sulphate terbutaline: effect of inhaled flow rate on drug deposition and efficacy. Int J Pharm 1991; 74:209-213.

(58) Olsson B, Borgström L. Author's response. J Aerosol Med 1996; 9:281-283.

(59) Borgström L, Bondesson E, Morén F, Trofast E, Newman SP. Lung deposition of budesonide inhaled via Turbuhaler: a comparison with terbutaline sulphate in normal subjects. Eur Respir J 1994; 7(1):69-73.

(60) Pedersen S. How to use a Rotahaler? Arch Dis Child 1986; 61:11-14.

(61) Dolovich M, Vanziegelheim M, Hidinger K. Influence of inspiratory flow rate on the response to terbutaline sulphate inhaled via the Turbuhaler. Am.Rev.Respir.Dis. 137, 433. 1988.

(62) Pedersen S, Hansen OR, Fuglsang G. Influence of inspiratory flow rate upon the effect of a Turbuhaler. Arch Dis Child 1990; 65(3):308-310.

(63) Pauwels R, Derom E. Deposition and pharmacodynamics of terbutaline inhaled via Turbuhaler. J Aerosol Med 1991; 4:A187.

(64) Borgström L, Derom E, Ståhl E, Wåhlin-Boll E, Pauwels R. The inhalation device influences lung deposition and bronchodilating effect of terbutaline. Am J Respir Crit Care Med 1996; 153(5):1636-1640.

(65) Derom E, Van Schoor J, Borgström L, Löfroos AB, Pauwels R. Lung deposition and clinical effect of Terbutaline delivered from pMDI and turbuhaler on airway responsiveness in patients with asthma. Am.J.Respir.Crit.Care Med. 159, A859. 1999.

(66) Svedmyr N, Löfdahl C, Svedmyr K. The effect of powder aerosol compared to pressurized aerosol. Eur J Respir Dis 1982; 63 Suppl. 119:81-88.

(67) Duncan D, Paterson I, Harris D, Crompton G. Comparison of the effects of salbutamol inhaled as a dry powder and by conventional pressurized aerosol. Br J Clin Pharmacol 1977; 4:669-671.

(68) Hetzel MR, Clark TJ. Comparison of salbutamol Rotahaler with conventional pressurized aerosol. Clin Allergy 1977; 7(6):563-568.

(69) Carlsson LG, Arwestrom E, Friberg K, Källén A, Lunde H, Löfdahl CG. Efficacy of cumulative doses of salbutamol administered via Turbuhaler or Diskhaler in patients with reversible airway obstruction. Allergy 1998; 53(7):712-715.

(70) Löfdahl CG, Andersson L, Bondesson E, Carlsson LG, Friberg K, Hedner J et al. Differences in bronchodilating potency of salbutamol in Turbuhaler as compared with a pressurized metered-dose inhaler formulation in patients with reversible airway obstruction. Eur Respir J 1997; 10(11):2474-2478.

(71) Goodman DE, Israel E, Rosenberg M, Johnston R, Weiss ST, Drazen JM. The influence of age, diagnosis, and gender on proper use of metered-dose inhalers. Am J Respir Crit Care Med 1994; 150(5 Pt 1):1256-1261.

(72) Böllert FG, Matusiewicz SP, Dewar MH, Brown GM, McLean A, Greening AP et al. Comparative efficacy and potency of ipratropium via Turbuhaler and pressurized metered-dose inhaler in reversible airflow obstruction. Eur Respir J 1997; 10(8):1824-1828.

(73) Gross G, Thompson PJ, Chervinsky P, Vanden Burgt J. Hydrofluoroalkane-134a beclomathasone dipropionate, 400 μg, is as effective as chlorofluorocarbon beclomethasone dipropionate, 800 μg, for the treatment of moderate asthma. Chest 1999; 115:343-351.

(74) Salzman GA, Pyszczynski DR. Oropharyngeal candidiasis in patients treated with beclomethasone dipropionate delivered by metered-dose inhaler alone and with Aerochamber. J Allergy Clin Immunol 1988; 81(2):424-428.

(75) Selroos O, Backman R, Forsen KO, Löfroos AB, Niemisto M, Pietinalho A et al. Local side-effects during 4-year treatment with inhaled corticosteroids: a comparison between pressurized metered-dose inhalers and Turbuhaler. Allergy 1994; 49(10):888-890.

(76) Borgström L. Local versus total systemic bioavailability as a means to compare different inhaled formulations of the same substance. J Aerosol Med 1998; 11:55-63.

(77) Thorsson L. Studies on the deposition, bioavailability and systemic activity of glucocorticoids in man. Lund University, Lund, Sweden, 1998.

(78) Morgan DJ, Paull JD, Richmond BH, Wilson-Evered E, Ziccone SP. Pharmacokinetics of intravenous and oral salbutamol and its sulphate conjugate. Br J Clin Pharmacol 1986; 22(5):587-593.

(79) Roy A, Leblanc C, Paquette L, Ghezzo H, Cote J, Cartier A et al. Skin bruising in asthmatic subjects treated with high doses of inhaled steroids: frequency and association with adrenal function. Eur Respir J 1996; 9(2):226-231.

(80) Brown PH, Matusiewicz SP, Shearing C, Tibi L, Greening AP, Crompton GK. Systemic effects of high dose inhaled steroids: comparison of beclomethasone dipropionate and budesonide in healthy subjects. Thorax 1993; 48(10):967-973.

(81) Pedersen S, Steffensen G, Ohlsson SV. The influence of orally deposited budesonide on the systemic availability of budesonide after inhalation from a Turbuhaler. Br J Clin Pharmacol 1993; 36(3):211-214.

(82) Borgström L. In vitro, ex vivo, in vivo veritas. Allergy 1999; 54:88-92.

(83) Derom E, Pauwels R. Bioequivalence of inhaled drugs. Eur Respir J 1995; 8(10):1634-1636.

(84) Miller MR, Bright P. Differences in output from corticosteroid inhalers used with a volumatic spacer. Eur Respir J 1995; 8(10):1637-1638.

(85) Barry PW, O'Callaghan C. In vitro comparison of the amount of salbutamol available for inhalation from different formulations used with different spacer devices. Eur Respir J 1997; 10(6):1345-1348.

(86) Borgström L, Bengtsson T, Derom E, Pauwels R. Variability in lung deposition of inhaled drug, within and between asthmatic patients, with a pMDI and a dry powder inhaler, Turbuhaler(R). Int J Pharm 2000; 193(2):227-230.

(87) Borgström L, Asking L, Beckman O, Bondesson E, Källén A, Olsson B. Discrepancy between in vitro and in vivo dose variability for a pressurized metered dose inhaler and a dry powder. J Aerosol Med 1998; 11:S59-S64.

(88) Meakin BJ, Cainey J, Woodcock PM. Effect of exposure to humidity on terbutaline delivery from turbuhaler dry power inhalation devices. Eur Respir J 1993; 6(5):760-761.

(89) Everard ML, Devadason SG, Summers QA, LeSoüef PN. Factors affecting total and "respirable" dose delivered by a salbutamol metered dose inhaler. Thorax 1995; 50(7):746-749.

(90) Wildhaber JH, Devadason SG, Eber E, Hayden MJ, Everard ML, Summers QA et al. Effect of electrostatic charge, flow, delay and multiple actuations on the in vitro delivery of salbutamol from different small volume spacers for infants. Thorax 1996; 51(10):985-988.

(91) Barry PW, O'Callaghan C. The effect of delay, multiple actuations and spacer static charge on the in vitro delivery of budesonide from the Nebuhaler. Br J Clin Pharmacol 1995; 40(1):76-78.

10.0 AFFILIATIONS

Prof. Eric Derom, Ph.D., M.D., works at the Department of Respiratory Diseases, Ghent University Hospital, De Pintelaan 185, B-9000 Gent, Belgium

P.ZANEN MD, PhD

TARGETTING AEROSOLS TO DISEASE AREAS

The Role of Monodisperse Aerosols

1. INTRODUCTION

Aerosolised drugs are nowadays cornerstones in the treatment of pulmonary diseases. The metered dose inhaler (MDI) was the first apparatus, which is both reliable and practical. The use of MDI's resulted in a decreased need for systemically administered bronchodilators or other anti-asthma drugs. This meant a considerable reduction of side-effects and an improvement of the quality of therapy[1]. The growing experience with MDI's revealed some drawbacks of the system: it appeared that a considerable number of patients were not able to use a MDI correctly[2]. These mistakes cause a suboptimal delivery of drugs to the airways and thereby a reduced therapeutic efficacy.

It became also known that only a small fraction (10-20%) of the emitted dose reaches the lower airways. The remainder deposits in the extrathoracic and upper airways, is swallowed and subsequently absorbed in the gastrointestinal tract[3]. This low availability is related to the formulation of the aerosols. All aerosol generators produce particles (or droplets) with a wide range of diameters. Such aerosols are defined as polydisperse, opposite to monodisperse aerosols, of which the diameter range is small. Large particles tend to deposit in the extrathoracic and upper airways; size measurements of the particles produced by all current generators reveal that the majority of the particles generated are these large particles[4].

To increase the efficacy of the administration, it is evident that the formulations of these aerosols have to be improved to increase lung deposition and to decrease extrathoracic and upper airway deposition. In general, the smaller a particle is the lower the extrathoracic and upper airway deposition will be and the deeper it penetrates the lower airways. However, the tendency to deposit declines as particle size decreases. Therefore, a particle may not be too large or too small.

A confounding parameter is the diameter of the airways. In small tubes a smaller particle shall penetrate further due to its lower deposition tendency, while in larger tubes such a low tendency proves to be the wrong choice. An absence of deposition is equivalent to a lack of therapeutic effect. Transferred to therapeutic aerosols in asthmatic patients the ideal formulation may depend on the degree of constriction of the airways. The localisation of receptors may also be of influence. Barnes showed that more β_2-receptors are present in small airways compared to large airways[5].

L. Gradoń and J. Marijnissen (eds.), Optimization of Aerosol Drug Delivery, 243–276.

Hence it may be hypothesized that β_2-mimetic aerosols should be directed towards the periphery of the lung and therefore should exhibit a small particle size.
This review will start with a summary of existing knowledge on the theoretical and physical backgrounds of deposition of aerosols. Subsequently the outcome of experiments to define the optimal aerosol formulation will be explained in some detail together with a short review on desired deposition patterns.

2. DEPOSITION PHYSICS

In 1935 Findeisen published his 'Uber das Absetzen kleiner in der Luft suspendierter Teilchen in der menschlichen Lunge bei der Atmung', which was the first mathematical attempt to reveal the deposition patterns of inhaled particles over the airways[6]. Findeisen defined four deposition-mechanisms, which are still used nowadays: a] *impaction*, b] *sedimentation*, c] *Brownian movement* and d] *the so called 'rim-effect'*.

a] Impaction. Along its travel through the airways, a flow of air changes direction often and suddenly. Due to their inertia, particles shall not always be able to follow these changes, causing them to collide with the wall of the airways. The general character of deposition by impaction is: large particles travelling at high speed exhibit the highest probability of deposition. In the human airways the velocity of the air is highest in the upper airways, therefore impaction is an important deposition mechanism in these parts. In other words, large particles are filtered out of the air in the upper airways.

b] Sedimentation. Due to gravity, particles will start to fall. At the moment the force of gravity equals the air resistance, the speed becomes constant: the so called terminal velocity. High terminal velocities are equivalent to high deposition-probabilities, because such particles travel larger distances in an unit of time. These higher velocities are found with the larger particles. Because the residence time is too short, sedimentation is of lesser importance in the upper airways. Sedimentation therefore causes particles, who 'escaped' impaction, to deposit in the smaller airways. An important aspect of sedimentation is, when particle-size reduces, the terminal velocity reduces too. So small particles need more time to reach the walls of the bronchi and when that time exceeds one complete breathing-cycle, the particles are exhaled. Increasing the volume of air inhaled and breath-holding are both important, because they both increase the residence time.

c] Brownian movement. Due to collisions with air molecules, very small particles (<0.01 μm) start to move in a random manner over random distances and a chance is present of a collision with the walls of the bronchi. It will be clear that only very small particles deposit efficiently by Brownian movement.

d] Rimeffect. By pure chance it is possible that a particle, in close vicinity of the wall touches it and deposits. Findeisen called that the 'rim-effect'. In the calculations the contribution of this effect to the total deposition turned out to be negligible. To keep the calculations by Findeisen at a practical level, the anatomy of the human airways had to be simplified considerably: it was restricted to only 9 airway generations. Findeisen also restricted his calculations to just one breathing volume and inhalation-flow. The outcome is presented in Fig. 1. The conclusions are clear: large particles deposit entirely (in the upper airways). Smaller particles do penetrate further, but the tendency to deposit reduces as they decrease in size. The deposition of 0.1-0.3 μm particles is minimal. These particles do not deposit by impaction any more, their terminal velocity is very low, while they are still too heavy for Brownian movement deposition.

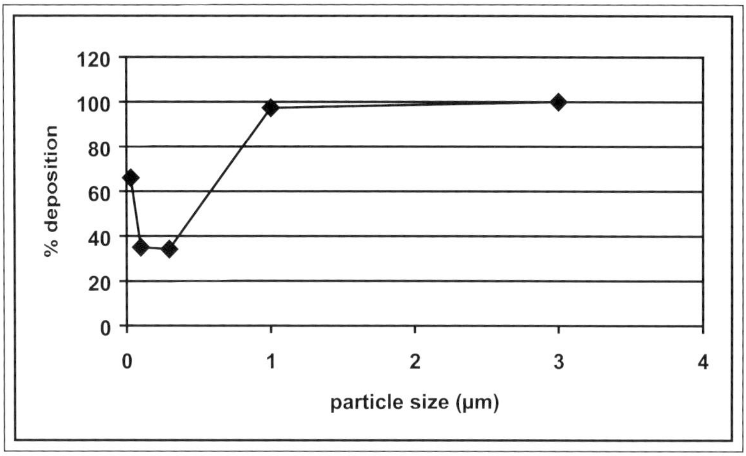

Fig. 1 Total lung deposition (= particles entering minus particles leaving the airways) as calculated by Findeisen[6]

In later years this first and crude model was refined many times. Landahl increased the number of bronchioli, alveolar ducts and alveoli considerably, redefined the deposition-equations and added to the model the influence of increasing breathing volumes[7]. A second improvement by Beeckmans was the incorporation of the Weibel-A lung-model[8 9].

In 1966 a significant contribution to the modelling was published by the Task Group of the International Committee on Radiological Protection (IRCP)[10]. This Task Group had to formulate a model to estimate the deposition and especially the clearance from the lungs of radioactive dust. One did not design a new model, but modified/improved the Findeisen/Landahl model. First, one defined three anatomical compartments: 1] nasopharyngeal, 2] tracheobronchial and 3] alveolar or pulmonary. Compartment 1] and 2] possess ciliated epithelium, which can remove insoluble deposited particles quickly, while the alveolar compartment can not. A second and major improvement is the use of the so called *aerodynamic particle size*. In all calculations up to now, spherical particles were 'used', but these are rare in nature.

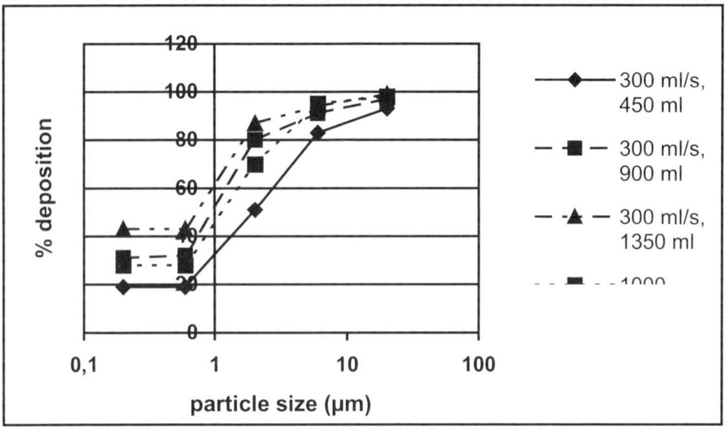

Fig. 2 Total deposition as function of inhalationflow and -volume, calculated by Landahl[7]

The terminal velocity of a particle also depends on its shape: needle shaped crystals show higher velocities. The definition of the aerodynamic size of a particle is: *the size of a spherical particle with a mass of 1, having the same terminal velocity of said particle*. In other words, every irregular shaped particle is transformed to a spherical one by comparing terminal velocities. The aerodynamic particle size is now standard in aerosol-sizing. The third change was the implementation of polydisperse aerosols in modelling. Natural aerosols exhibit a range of sizes (polydisperse) and as result, deposition therefore is expected to be widespread in the airways. In Table 1 results on the deposition of polydisperse aerosols are presented. From the data it is clear that the increasing polydispersity of 'large' aerosols poses a threat in terms of alveolar deposition. Large aerosols were not supposed to reach the alveoli and therefore they could not damage alveolar tissues. This is, however, only correct as long as the aerosols are monodisperse (=show a narrow size-distribution).

Table 1 Percentual deposition of polydisperse aerosols, according to the IRCP Task Group[10]. *MAD = median aerodynamic diameter, σ = geometric standard deviation*

MAD (μm)	σ	nasopharyngeal deposition	tracheobronchial deposition	pulmonary deposition
0.2	1.2	0.00	2.06	36.4
0.2	1.5	0.01	2.37	39.1
0.2	2.0	0.78	2.91	41.2
0.2	2.5	2.36	3.61	42.3
0.2	3.0	4.09	4.24	42.8
2	1.5	51.1	4.70	27.2
2	2.0	50.7	4.30	23.6

MAD (μm)	σ	nasopharyngeal deposition	tracheobronchial deposition	pulmonary deposition
2	2.5	50.4	3.90	21.8
2	3.0	50.2	3.61	21.0
2	4.0	50.1	3.57	20.6
20	1.2	99.9	0	0
20	2.5	97.2	0.81	1.70
20	3.0	95.6	1.03	2.60

2.1. Regional deposition

Up to here only 'total' deposition was dealt with. Total being defined as the difference in number of inhaled minus exhaled particles. This gives no information where particles in the airways deposit. Findeisen attempted to calculate the deposition per individual airway-generation, but his rather crude lung-model prevented accurate conclusions. Based on the Landahl equations and the Weibel-A model Gerrity made a second attempt[11]. In Fig. 3 the results are presented. In general one can say that large particles deposit in the proximal airways and small ones in distal. Important to note is that deposition patterns are always widespread. Large particles do reach the distal airways in low numbers. So when an abundant number of large particles are inhaled, the burden to the alveoli can still be significant.

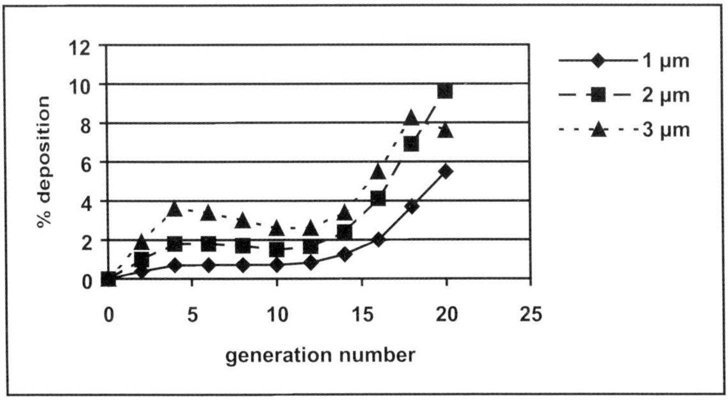

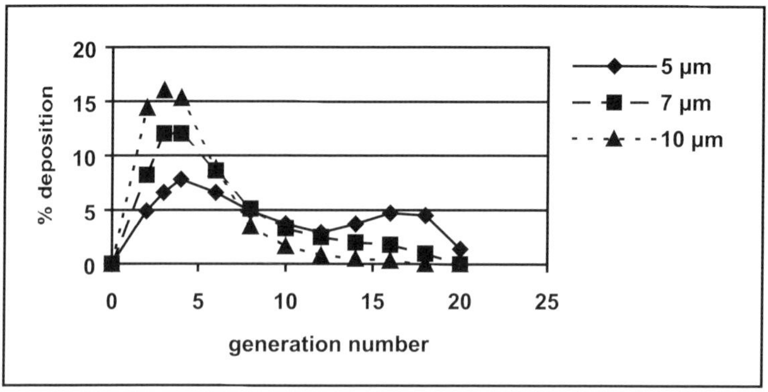

Fig. 3 Calculated regional deposition of 1 µm, 2 µm, 3µm, 5 µm, 7 µm and 10 µm non-hygroscopic particles[11]

2.2. In-vivo measurements of lung deposition

The group of Heyder measured deposition in healthy volunteers in a highly controlled way, so the results of these experiments are of value. Their main objective was a tight control of the inhalation manoeuvre: they changed the flow, keeping the volume constant or the volume, keeping the flow constant (breath-holding periods were absent)[12]. In general it was found that the deposition increases with larger inhaled volumes. An example: an increase of the inhaled volume from 250 ml to 2000 ml (flow 250 ml/s) causes 30% of all 0.5 µm particles to deposit in stead of 5%, while the deposition of 3 µm particles increases from 25% to 85%. The explanation offered is that these particles mainly deposit by sedimentation, the larger particles of course to a higher extent than the smaller ones. The increase in inhaled volume actually means an increase in residence time, favouring deposition. Contrary, an increase in inhalation flow leads to a decrease in deposition, because residence time reduces. So changing flow and volume in such a way that residence time is kept constant, would not change the deposition. In Table 2 the results of this experiment are presented, showing the validity of this thought. It means that the principal deposition mechanism for particles <3 µm is sedimentation, inertial impaction plays a less important role[13].

Table 2 Deposition as function of particle size, inhaled volume and breathing frequency as determined by Heyder[13]. The residence time comprises the inhalation and exhalation and is calculated as volume divided by flow multiplied by two.

inhaled volume (ml)	flow (ml/s)	residence time (s)	particle size (μm)				
			0.2	0.62	1	2	3
250	250	2	8	6	7	13	21
1000	1000	2	8	7	8	24	44
500	250	2	14	11	14	27	42
1000	500	2	14	11	15	34	58
1000	250	4	33	25	40	69	79
2000	500	4	34	29	44	75	82

In a next experiment the group used the technique described above of immediate and 24 hours measurements of the radioactive count after administering insoluble radioactive aerosols to estimate the total, tracheobronchial and alveolar deposition. Again, one controlled the inhalation manoeuvre tightly and the results clearly indicate that the alveolar deposition (by sedimentation) of the smaller particles is very high[14]. For particles <2 μm the total and alveolar deposition are the same. Later an extension of these investigations was published: the total and regional deposition of a larger range of particles became known (Fig. 4).[13]

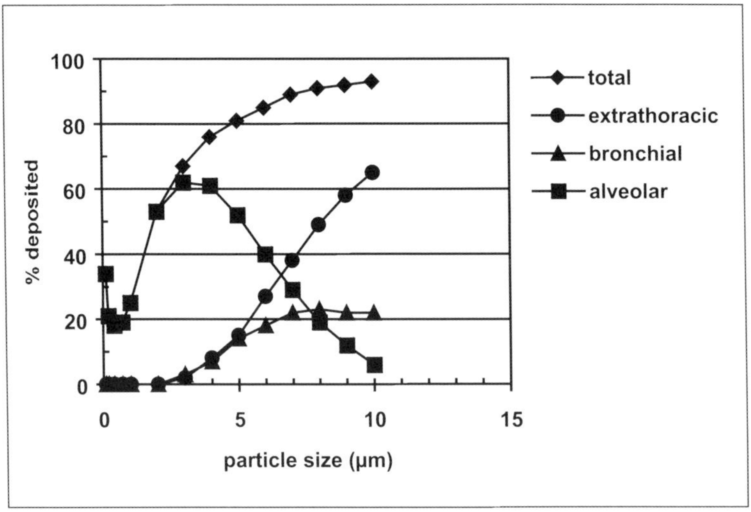

Fig. 4 In-vivo deposition of particles in the human airways[13].
(flow rate : 250 ml/s, tidal volume: 1000 ml)

In another experiment by the same group, one was able to prove that the total deposition is strongly governed by the aerodynamic particle size. Two types of aerosols were administered: sebacate and iron oxide. Sebacate has a lower density than iron oxide and 23% of the 1 μm sebacate particles deposited compared to 50% of the heavier iron oxide particles. The aerodynamic particle size of iron oxide is by definition larger and when one corrected the deposition for these differences, the differences in deposition disappeared completely. Similar 'corrections' for the flow or residence time were calculated and the group was able to construct a so-called deposition parameter (X_m). This parameter normalises the total deposition in terms of particle size, particle density, residence time and inhalation flow. In other words, besides particle size, particle density, residence time and inhalation flow, there are no other factors governing total deposition[15].

$$X_m = \left(\lg\frac{Q}{Q_0} - 1.43\right)\lg\left[\frac{\rho}{\rho_0}\left(\frac{d}{d_o}\right)^2\left(\frac{\tau}{\tau_0}\right)^{24/\sqrt{Q/Q_0}}\right]$$

Eqn 1

in which: d= particle diameter (μm), d_0= 1 μm,
ρ= particle density (g/cm^3), ρ_0= 1 g/cm^3
V= inhalation flow (cm^3/s), V_0= 1 cm^3/s
F= breathing frequency (min^{-1}), F_0= 1 min^{-1}

3. THE TARGETS FOR INHALED AEROSOLS

From the preceding paragraphs it will be clear that to a certain extent it is possible to target aerosols to a specific area of the airways by controlling particle size, inhalationflow and -volume. So optimising aerosol therapy starts amongst others with defining what is the best region to target to. A short review for bronchodilators and steroids is given.

3.1. Bronchodilators

Bronchodilators relax smooth muscle by stimulating β_2- or inhibiting cholinergic receptors. Those areas in the airways where both receptors and muscle are abundant are (intuitively) the areas of choice for targeting.

Ebina showed in patients not dying from respiratory causes that towards the periphery of the airways the smooth muscle layer decreases in thickness, but the rate at which this occurs is less than the reduction of airway diameter and therefore concludes that in healthy lungs the peripheral bronchi contain a relative excessive

amount of smooth muscle[16]. Hossain compared the smooth muscle area in airway walls in asthmatics and healthy controls and found a decrease in all groups. However, in asthmatics more muscle was present in all airways and the decrease towards the periphery was less strong[17].
Another study showed that based, on muscle distribution, that a distinction can be made between asthmatics. Type 1 shows hypertrophy only in the large airways and type 2 in all airways[18].

Table 3 Thickness (μm) of airway smooth muscle in various diseases and controls[18]

group	large bronchi 2<R<3 mm	small bronchi 0.3<R<0.4 mm
controls	12.4	5.6
asthma, type 1	64.6	5.6
asthma, type 2	32.8	11.5
emphysema	26.8	6.6

Carroll found that the smooth muscle area in large (>4 mm diameter) airways was significantly increased in fatal asthmatics. Between non-fatal and controls no differences exist in these large airways, but in the small 2-4 mm airways the smooth muscle area was somewhat increased. In airways <2 mm again no differences were found[19].
Studies in ferret lungs showed that ß-receptors are located near airway smooth muscle and that their number increases in small airways, while in alveoli large numbers are present[20]. Later this study similar was found for α- and cholinergic receptors. Summarising, higher numbers of α- and ß-receptors towards the periphery of the airways and a decrease of the cholinergic receptors[21].
In human lungs a similar pattern is present (see Table 4), but one has to note that the ß-receptor change towards $ß_1$-recpetor in the periphery[22].

Table 4 Localisation of ß-receptors over human airways[22]

	grain count/1000μm^2	$ß_1$ (%)	$ß_2$ (%)
bronchus epithelium	213	0	100
bronchioli epithelium	129	0	100
bronchus smooth muscle	28	0	100
bronchioli smooth muscle	92	10	90
alveolar wall	513	30	70

3.2. Corticosteroids

Asthma is an eosinophilic inflammation, so one must know the distribution of eosinophils in the asthmatic airway.

Saetta studied six patients, dying from an asthmatic attack. They were compared to a group non-asthmatics, who died from non-pulmonary causes. Both eosinophils and mononuclear cells were more abundant in non-respiratory bronchioles, less than 2 mm in diameter. The inflammatory infiltrate occupied 60% of the wall, opposed to 30% in the 'healthy' subjects[23]. Unfortunately the distribution of the inflammation over the airways did not become known due to one point sampling technique. In five patients, rapidly dying after onset of their asthmatic attack, proximal and distal airways were sampled and cellular content determined. The results clearly show a stronger cellular infiltrate of the proximal airways, which might indicate a centrally located inflammation in these typical cases[24].

Table 5 Number of cells ($x10^4$ per μm^2) present in proximal or distal airways of asthmatics patients dies within 1 hour after onset of symptoms[24]

	proximal airways	distal airways
T-cells	10.9	7.7
macrophages	9.7	3.4
eosinophils	1.7	0.3
activated eosinophils	1.14	0.3

Morphology of 0.5 to 1.5 cm intrapulmonary bronchi of patients, not dying quickly, showed higher number of leucocytes, T-cells and eosinophils. However, a correlation the number of eosinophils and the duration of the terminal episode exists (a longer the terminal period, the higher the eosinophilic count)[25].

Table 6 Number of cells/mm lenght of epithelial basement membrane present in intrapulmonary bronchi of asthmatics patients and controls[25]

	asthma	controls
leucocytes	96.9	41
T-cells	72.8	27.9
B-cells	12.4	13.0
eosinophils	62.7	1.0
activated eosinophils	35.7	0

Holgate counted and compared mast cells, activated and resting eosinophils, monocytes/macrophages, neutrophils and $CD3^+$/$CD8^+$ T-cells in large (diameter >2 mm) and small (diameter <2 mm) airways. In small (<2 mm) airways cell count did not differ, but in the airway wall of the large larger number of eosinophils were present. $CD3^+$ T-cell count, on the other hand was lower in large airway epithelium[26].
The above studies mostly indicate at a heavier eosinophilic infiltrate in large central airways in fatal asthma. Airways obtained from surgical asthmatics were stratified into <2 mm and >2 mm groups (small and large) and mast cells, activated and resting eosinophils, macrophages, neutrophils and T-cells were counted. All airways

showed a higher cellular content, but not for macrophages and neutrophils. Mast cells were only more abundant in the large airways. Comparing large and small airways revealed a small increase of activated, but not total eosinophils in the small airways ($p<0.05$)[27]. Carroll, defined airways as <6 mm, 6-16 mm and >16 mm. In these airways eosinophils and lymphocytes were counted in asthmatics dying of their asthma or not and in controls. Lymphocytes decreased rapidly in the smallest airways and between fatal and non-fatal asthmatics no big differences were noted. In fatal asthma eosinophils were abundant in the large airways, in non-fatal asthma only a slight difference between small and large airways was found[28].
In nocturnal asthma Kraft showed that the number of eosinophils was slightly higher in the alveoli. Interesting is that in the nocturnal patients the eosinophils increased during the night, while their lung function deteriorated[29].
A second approach, quite similar to bronchodilators, is to locate the glucocorticosteroid receptors. Data are limited, however. In 3 asthmatics and 15 controls human Adcock measured the amount of glucocorticosteroid receptor mRNa and protein. The highest concentration mRNA was recovered from alveolar walls, endothelium and smooth muscle cells of bronchial and pulmonary vessels. Lower amount from airway epithelium and smooth muscle. No differences existed between asthmatics and controls. Antibody staining of glucocorticosteroid receptor protein revealed that most mass was found in alveolar walls, endothelium and smooth muscle cells of vessels. Again no difference was found between asthmatics and controls[30]. Vachier confirmed these results[31].

4. RELATING AEROSOL PARAMETERS TO CLINICAL EFFECT

In this section an overview is given of experiments relating aerosol parameters (mostly particle size) and the clinical efficacy. As known the deposition patterns are governed by few parameters (inhalation flow/volume and particle size). When one standardises flow and volume only size remains as a governing parameter.

4.1. Bronchodilators

Ruffin administered 3.3 μm and 1.5 μm isoproterenol aerosols and noted that the cumulative dose-response curve of the 1.5 μm aerosol was shifted to the right, indicating a lower potency. Whether this shift was significant, is not mentioned, however[32].
Rees milled terbutaline crystals down to three different size ranges. The mass median diameters were 5.6 μm, 9.1 μm and 13.6 μm respectively, but the distribution showed strong polydispersity with considerable overlap. These crystal sizes were incorporated into standard MDI's and it was shown that only the smallest aerosol improved the FEV_1, sG_{aw} and MEF_{50} significantly. The two remaining aerosols had no dilating effect at all and the investigators concluded that effective particles should be smaller than 5 μm[33]. Using three different nebulisers, Clay generated terbutaline aerosols with a mass median diameter of 1.8 μm, 4.6 μm and 10.3 μm. All aerosols elicited significant changes with respect to baseline. Except in case of the $MEF_{50/25}$,

where the smallest aerosol was most potent, none of the aerosols differed significantly amongst each other[34]. A year later Clay showed that 80% of the 1.8 μm aerosol deposited in the lung compared to 60% and 44% of the 4.6 μm and 10.3 μm aerosols[35].
Again using different types of nebulisers, Mitchell generated 1.4 μm and 5.5 μm salbutamol aerosols. The lung deposition was studied by radiolabelling and no significant difference was found between the deposition patterns. Mitchell constructed cumulative dose-FEV_1 curves and these curves showed to be superimposable[36]. These conclusions, which are only based on the FEV_1-improvement, are in line with those of Clay, t.i. no clear cut relationship between particle size distribution and clinical efficacy.
The data above suggest that the clinical effect of bronchodilator aerosols is not strongly influenced by the size distribution, but some evidence emerged that particles below 5 μm are more potent. The importance of that mass <5 μm was shown by Persson, who modified the mouthpiece of terbutaline Turbuhalers in such a way that the inhalers delivered either 90 μg, 40 μg or 5 μg as <5 μm particles. The increase in FEV_1 with the 5 μg mouthpiece was significantly lower than with the higher dose mouthpieces. The latter two did not differ from each other[37].
Johnson reported that 3.3 μm particles were more potent than 7.7 μm particles[38]. Two different nebulisers generated these particles and both for ipratropium bromide and salbutamol cumulative dose-response curves were constructed. The ipratropium curves were not significantly different for all lung function parameters measured, while the salbutamol curves did show a higher potency for the smallest aerosol. Radiolabelling studies did not show significant changes in deposition patterns of the aerosols, but a higher total lung dose of smaller aerosol. This indicates that the oropharyngeal passage of the 3.3 μm aerosol is better, which is apparently the explanation for the higher potency. Patel and co-workers used a spinning top generator to manufacture monodisperse 2.5 μm and 5 μm isoproterenol aerosols[39]. The advantage of this approach is that it has an higher discriminative power: nebuliser aerosols do generate highly polydisperse aerosols and the overlap in the particle size distributions tend to obscure differences in size-effects. The 2.5 μm dose-response curves of Patel for all parameters measured were shifted to the left, indicating a higher potency. He calculated that on a weight basis the 2.5 μm aerosol was 2 to 4 times more potent compared to the 5 μm aerosol. This conclusion is however questionable because one did not control the inhaled dosage accurately.
The parallel designed study by Hultquist comparing 1.5 μm and 4.8 μm particles generated by two different nebulisers could not show an advantage for the smaller aerosol[40]. Neither the cumulative FEV_1 dose response curves, nor parameters for gas-exchange did show significant differences.

The data on a relationship between particle size and clinical efficacy, as reviewed above are hard to interpret. Results of bronchodilators studies are contradictory: early trials suggest that particles simply should be smaller than 5 μm and that the actual particle size below that 5 μm is irrelevant[33 34 36]. Other researchers, however, do show differences in the degree of bronchodilation in the <5 μm size range[32 38 39].

Unfortunately, an optimal particle size for a bronchodilator aerosol cloud can not be derived from the latter experiments. Frequently one compared just two particle sizes *and* amongst the studies the range of sizes varied.
The contradictions on the importance of the below 5 μm size may find their origin in the experimental set up of the studies. Most experiments show a common weakness: in order to generate the aerosols, one uses different nebulisers. The nebulisers generate smaller of larger *polydisperse* aerosols, which renders it hard to define the effect of a changing particle size due to the considerable overlap in size distributions[41 42]. The stronger that overlap, the lower the chance to find a particle size effect. So, whether one reports a significant size/efficacy relationship or not, seems to be nebuliser dependent. Studies, using nebulisers with narrow particle size distributions, have the highest probability is to find a particle size efficacy relationship. Information on that distribution is lacking in the reports.
Second confounders are the differences in nebuliser output. It is well known that the different types nebulisers used show widely different mass outputs[41 42 43]. It is conceivable that a high mass output of a large particle nebuliser counterbalanced the low particle size efficacy. This effect of course lowers the discriminative power of the study, favouring the conclusion of non-significant differences. So an accurate estimation of the mass output is highly advisable, but most experiments did not control this confounder.
Taking these problems into account we designed studies, which avoided as much as possible dose bias and overlapping size distributions. We produced monodisperse aerosols (geometric sd <1.2) were produced by a spinning top generator[44]. A spinning top generator consists of a compressed air driven disk, rotating at 12000 rpm. Liquid is fed to the centre of the disk by a pump identical sized droplets are expelled from the rim of the disk.
By blowing hot air over the disk these droplets are dried and subsequently transported to a small tank, from which the volunteers inhale. The concentration of the drug in and the viscosity of the solution govern the diameter of the resulting dry particles. Bronchodilator solutions (50% water/50% ethanol) was used to obtain aerosol with a MMAD of 1.5 μm, 2.8 μm and 5 μm. Each time a volunteer was due to start the aerosol inhalation, the mass of drug per litre of air and the particle diameter of the dry aerosol particles in the tank were measured by an Aerodynamic Particle Sizer 33 (TSI, St. Paul, Min). This is a so called time of flight particle sizer and gives information on both the particle size distribution and the mass contained in the aerosol in μg per litre of air. To calculate the volume of air to be inhaled, we divided the dose to be administered by the mass of drug per litre of air rendering the litres to be inhaled. If sufficient dosage (expressed as μg delivered to the mouth) had been inhaled, the aerosol inhalation was discontinued by switching over to non-aerosol containing air. The inhalation manoeuvre in all experiments was standardized and consisted of inhalation of the slow vital capacity with a flow of 40-60 L/min, followed by a breath-holding period of 10 seconds and a slow exhalation.
Using this set up, we conducted several experiments. In the first one we tried to define the optimal size of salbutamol in stable asthmatics[45]. Eight mild asthmatic patients participated in this trial. The average age (sd) was 40 (10) years, the mean

FEV_1 (sd) was 72.3 (6.8) percent of the predicted value. In all patients a bronchodilator response of >15% after inhalation of 200 µg salbutamol had been measured just before the trial. None of the patients were smokers. Each patient was studied four times at the lung function laboratory: placebo, 1.5 µm, 2.8 µm and 5 µm aerosols were randomised. First, 5 µg salbutamol was administered, followed by 5 µg, 10 µg and 20 µg during the second, third and fourth cycle, respectively, resulting in cumulative doses of 5 µg, 10 µg, 20 µg and 40 µg salbutamol. (All dosages are expressed as µg delivered to the mouth). In Fig. 5 the dose-response curve for the FEV_1 and peak flow are represented. In any of the parameters no change was measured during the inhalation of placebo. The evaluation of the dose effects demonstrated that for all lung function parameters statistically significant differences existed between dosages, with the higher dosages causing a stronger bronchodilation ($p<0.05$). The interaction between the dose and the effects of the three types of salbutamol aerosols was non-significant for all lung function parameters ($p>0.1$), which indicates that the dose-response curves run parallel.
In evaluating the aerosol-size effect, the analysis of variance demonstrated significant differences with reference to placebo for the FEV_1 ($p<0.01$), the PEF ($p<0.01$), FVC ($p<0.01$), MEF_{75} ($p<0.01$), MEF_{50} ($p<0.01$) and MEF_{25} ($p<0.01$). This implies that all the dose-response curves of the salbutamol aerosols are located higher than the placebo curves. For the R_{tot} ($p=0.116$) and the VC ($p=0.068$) no significant differences due to the different aerosol-sizes were demonstrable.

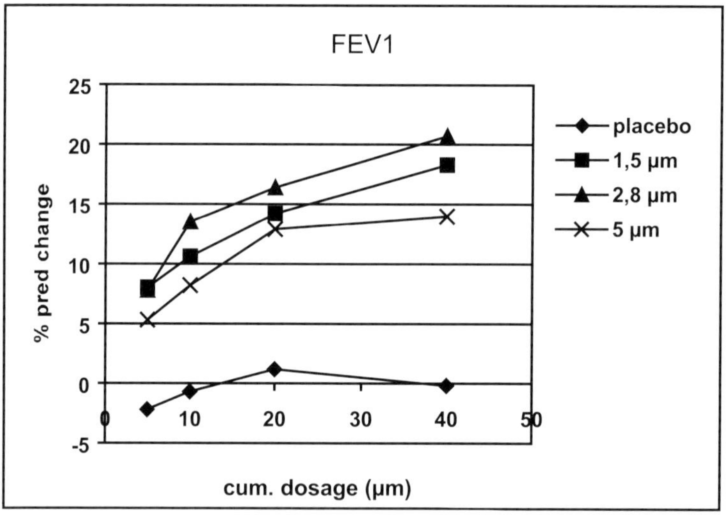

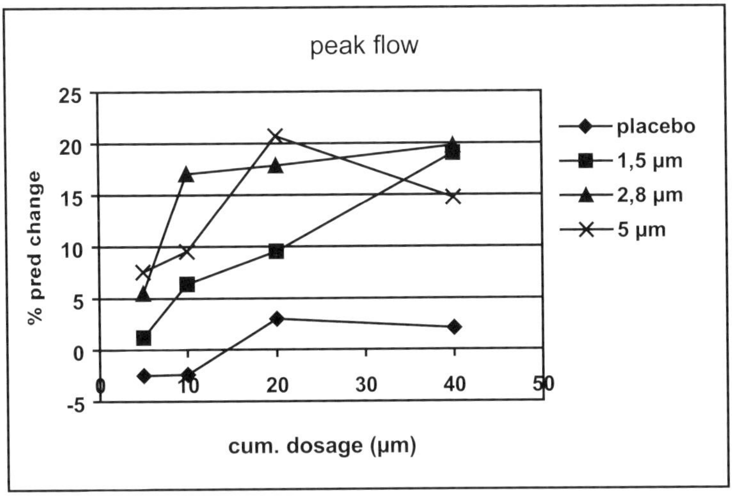

Fig. 5 Dose-response curves for the FEV_1 and peak flow as function of particle size

Table 7 Mean (sd) improvement in lung function (%predicted) over all four dosages per type of aerosol (NS= non significant)

lung function parameter	particle size of the aerosol (μm)			significant different aerosols
	1.5	2.8	5	
FEV1	12.8(4.3)	14.6(6.5)	10.1(4)	2.8 vs. 5
FVC	8.9(5.6)	9.6(9.3)	7.9(6.6)	NS
MEF25	12.1(6.6)	14.9(9.5)	9.2(7.9)	2.8 vs. 5
MEF50	15.2(6.7)	18.5(9.4)	12.1(6.2)	2.8 vs. 5
MEF75	17(4.4)	21.4(9)	14.6(5.6)	2.8 vs. 5
PEF	9(7.1)	15(7)	13.1(7.9)	2.8 vs. 1.5
Rtot	-27.6(93)	-84.3(92)	-52.7(62.1)	NS
VC	7(7.8)	8.1(9.4)	6(8.9)	NS

A significant difference with reference to 2.8 μm aerosol would occur in this experiment for the FEV_1 if the deviation between the means exceeds 2.9%, for the PEF this deviation should be at least 5.9%, for the $MEF_{75/50/25}$ at least 5.4%, 4.3% and 4.9%, respectively, and, finally, for the FVC at least 9.4%. In the case of the FEV_1 and the $MEF_{75,50,25}$ a statistically significant difference occurred between the 5 μm aerosol and the 2.8 μm aerosol. For the PEF a significant difference was found between the 1.5 μm and the 2.8 μm aerosol. For the FVC the differences were too small to be significant.

Table 8 lists the mean improvements as %predicted with 95% confidence intervals after administration of 40 μg salbutamol for all lung function parameters for the 2.8 μm aerosol. Despite the low dosage of 40 μg a significant improvement of the lung function can be observed.

Table 8 Mean improvement in lung function (%predicted) following 40 μg salbutamol administered as a 2.8 μm aerosol

lung function parameter	improvement (% predicted)	95% confidence interval	
FEV_1	20.7	13.5	28.0
FVC	12.6	2.4	22.7
MEF_{25}	26.9	20.6	36.9
MEF_{50}	26.7	18.5	34.8
MEF_{75}	28.8	06.4	37.6
PEF	-102.2	-29.0	-175.0
R_{tot}	19.8	12.4	27.1
VC	9.3	-0.11	18.7

In a second experiment we did similar but now for ipratropium bromide[46]. We formed a hypothesis that ipratropium bromide would favour a central deposition pattern. A change with the previous experiment we only administered 8 μg of ipratropium bromide. In evaluating the aerosol-size effect, the analysis of variance demonstrated significant differences with reference to placebo for the FEV_1 ($p<0.01$), the PEF ($p=0.001$), the FVC ($p=0.034$), the MEF_{50} ($p<0.001$) and the MEF_{25} ($p=0.001$). For the R_{tot} ($p=0.202$) and the VC ($p=0.052$) no significant differences were demonstrable.

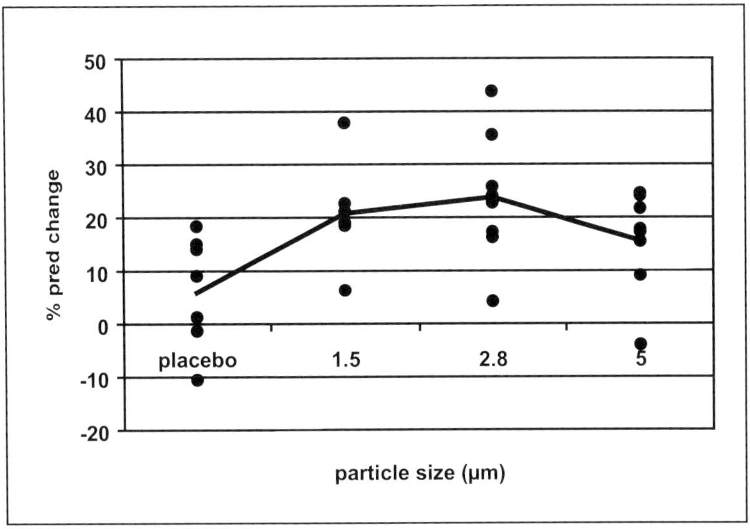

Fig. 6 Scatterplot of FEV_1 improvement after administration of the different aerosols, the mean improvement is depicted by the solid line

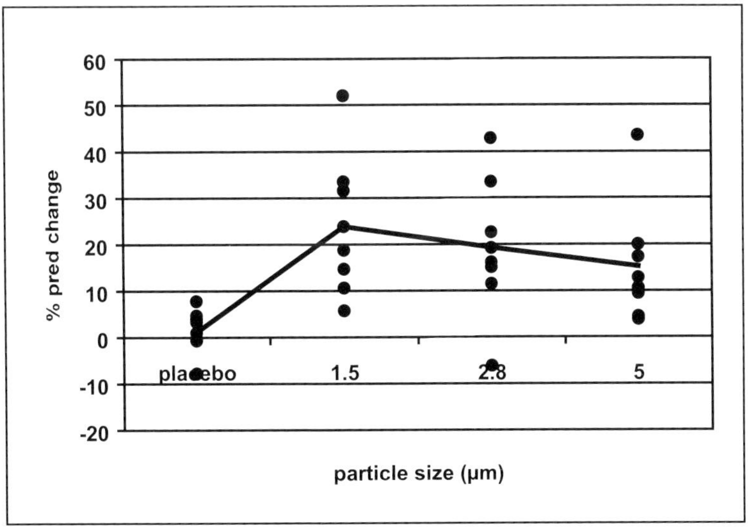

Fig. 7 Scatterplot of MEF_{25} improvement after administration of the different aerosols, the mean improvement is depicted by the solid line

Table 9 Mean (sd) improvement in lung function (% predicted) after inhalation of ipratropium bromide aerosols with different particle sizes (NS = non-significant)

lung function parameter	particle size of the aerosol			significant different aerosols
	1.5 μm	2.8 μm	5 μm	
VC	13.6(13)	11.5(12.5)	7.7(7.8)	NS
R_{tot}	-74.4(73.2)	-114.8(97.1)	-104.5(135.9)	NS
FVC	12.4(12.4)	13(12.5)	9.4(10.8)	NS
FEV_1	20.3(8.7)	23.7(12.1)	15.6(9.8)	1.5/2.8 vs. 5
MEF_{25}	23.7(15)	19.3(14.7)	15.3(12.7)	1.5/2.8 vs. 5
MEF_{50}	24.7(10.7)	24.7(12.1)	17.7(10.3)	1.5/2.8 vs. 5
PEF	21(7.2)	22(13.2)	16.6(12.4)	NS

As said earlier the baseline lung function or better said the degree of airway obstruction can influence the deposition patterns and thereby the efficacy of inhaled drugs. As direct consequence we repeated the above experiments in more severe asthmatics with a mean FEV_1 (sd) of 37.9 (7.3) percent of the predicted value[47]. In all patients a more than 15% increase of baseline FEV_1 after inhalation of 200 μg salbutamol had been measured just before the trial.
No significant change was measured in any of the lung function parameters during the inhalation of placebo. Nor could we show significant differences between the bronchodilator response of salbutamol and ipratropium bromide. Comparing identical particle sizes there were no significant differences between salbutamol and ipratropium bromide.

Table 10 Mean improvement in lung function (% predicted) after inhalation of aerosols with different particle sizes or placebo. The improvements shown reflect the net effects of the different particle sizes, not confounded by the type of bronchodilator used. An asterisk indicates a significant difference versus the 2.8 μm aerosol.

lung function parameter	placebo	particle size of the aerosol		
		1.5 μm	2.8 μm	5 μm
sG_{aw}	-2.88*	-14.46	-21.93	-12.86
FEV_1	0.08*	3.57*	8.87	3.84*
MEF_{75}	-0.43*	1.44*	4.85	1.95
MEF_{50}	-0.30*	0.58*	4.23	1.17*

Table 11 Mean improvement in lung function (sd, as % predicted) after inhalation of a 2.8 μm salbutamol or ipratropium bromide aerosol.

lung function parameter	improvement (sd)	
	after 20 μg salbutamol	after 8 μg ipratropium bromide
sG_{aw}	-22.4% (17.3%)	-21.4% (16.9%)
FEV_1	8.5% (7%)	9.3% (4.1%)
MEF_{75}	3.4% (4.6%)	6.3% (4.1%)
MEF_{50}	3.4% (4.8%)	5% (3.2%)

With respect to the effects of different particle sizes of the aerosol administered, we found significant differences in case of the FEV_1 ($p<0.001$), the MEF_{75} ($p=0.02$), the MEF_{50} ($p<0.001$) and sG_{aw} ($p=0.01$) (see Fig. 8). In Table 10 we depict for each lung function parameter between which particle sizes these significant differences were present. In

Table 11 we list the improvements of various lung function parameters after administration of 20 μg salbutamol and 8 μg ipratropium bromide respectively as 2.8 μm aerosols.

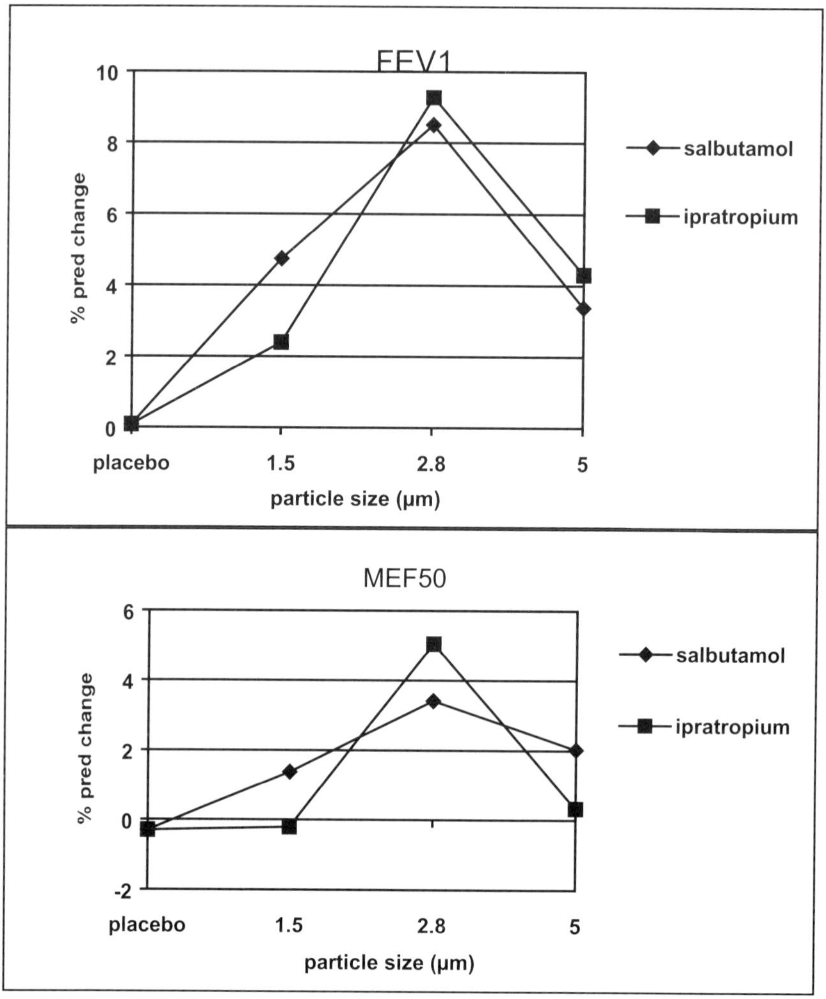

Fig. 8 Improvement of the FEV_1, $MEF_{75/50}$ due to placebo or salbutamol and ipratropium bromide aerosols of increasing particle sizes. The improvement is presented as percentage change of the predicted value.

A last experiment was that we compared adverse effects of the monodisperse aerosols to those after administration through a metered dose inhaler. Before we give the results on this experiment it is necessary to know that we showed earlier that it is possible to obtain similar bronchodilating effects after administration of 8 μg of ipratropium bromide as 2.8 μm monodisperse aerosol or after 40 μg from a standard MDI[48]. This strongly indicates that ±80% of a MDI aerosol in not contributing much to the effect, although this 'overdose' still consists of active drug with a potential to raise adverse effects. To unravel the role of this 'overdose' in eliciting adverse effects, we compared the adverse effects of a standard fenoterol MDI and a 2.8 μm monodisperse aerosol in a 80% lower dose (800 μg versus 160 μg) in healthy volunteers[49]. As measure for adverse effects changes in serum potassium and finger tremor (depicted as a voltage from a displacement meter) were taken. Bronchodilation was measured through sGaw determinations.

The differences in serum potassium changes between the monodisperse and MDI-aerosol on the one hand and placebo on the other were highly significant ($p=0.004$ resp. $p<0.001$), but also between the two active formulations ($p=0.001$). The monodisperse aerosol elicited a mean (sd) fall of 0.27 (0.27) mmol/l, compared to 0.67 (0.18) mmol/l after the MDI-aerosol. The tremor response showed an identical picture: between placebo and actives a significant difference ($p=0.001$ resp. $p=0.01$) and between the actives ($p=0.029$). The monodisperse aerosol elicited a mean (sd) increase of 0.07 (0.07) volt, compared to 0.29 (0.34) volt after the MDI-aerosol. The changes in pulse pressure and heart rate did not differ significantly from placebo or between the actives. The specific airway conductance changes were significantly different between placebo and the actives ($p<0.001$), but not between the two drug aerosols ($p=0.87$). An 0.31 (0.24) resp. 0.29(0.49) $kPa^{-1}sec^{-1}$ increase was noted after the monodisperse or MDI-aerosol.

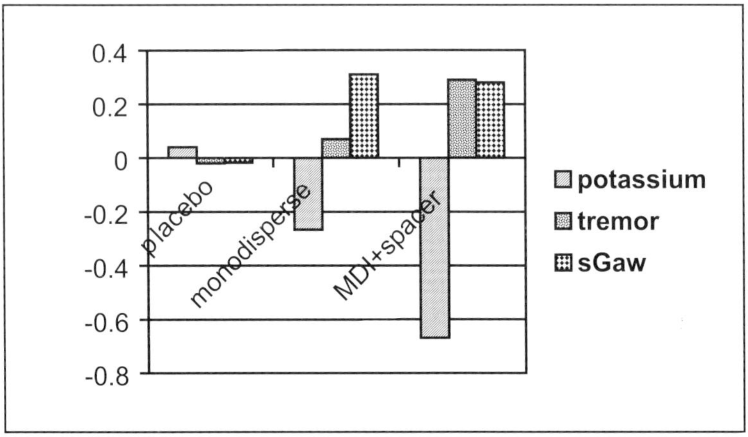

Fig. 9 Changes in serum potassium (mmol/l), tremor response (volt) and sGaw ($kPa^{-1}s^{-1}$) due to placebo, 160 μg fenoterol as 2.8 μm monodisperse aerosol or 800 μg fenoterol from a MDI + spacer.[49]

The first main conclusion from all these experiments is that in patients with mild and severe airflow limitation, both for $ß_2$-mimetics and parasympathicolytic drugs the optimal particle size is ±3 μm. It also appears is that administration of these optimal sized aerosols can lead to a significant reduction of adverse effects.
The first conclusion can be explained by the existence of an upper airway filter. The extrathoracic and upper airways function as an effective filter for large particles, protecting the lower airway against a high load of dust or other possibly harmful materials. Administration of large particles is therefore inefficient, because these are lost in the upper airways filter leading to low dosages in the lower airways. This may result in a low therapeutic effect. This phenomenon occurs both for $ß_2$-mimetics and parasympathicolytic aerosols and offers an explanation for the lower efficacy of the 5 μm aerosols, which we found in all experiments. In this respect the calculations of the regional deposition by Gerrity et al. are highly informative[11]. For 5 μm particles they show a predominant upper airway deposition, while the lower airway deposition is lower compared to the 2 and 3 μm particles. The experimental data by Svartengren et al. on the cut-off characteristics of the oropharynx underline the importance of the upper airway filter[50]. Moreover, it has to be taken into account that $ß_2$-mimetics and parasympathicolytic aerosol particles will be hygroscopic to a certain extent. In a moisture environment, their size will grow and thereby the upper airway deposition of the larger particles. 5 μm particles grow into much larger particles than 2.8 μm, favouring their upper airway deposition.

Having explained the reasons for the low 5 μm efficacy, the differences between the 2.8 μm and the 1.5 μm aerosols need explanation. We expected that for $ß_2$-mimetics and parasympathicolytic aerosols, due to the differences in receptor distributions, different optimal particle sizes would show. A more centrally orientated deposition pattern for parasympathicolytic drugs seemed logical and our results confirm this hypothesis (2.8 μm more potent than 1.5 μm). Maybe the deposition pattern of 5 μm particles would be better, in the sense of an even better fit to the receptor distribution. However, if this would be the case, the low penetration and dose apparently outweighs the closer fit to the receptor distribution. In other words, an optimal fit between deposition pattern is of relatively low value when the deposited mass is (very) low. An ill-fitting pattern with higher local masses is then preferable.
For $ß_2$-mimetics it might have been expected or argued that 1.5 μm aerosols are a better choice, because of a better fit to the receptor distribution. We feel that a combination of effects can explain the $ß_2$-mimetics results. First, more smaller particles are exhaled than larger particles, even in asthmatics. The experiments by Schiller-Scotland et al. clearly show an almost twice as high deposition for 3 μm particles compared to 1 μm particles. The differences in deposition rates are less pronounced in asthmatics compared to healthy volunteers, but they are present[51]. So it is expected that 2.8 μm aerosols are equivalent to higher dosages and hence more potent. Apparently, the dose outweighs the deposition pattern.
It has to be realised that depositions patterns are never focused. The data by Gerrity et al. indicate that the deposition patterns of 1 μm, 2 μm and 3 μm particles in the lower airways do not differ strongly. The spread of particles is more or less

comparable, but the mass deposited is higher for the 3 μm particles. Therefore when the deposited dose is higher with comparable deposition patterns, the dose definitely outweighs the deposition patterns.

The results in patients with severe airflow limitations can be explained by the same principles. 5 μm particles hardly will enter the lower airways. The function of the upper airway filter will not be highly different from that of mild asthmatics or healthy subjects. One might have expected that the penetration of 1.5 μm particles into constricted airways would be the most optimal, but this proved not to be true. The deposition of 1.5 μm particles will always be lower than that of 2.8 μm particles and apparently the latter reach the β_2-receptor in sufficient numbers. Again dose outweighs the deposition pattern.

In the explanations above the nature of the drug inhaled hardly plays a role. Inhaled particle deposition is a physical process, not linked to the chemical nature of the drug. The aerodynamic particle size is not directly influenced by the chemical structure, except for the density and hygroscopy. Density only plays a minor role, because a density increase by a factor of eight increases the aerodynamic size only by a factor two. So the above relationship between size and efficacy is most probably true for all inhaled drugs, the most conceivable exceptions being receptor distributions peaking in the upper airways or in the alveoli.

Metered dose and dry powder inhalers both generate polydisperse aerosol clouds. Therefore it is predicted that the lower airway deposition (as percentage of the emitted dose) is low and that the dose can be reduced substantially if the deposition patterns can be improved. Most of the larger particles deposit in the upper airway filter and unfortunately is absorbed from the local mucosa or from the gastro-intestinal tract after being swallowed, which may result in systemic adverse effects. Elimination of the large particles reduces the systemic load and adverse effects and offers a reasonable explanation for our findings.
However, any drug deposited in the lower airways is also prone to absorption o the systemic circulation, which may lead to adverse effects. The severity and frequency of adverse effects will be lower with optimal formulated aerosols. Nevertheless, rapid deactivation of absorbed drug is necessary to eliminate this contribution to systemic adverse effects. Such deactivation must occur before the drug occupies receptors in the body, which may result in requirements of a deactivation within seconds, regarding the circulation time of the blood.

4.2. Corticosteroids

Measuring and comparing the beneficial (and adverse effects) of inhaled corticosteroids proved to be very difficult, because they develop their full effect over weeks to months. Next to this, the effects are highly variable and studies are generally not very powerful[52]. To add to this, studying the relationship between aerosol size and clinical effect is technically demanding because of the complicated nature of monodisperse aerosol generators. These systems are not mass produced nor a they suitable for use at home, which makes daily aerosol delivery for long periods not practical. However, the occurrence of beclomethasone MDI-formulations, based on HFA-propellants and dissolution of beclomethasone in the HFA and ethanol, gives some insight in the size-effect relationship[53]. These data are reviewed in the following sections.

The HFA/ethanol formulation is a solution aerosol compared to the well known suspension aerosol and this affects (as expected) the particle size distribution. In vitro cascade particle sizing showed that the MMAD is much finer. The bulk of the mass of the solution aerosol consists of 1.1-2.1 μm particles compared to 3.3-4.7 μm particles of the CFC MDI[54]. Such a change in particle size distribution must affect the deposition patterns within the airways and shift to the periphery may be expected. This hypothesis was confirmed by the experiments of Leach. He labelled a CFC and HFA-beclomethasone MDI (suspension versus solution) and determined the lung and oropharynx deposition. For the HFA MDI 55-60% of the dose was delivered to the lungs and 29-33% to the oropharynx. The CFC MDI delivered only 4-7% and 90% to resp. the lungs and oropharynx.[55] Rephrased, the solution aerosol delivered approx. eight times more drug to the airways than the suspension aerosol did. The shift in MMAD must result in a higher alveolar deposition. Beclomethasone is absorbed efficiently from the lungs, but not from the gastrointestinal tract. Soria showed that beclomethasone plasmalevels after an oral dose of 2 mg equals those of an 0.8 mg inhaled dose from a solution aerosol (the relative bioavailability of the inhaled dose is 250%)[56]. One might expect that absorption from the alveolar compartment is very efficient: Harrison showed that the area under the curve (AUC) of beclomethasone (in the form of BOH) after 200 μg of the HFA formulation equals that of 400 μg from the CFC formulation[57]. In line with the concept of a swift absorption the C_{max} was higher and T_{max} In Table 12 a summary is given.

Table 12 Summary of pharmacokinetics[56]

	200 μg HFA-BDP	400 μg HFA-BDP	400 μg CFC-BDP
C_{max} (pg/ml)	590	1191	410
T_{max} (h)	3	3	4
AUC (pg/ml*h)	2339	4962	2092

A comparison of the therapeutic efficacy of the HFA- and CFC-formulation showed subsequently that both preparations can be considered equipotent[58]. The design of this study started with a course of oral prednisolone (30 mg/day) for 7-13 days, after which the subject were randomised to use the HFA- or CFC-preparation. (The oral course was inserted to test steroid responsiveness). Peakflow and FEV_1 were measured at regular intervals after the switch and in both groups appeared to decline gradually in a similar way to a lower plateau. Statistical analysis showed equivalence of the two preparations. Another study was able to show similar or somewhat lower adverse effects of the HFA formulation[59]

This change of formulation, size distribution leads to several questions. An yet unexplained problem is that the eightfold increase in deposition does not lead to an equally strong increase in effect. The HFA-formulation is marketed with a 50% reduced dose, so even then (compared to the CFC-formulation) a fourfold increase in lung mass is present. One would expect a substantial increased clinical effect from that increase of mass or dosage, which isn't hard to shown. More research is needed to unravel the contradictions.

5. NEW APPROACHES

5.1. Porous particles

Several times above it was stressed that absorption of inhaled products is fast, but bioavailability is low. The latter caused by the problems with particle size distributions, while rapid absorption may be modified by a slow dissolution (of very small particles).

Recently a few publications emerged on porous particles: these particles have a very low density (due to her hollow structure) while their actual size is large. The low density allows that large physical size because the aerodynamic diameter of such particles equals that of much smaller solid particles. The deposition patterns in the lungs of a 9 μm low density particle will therefore be the same as a 3 μm solid particle: the porous particles can be targeted to the same regions. The nice feature is that large low density particles are easier to handle: they do not, like micronised particles, agglomerate and flow better. In dry powder inhalers this would mean an advantage because the separation of these particles from carriers is much better and it might even be so that carriers aren't needed anymore. The bioavailability of these porous particle formulation increases dramatically. In animal experiments with insulin it appears that a conventional dry powder showed a relative bioavailability of 12%, but the porous formulation of 87%.[60]

Other experiments indicate that the length of action of these particles increase: porous salbutamol were administrated to guinea pigs who were repeatedly challenged with carbachol. Over a period of 36 hours, airway resistance was determined in response to the repeated challenges. Solid salbutamol particles protected for up to 5 hours, but the porous particles for at least 16 hours. [61] Similar findings were found for insulin.[60] The first explanation given was the lower phagocytosis of large particles by macrophages, apparently this must lead to reduced elimination of the drug from the airways.

5.2. New atomisation techniques

A possible new way to obtain better defined aerosols is electro hydrodynamic atomisation (EHDA).[62 63 64] The energy source for EHDA is an electric field (see Fig. 10): a liquid is supplied to a nozzle and an electric field is generated between the nozzle and a counter electrode. Normally droplets emerge from a nozzle, when a fluid is slowly pumped through, but now the electrical stress forces another shape: a cone. In that cone the fluid is accelerated greatly and from the tip a thin jet emerges. The jet breaks up into monodisperse droplets.

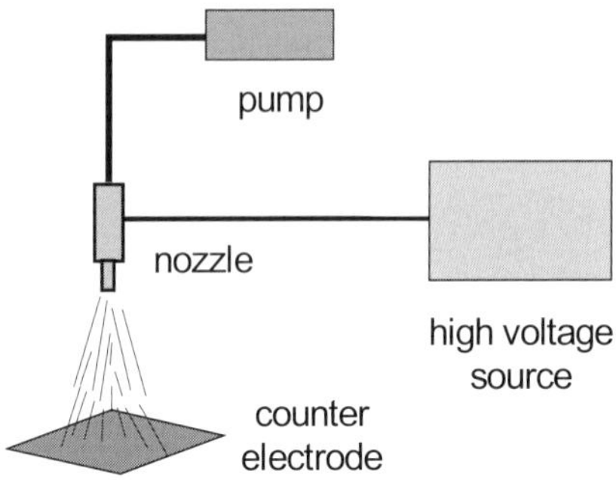

Fig. 10 Basic set-up used for EHDA

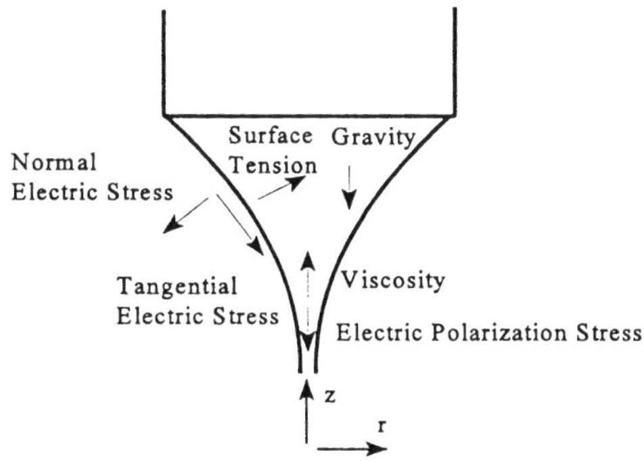

Fig. 11 Stresses acting on the cone

Fig. 12 Photograph of EHDA

The droplet size depends on liquid flow rate and properties (i.e. conductivity, surface tension and density) and Eqn 2 predicts the primary droplet size D:

$$D = 0.767c\left(\frac{\rho\varepsilon_0}{\gamma K}\right)^{1/6} Q^{1/2}$$

Eqn 2

where ρ = density of the liquid ($kg.m^{-3}$)
ε_0 = permittivity of vacuum ($8.85.10^{-12}$ $C^2.N^{-1}.m^{-2}$)
Q = applied flow rate of the liquid ($m^3.s^{-1}$)
K = conductivity of the liquid ($S.m^{-1}$)
γ = surface tension of the liquid ($N.m^{-1}$)
c = 2.05.[64]

The equation predicts that the flow rate at which fluid is pumped through the nozzle is an important parameter in controlling the droplet size: the higher that is the larger the droplets will be. The concentration of the drug in solution is of course a second parameter. Evaporation of the solvent leaves smaller dry particles (size depending on the concentration of the dissolved drug).
The droplets produced need to be neutralised, because the generation process guarantees highly charged droplets and a corona discharge is used. (A corona discharge occurs when between a sharp pointed and earthed needle and the nozzle a high electric field exists. The needle is placed opposite the nozzle and acts as counter electrode). A corona produces free electrons forming negative ions, which discharge the positively charged droplets.[65] We use this set up to produce monodisperse beclomethasone dipropionate aerosols from an ethanol solution, below we give some data on the characteristics.

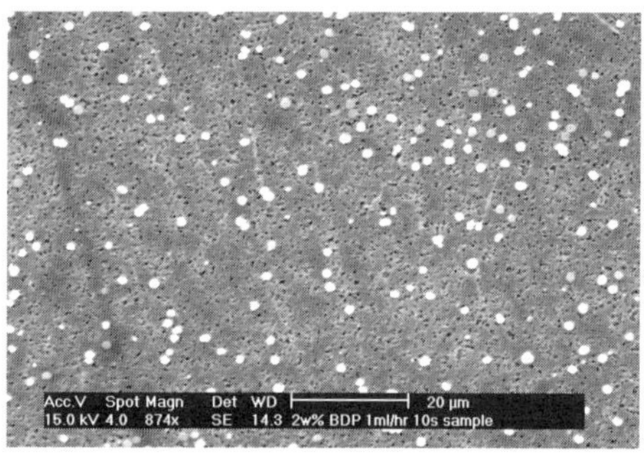

Fig. 13 SEM picture of monodisperse beclomethasone dipropionate captured on a filter

Table 13 Influence of the beclomethasone dipropionate concentration in ethanol and flow rate on the MMAD and geometric SD. The data depicted represent the mean (SD) of 10 consecutive measurements at 1 min intervals.[66]

BDP in ethanol (w/w)	flow rate	mmad (µm)	gsd
0.5 %	1 ml.hr^{-1}	1.58 (0.05)	1.18 (0.03)
2 %	1.5 ml.hr^{-1}	2.51 (0.03)	1.2 (0.03)
3 %	3 ml.hr^{-1}	4.6 (0.06)	1.25 (0.06)

5.3. liposomes

Fast disappearing effects of bronchodilators, insulin etc etc has raised (quite predictable) an interest in slowing down the absorption of inhaled drugs. Currently long acting bronchodilators gained a well defined place in asthma therapy. These long acting compounds act longer because they either adhere strongly to the receptors or to neighbouring structures: the residence time at the receptors is prolonged due to their physico-chemicals properties. An obvious alternative route is to delay absorption and disappearance of inhaled drugs. Liposomes are a possible approach.

The kinetics of sodium cromoglycate (SCG) was studied in healthy volunteers, who inhaled SCG 20 mg as a solution or encapsulated in dipalmitoyl phosphatidylcholine/cholesterol (1:1) liposomes. After administration of liposomal SCG detectable plasmalevels were present 24 and 25 hr after inhalation, while after the solution no SGC was detectable in the 24-hr samples. C_{max} levels were more than sevenfold greater for the solution. It is no surprise that the absorption rate constants differed strongly, indicating *slow drug release* from the liposomes.[67] Soon hereafter it was shown that single doses of beclomethasone dipropionate-dilauroylphosphatidylcholine (DLPC) aerosol in volunteers were safe to administer and that 82% ± 5.9 of the total pulmonary dose of ^{99m}Tc-labeled liposome suspension that at 4 hours after administration was still present in the lungs of mild asthmatic patients, while in severe asthma this was 69% ± 10.9.[68, 69]

The clearance of labelled beclomethasone dipropionate dilauroylphosphatidylcholine (DLPC) and dipalmitoylphosphatidylcholine (DPPC) liposomes was measured in 11 healthy volunteers. 24 h after inhalation 79% of the DLPC and 83% of the of DPPC liposomes was still present in the lungs. The results indicate liposome encapsulation leads to sustained release in the lower airways.[70] Apparently the liposomes escape the mucociliairy transport and remain in the lungs for longer times. That feature combined with slow release of drugs from the liposomes opens the opportunity to formulate inhaled slow-release preparations.

6. LITERATURE

[1] Timsina M.P., Martin G.P., Marriott C., Ganderton D. and Yianneskis M. Drug delivery to the respiratory tract using dry powder inhalers. Int J Pharm 101: 1-13 1994.

[2] Pedersen S., Frost L. and Arnfred T. Errors in inhalation technique and efficiency in inhaler use in asthmatic children. Allergy 41: 118-124 1986.

[3] Biddiscombe M.F., Melchor R., Mak V.H.F., Marriott R.J., Taylor A.J., Short M.D. and Spiro S.G. The lung deposition of salbutamol, directly labelled with technetium-99m, delivered by pressurised metered dose and dry powder inhalers. Int J Pharm 91: 111-121 1993.

[4] Bouchikhi A., Becquemin M.H., Bignon J., Roy M. and Teillac A. Particle size study of nine metered dose inhalers and their deposition probabilities in the airways. Eur J Respir 1: 547-552 1988.

[5] Barnes P.J., Basbaum C.B., Nadel J.A. and Roberts J.M. Localization of ß-adrenoreceptors in mammalian lung by light microscopic autoradiography. Nature 299: 444-447 1982.

[6] Findeisen W. Uber das Absetzen kleiner in der Luft suspendierten Teilchen in der menschlichen Lunge bei der Atmung. Phlügers Arch Ges Physiol 236: 367-379 1935.

[7] Landahl H.D. On the removal of airborne droplets by the human respiratory tract. I The lung. Bull Math Biophys 12: 43-56 1950.

[8] Beeckmans J.M. Correction factor for size-selective sampling results, based on a new computer alveolar deposition curve. Ann Occup Hyg 8: 221-231 1965.

[9] Weibel E.R. Morphometry of the human lung. Springer Verlag 1963.

[10] Task Group on Lung Dynamics. Deposition and retention models for internal dosimetry of the human respiratory tract. Health Phys 12: 173-207 1966.

[11] Gerrity T.R., Lee P.S., Hass F.J., Marinelli A., Werner P. and Lourengo R.V. Calculated deposition of inhaled particles in the airway generations of normal subjects. J Appl Physiol 47: 867-873 1979.

[12] Heyder J., Armbuster L., Gebhart J., Grein E. and Stahlhofen W. Total deposition of aerosol particles in the human respiratory tract for nose and mouth breathing. J Aerosol Sci 6: 311-328 1975.

[13] Heyder J., Gebhart J., Rudolf G., Schiller C.F. and Stahlhofen W. Deposition of particles in the human respiratory tract in the size range 0.005-15 μm. J Aerosol Sci 17: 811-825 1986

[14] Stahlhofen W., Gebhart J. and Heyder J. Experimental determination of the regional deposition of aerosol particles in the human respiratory tract. Am Ind Hyg Ass J 41: 385-398 1980.

[15] Heyder J., Gebhart J., Rudolf G. and Stahlhofen W. Physical factors determining particle deposition in the human respiratory tract. J Aerosol Sci 11: 505-551 1980.

[16] Ebina S., Yaegashi H., Takahashi T., Motomiya M. and Tanemura M. Distribution of smooth muscle along the bronchial tree. Am Rev Respir Dis 141: 1322-1326 1990

[17] Hossain S. Quantitative measurement of bronchial muscle in men with asthma. Am Rev Respir Dis 107: 99-109 1973

[18] Ebina S., Yaegashi H., Chiba R., Takahashi T., Motomiya M. and Tanemura M. Hyperreactive site in the airways tree of asthmatic patients revealed by thickening of bronchial muscles. Am Rev Respir Dis 141: 1327-1332 1990

[19] Carroll N., Elliot J., Morton A. and James A. The structure of large and small airways in nonfatal and fatal asthma. Am Rev Respir Dis 147: 405-411 1993

[20] Barnes P.J., Basbaum C.B. and Nadel J.A. and Roberts J.M. Localization of ß-adrenoreceptors in mammalian lung by light microscopy autoradiograhpy. Nature 299: 444-447 1982

[21] Barnes P.J., Basbaum C.B. and Nadel J.A. Autoradiographic localization of autonomic receptors in airway smooth muscle. Am Rev Respir Dis 127 : 758-762 1983

[22] Carstairs J.R., Nimmo A.J. and Barnes P.J. Autoradiographic visualization of beta-adrenorecptor subtypes in human lung. Am Rev Respir Dis 132: 541-547 1985

[23] Saetta M., Di Stefano A., Rosina C., Thiene G. and Fabbri L.M. Quantitative structural analysis of peripheral airways and arteries in sudden fatal asthma.. Am Rev Respir Dis 143: 138-143 1998

[24] Faul J.L., Tormey V.J., Leonard C., Burke C.M., Farmer J., Horne S.J. and Poulter L.W. Lung immnopathology in cases of sudden death. Eur Respir J 10:301-307 1997

[25] Azzawi M., Johnston P.W., Majumdar S., Kay A.B. and Jeffrey P.K. T lymphocytes and activated eosnophils in airway mucosa in fatal asthma and cystic fibrosis. Am Rev Respir Dis 145: 1477-1482 1992

[26] Synek M., Beasley R., Frew A.J., Goulding D., Holloway L., Lampe F.C., Roche W.R. and Holgate S.T. Cellular infiltration of the airways in asthma of varying severity. Am J Respir Crit Care Med 154: 224-230 1996

[27] Hamid Q., Song Y., Kotsimbos T.C., Minshall E., Bai T.R., Hegele R.G. and Hogg J.C. Inflammation of small airways in asthma. J Allergy Clin Immunol 100: 44-51 1997

[28] Carroll N., Cooke C. and James A. The distribution of eosinophils and lymphocytes in the large and small airways of asthmatics. Eur Respir J 10: 292-300 1997

[29] Kraft M., Djukanovic R., Wilson S., Holgate S.T. and Martin R.J. Alveolar tissue inflammation in asthma. Am J Respir Crit Care Med 154: 1505-1510 1996

[30] Adcock I.M., Gilbey T., Gelder C.M., Chung K.F. and Barnes P.J. Glucocorticoid receptor localization in normal and asthmatic lung. Am J Respir Crit Care Med 154:L 771-782 1996

[31] Vachier I., Chiappara G., Vignola A.M., Gagliardo R., Altieri E., Terouanne B., Bousquet J., Godard P. and Chanez P. Glucocorticoid receptors in bronchial epithelial cells in asthma. Am J Respir Crit Care Med 158: 963-970 1998

[32] Ruffin R.E., Dolovich M.B., Oldenburg F.A. and Newhouse M.T. The preferential deposition of inhaled isoproterenol and propanolol in asthmatic patients. Chest 80: 904-907 1981.

[33] Rees P.J., Clark T.J.H. and Morén F. The importance of particle size in response to inhaled bronchodilators. Eur J Respir Dis suppl 119: 73-78 1982.

[34] Clay M.M., Pavia D. and Clarke S.W. Effect of aerosol particle size on bronchodilation with nebulised terbutaline in asthmatic subjects. Thorax 41: 364-368 1986.

[35] Clay M.M. and Clarke S.W. Effect of nebulised aerosol size on lung deposition in patients with mild asthma. Thorax 42: 190-194 1987.

[36] Mitchell D.M., Solomon M.A., Tolfree S.E.J., Short M. and Spiro S.G. Effect of particle size of bronchodilator aerosols on the lung distribution and pulmonary function in patients with chronic asthma. Thorax 42: 457-461 1987.

[37] Persson G. and Wirén J.E. The bronchodilator response from inhaled terbutaline is influenced by the mass of small particles: a study on a dry powder inhaler (Turbuhaler). Eur Respir J 2: 253-256 1989.

[38] Johnson M.A., Newman S.P., Bloom R., Talaee N. and Clarke S.W. Delivery of albuterol and ipratropium bromide from two nebulizer systems in chronic stable asthma. Chest 96: 1-10 1996.

[39] Patel P., Mukai D. and Wilson A.F. Dose-response effects of two sizes of monodisperse isoproterenol in mild asthma. Am Rev Respir Dis 141: 357-360 1990.

[40] Hultquist C., Wollmer P., Eklund G. and Jonson B. Effect of inhaled terbutaline sulphate in relation to its deposition in the lungs. Pulm Pharm 5: 127-132 1992.

[41] Loffert D.T., Ikle D. and Nelson H.S. A comparison of commercial jet nebulizers.Chest 106: 1788-1793 1994.

[42] Smith E., Dnyer J. and Kendrick A.H. Comparison of twenty three nebulizer/compressor combinations for domiciliary use. Eur J Respir 8: 1214-1224 1995.

[43] Thomas S.H.L., O'Doherty M.J., Page C.J. and Nunan T.O. Variability in the measurement of nebulized aerosol deposition in man. Clin Sci 81: 767-775 1991.

[44] Cheah P.K.P. and Davies C.N. The spinning top aerosol generator-improving the performance. J Aerosol Sci 15: 741-751 1984.

[45] Zanen P., Go L.T. and Lammers J.W. The optimal particle size for beta-adrenergic aerosols in mild asthmatics. Int J Pharm 107:211-217 1994

[46] Zanen P., Go L.T. and Lammers J.W. The optimal particle size for parasympathicolytic aerosols in mild asthmatics Int J Pharm 114:111-115 1995

[47] Zanen P., Go L.T. and Lammers J.W. Optimal particle size for beta 2 agonist and anticholinergic aerosols in patients with severe airflow obstruction. Thorax. 51: 977-80 1996.

[48] Zanen P., Go L.T. and Lammers J.W. The efficacy of a low-dose, monodisperse parasympathicolytic aerosol compared with a standard aerosol from a metered-dose inhaler. Eur J Clin Pharmacol. 54: 27-30 1998.

[49] Zanen P. and Lammers J.W. Reducing adverse effects of inhaled fenoterol through optimization of the aerosol formulation. J Aerosol Med.12: 241-7 1999.

[50] Svartengren M., Anderson M., Bylin G., Philipson K. and Camner P. Mouth and throat deposition of 3.6 μm radiolabelled particles in asthmatics. J Aerosol Med 4: 313-321 1991.

[51] Schiller-Scotland C.F., Gebhart J., Hochrainer D. and Siekmeier R. Deposition of inspired aerosol particles within the respiratory tract of patients with obstructive lung disease. Toxicology Letters 88: 255-261 1996.

[52] Zanen P. and Lammers J.W. Sample sizes for comparative inhaled corticosteroid trials with emphasis on showing therapeutic equivalence.Eur J Clin Pharmacol 48: 179-84 1995.

[53] Steckel H. and Müller B.W. Metered dose inhaler formulations with beclomethasone-17, 21-dipropionate using the ozone friendly propellant R134a. Eur J Pharm Biopharm 46; 77-83 1998

[54] June D. Achieving the change: challenges and successes in the formulation of CFC-free MDI's. Eur Respir Rev 41: 32-34 1997

[55] Leach C.L., Davidson P.J. and Boudreau R.J. Improved airway targeting with the CFC-free HFA-beclomethasone metered dose inhaler compared with CFC-beclomethasone. Eur Respir J 12: 1346-1353 1998

[56] Soria I., Harrison L.I., Machacek J.H., Cline A.C. and Stampone P.A. Beclomethasone relative availability of oral versus inhaled beclomethasone dipropionate from an HFA-134a metered dose inhaler. Biopharm Drug Dispos 19: 297-302 1998

[57] Seale J.P. and Harrison L.I. Effect of changing the fine particle mass of inhaled beclomethasone dipropionate on intrapulmonary deposition and pharmacokinetics. Resp Med 92 (suppl A) 9-15 1998

[58] Davies R.J., Stampone P. and O'Connor B.J. Hydrofluoroalkane-134a beclomethasone dipropionate extrafine aerosol provides equivalent asthma control to chlorofluorocarbon beclomethasone dipropionate at approximately half the total daily dose. Resp Med 92 (suppl A) 23-31 1998

[59] Thompson P.J., Davies R.J., Young W.F., Grossman A.B. and Donnell D. Safety of hydrofluoroalkane-134a beclomethasone dipropionate extrafine aerosol. Resp Med 92 (suppl A) 33-39 1998

[60] Edwards D.A., Ben-Jebria A. and Langer R. Recent advances in pulmonary drug delivery using large, porous inhaled particles. J Appl Physiol 85: 379-385 1998

[61] Ben-Jebria A., Chen D., Eskew M.L., Vanbever R., Langer R. and Edwards D.A. Large porous particles for sustained protection from carbachol-induced bronchoconstriction in guinea pigs. Pharm Res 16: 555-561 1999

[62] Cloupeau M. and Prunet-Foch B. Electrohydrodynamic spraying functioning modes: a critical review J Aerosol Sci 25: 1021-1036 1994

[63] Grace J.M. and Marijnissen J.C.M. A review of liquid atomization by electric means. J Aerosol Sci 25: 1005-1019 1994

[64] Hartman R., Brunner D.J., Camelot D.M.A., Marijnissen J.C.M. and Scarlett B. Electrohydrodynamic atomization in the cone-jet mode physical modeling of the liquid cone and jet. J Aerosol Sci 30: 823-850 1999

[65] Meesters G.M.H., Vercoulen P.H.W., Maijnissen J.C.M. and Scarlett B. Generation of micron-sized droplets from the Taylor cone. J Aerosol Sci 23: 37-49 1992

[66] IJsebaert J.C., Geerse K.B., Marijnissen J.C.M., Lammers J.W and Zanen P. Electro hydrodynamic atomisation of drug solutions for inhalation purposes. J Appl Phys 91: 2735-41 2001

[67] Taylor K.M., Taylor G., Kellaway I.W. and Stevens J. The influence of liposomal encapsulation on sodium cromoglycate pharmacokinetics in man. Pharm Res 6: 633-6 1989

[68] Waldrep J.C., Gilbert B.E., Knight C.M., Black M.B., Scherer P.W., Knight V. and Eschenbacher W. Pulmonary delivery of beclomethasone liposome aerosol in volunteers Tolerance and safety. Chest 111: 316-323 1997

[69] Saari S.M., Vidgren M.T., Koskinen M.O., Turjanmaa V.M., Waldrep J.C, and Nieminen M.M. Regional lung deposition and clearance of 99mTc-labeled beclomethasone-DLPC liposomes in mild and severe asthma. Chest 113: 1573-1579 1998

[70] Saari M., Vidgren M.T., Koskinen M.O., Turjanmaa V.M. and Nieminen M.M. Pulmonary distribution and clearance of two beclomethasone liposome formulations in healthy volunteers. Int J Pharm 181: 1-9 1999

GIUSEPPE A. MARRARO, MD

TREATMENT OF ASTHMA PATIENTS UNDER SEVERE CONDITIONS

INTRODUCTION

Asthma is a chronic inflammatory disorder of the airways characterized by an obstruction of airflow. It affects people of all ages and is characterized by repeated attacks of difficulty in breathing. Airway inflammation may cause recurrent or persistent bronchospasm, wheezing, breathlessness, tightness in the chest and coughing, particularly at night, in the early morning, or after exercise.

Airway inflammation is associated with airway hyper-reactivity and/or bronchial hyper-responsiveness, with a tendency for the airways to narrow after exposure to allergens, environmental irritants, viruses, cold air, or exercise. In some patients with chronic asthma, airflow limitation may be only partially reversible because of airway remodeling (hypertrophy and hyperplasia of smooth muscle, sub epithelial fibrosis) that occurs with chronic untreated disease.

The mainstay of therapy for acute asthma is a drug inhaled via an aerosol and this therapy is highly effective in resolving the symptoms of a large number of patients.

Status asthmaticus is an emergency in which an acute exacerbation of asthma is unresponsive to bronchodilator therapy and conventional treatment (oxygen supplementation, humidification, etc.). Status asthmaticus can vary in form from mild to severe with severe bronchospasm, airway inflammation and mucous plugging that can cause breathing difficulties, CO_2 retention, hypoxemia, and severe respiratory failure which can lead to cardiac arrest.

L. Gradoń and J. Marijnissen (eds.), Optimization of Aerosol Drug Delivery, 277–293.

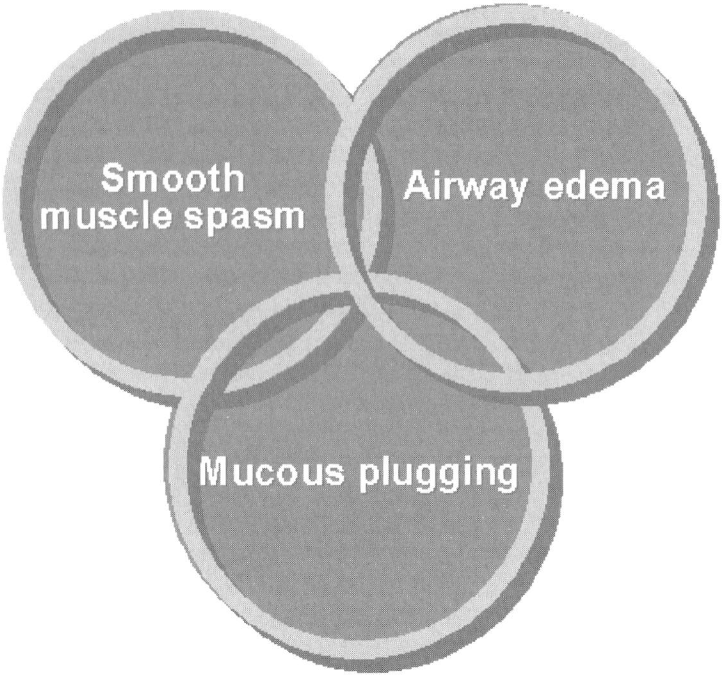

Figure 1. Characteristics of asthma as an inflammatory disease of airway.

Asthma affects up to 10% of the population of the United States of America. Worldwide its incidence is unclear, but it is estimated that there are about 20 million cases, 15% of who are children. A dramatic rise in the worldwide incidence of asthma has been described in the past 20 years. This has been accompanied by a significant rise in hospitalization and mortality. A significant increase in mortality has also been noted, especially amongst children between the ages of 0-4 and 9-16 years in recent years not only in the United States of America but also in Australia, Canada, the United Kingdom, Germany, and Switzerland. This has been attributed, in part, to pollution and industrialization. [1-4].

The complications associated with a severe asthmatic attack may include apnea, central nervous system (CNS) damage from hypoxia and toxicity from medications. The two common causes of death from asthma are hypoxic cardiac arrest and pneumothorax or pneumomediastinum. Mortality is higher in very young children and the elderly.

Patients with other preexisting conditions, such as restrictive lung disease, congestive heart failure, and chest deformities are at particular risk of death from status asthmaticus. Death may result from a single attack, when the patient does not respond to therapy, but occurs more frequently in long-term, uncontrolled asthma patients. In many cases, mortality and morbidity are related to cerebral hypoxia occurring prior to emergency department admission and/or consequent to lung barotrauma [5-9].

Pathophysiology

Asthma is an inflammatory lung disorder characterized by smooth muscle spasm, airway edema and mucous plugging. Different airway cells, including mast cells, epithelial cells, macrophages, and activated T lymphocytes are involved.
T lymphocytes play an important role in the regulation of airway inflammation through the release of numerous cytokines. Other airway cells, such as fibroblasts, endothelial cells, and epithelial cells, contribute to the chronicity of the disease. Additional factors, such as adhesion molecules (selectins and integrins), are critical in directing the inflammatory changes in the airway. Finally, cell-derived mediators influence smooth muscle tone and produce structural changes and remodel the airway.

This inflammation causes recurrent episodes of wheezing, dyspnea, and cough. Episodes are associated with obstruction in predominantly small-to-medium airways that reverses partially or completely, either spontaneously or with treatment.

Bronchospasm, mucus plugging, and edema in the peripheral airways are characteristics of the pathology and result in a non-uniform pattern of airflow obstruction and increased airway resistance. With a decreased ability to expel air, hyperinflation occurs, which can lead to severe ventilation/perfusion mismatch and increased dead space ventilation. This results in alveolar hypoventilation and atelectasis. The increased pleural and intra-alveolar pressures that result from obstruction and hyperinflation, together with the mechanical forces of the distended alveoli, can lead to a decrease in perfusion. The combination of atelectasis and decreased perfusion leads to ventilation perfusion mismatch within lung units and the resultant hypoxemia triggers an increase in minute ventilation. Over distention helps to maintain airway patency, thereby improving expiratory flow. However, it also leads to deterioration in pulmonary mechanics and increases the work of breathing and oxygen consumption.

Hyperinflation compensates for the airflow obstruction, but this compensation is limited when the tidal volume approaches the volume of the pulmonary dead space: the result is alveolar hypoventilation. Uneven changes in airflow resistance, the resulting uneven distribution of air, and alterations in circulation from increased intra-alveolar pressure due to hyperinflation all lead to ventilation-perfusion mismatch. Hypoxic vasoconstriction also contributes to this mismatch.

In the early stages of acute asthma, hyperventilation results in respiratory alkalosis. This is due to the fact that there are relatively fewer obstructed lung units (slow compartment) than there are unobstructed lung units (fast compartment). Hyperventilation allows CO_2 removal via the fast compartment. However, as the disease progresses and more lung units become obstructed, there is an increase in the slow compartments with decreased ability for CO_2 removal, eventually resulting in hypercarbia.

When ventilation-perfusion mismatch results in hypoxia, hypercarbia is prevented by the ready diffusion of carbon dioxide across alveolar capillary membranes. Thus, asthmatic patients who are in the early stages of an acute attack have hypoxemia in the absence of carbon dioxide retention.

Hyperventilation triggered by the hypoxic drive causes a decrease in $PaCO_2$ with consequent alkalosis. With worsening obstruction and increasing ventilation-perfusion mismatch, carbon dioxide retention appears. Respiratory failure leads to respiratory acidosis. Later, the increased work of breathing, increased oxygen consumption, and increased cardiac output result in metabolic acidosis

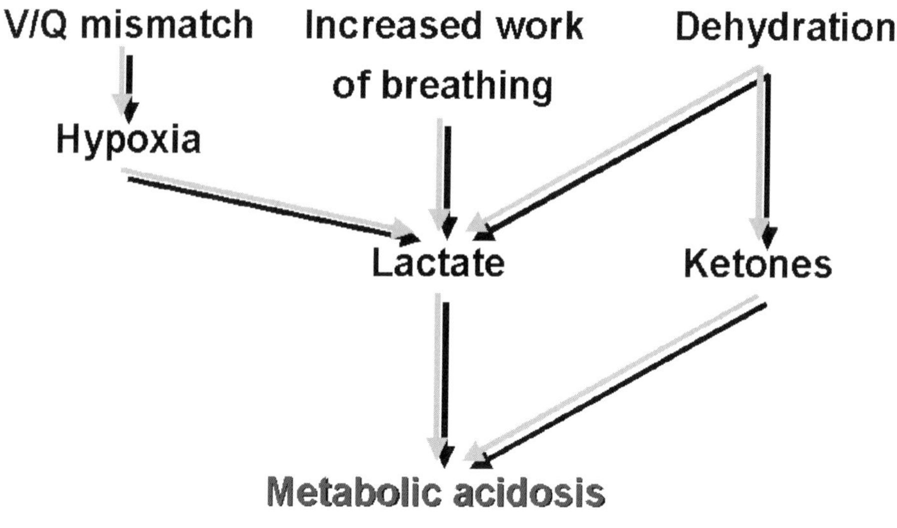

Figure 2. Origin of metabolic acidosis.

Lung hyperinflation and thoracic hyper-expansion cause reduced respiratory muscle efficiency and the flattened diaphragm is forced to contract with shortened muscle fibers.

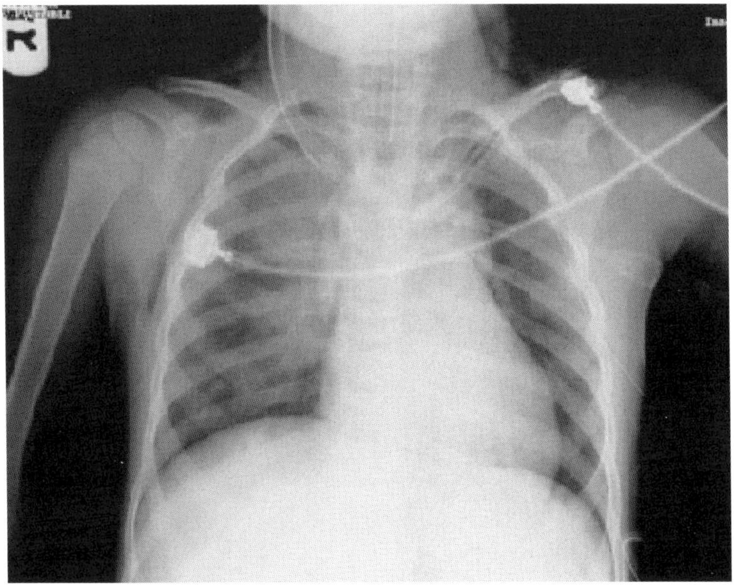

Figure 3. Chest x-ray of severe asthma in adolescent boy. Lungs are over inflated and hyper transparent, diaphragm and ribs are flattened, heart is reduced in volume due to external compression.

Airway rupture as subcutaneous emphysema, pneumodiastinum and pneumothorax can complicate the pathology.

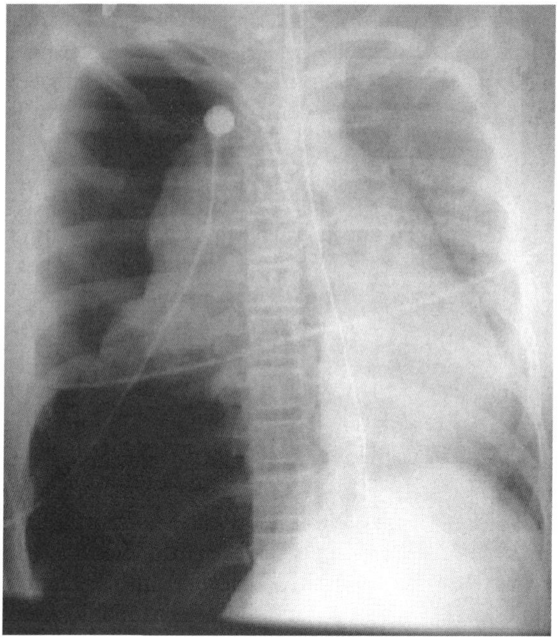

Figure 4. Dramatic right pneumothorax complicating asthma attack.

The greatly increased intra-thoracic pressure connected with lung over-inflation impedes venous return, thus decreasing right and left ventricular pre-load. Children breathing spontaneously with severe asthma have negative intra-pleural pressure (as low as -35 cmH_2O) during the entire respiratory cycle that can lead to left ventricular load. Negative intra-pleural pressure causes increased left ventricular after-load, resulting in risk of pulmonary edema. Hypoxic pulmonary vasoconstriction and lung hyperinflation lead to increased right ventricular afterload with right ventricular load. Tachycardia and hypotension related to reduction of cardiac output ensues and can lead to cardiac arrest.

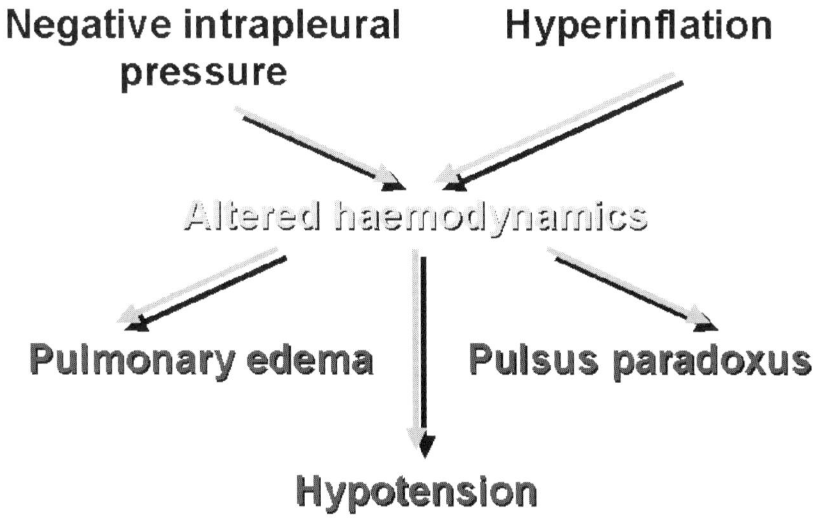

Figure 5. Cardiopulmonary interaction in asthmatic patients.

In the past decade, evidence has emerged for surfactant as a factor in the regulation of airway calibers and a modulator of allergic inflammation. Airway obstruction in asthma is commonly connected with smooth muscle constriction, mucosal edema and mucus presence in the small airway, may be due to a lack of pulmonary surfactant [10-12].

Indication for intubation and mechanical ventilation

Tracheal intubation and mechanical ventilation is frequently the last therapeutic option after the failure of standard therapy, i.e., β2–antagonists, aminophylline, steroids, epinephrine [13, 14]. The immediate indicators for initiating mechanical ventilation are respiratory arrest (apnea) and cardiac arrest [5, 15].

There is a hesitation to intubate the asthmatic patient. This is related to:

- the presence of the tube in the trachea can aggravate bronchospam;
- positive pressure ventilation increases risk of barotrauma and hypotension;
- over 50% of the morbidity and mortality during severe asthmatic attack occurs during or immediately after intubation.

Tracheal intubation must be applied promptly once it is deemed necessary. A delay in intubation and mechanical ventilation can increase morbidity and mortality.

In less severe conditions, arterial blood gas does not appear to be a good indicator of the need for mechanical ventilation because of delay in appearance of marked signs of worsening of ventilation. Apnea due to exhaustion which requires assisted ventilation often appears before gas exchange impairment. Arterial blood gases are often indicative of the efficacy of treatment [16]. In the initial phase of an asthma attack, $paCO_2$ can be reduced < 40 mmHg, but when status asthmaticus progresses $paCO_2$ rises, reflecting the severity of bronchospasm and difficulty of CO_2 elimination [17, 18].

PaO_2 falls precociously in an asthma attack because of a misdistribution of ventilation to alveolar perfusion. Oxygen supplementation is necessary to maintain normal PaO_2 but a fraction of inspired oxygen (FiO_2) > 0.60 is rarely needed to obtain SpO_2 >90 %. Oxygen supplementation may obscure the patient's need for ventilation for a long time.

According to arterial blood gases , mechanical ventilation is generally initiated when PaO_2 is < 60 mm Hg using oxygen concentration over 60-70% and/or $PaCO_2$ is > 50 -55 mmHg and rising more than 5 mmHg/hour.

The decision to intubate and mechanically ventilate a patient with severe status asthmaticus should preferably be based on the following clinical criteria rather than on gas analysis alone [19, 20]:
Exhaustion and apnea;
Significant hypoxemia poorly or not responsive to supplemental oxygen therapy alone;
Severe respiratory muscle fatigue and increased effort to breath;
Continued rise in PCO_2 despite therapy with marked decrease in airflow;
Markedly decreased air entry into the lungs and limited thoracic expansion;
Presence of pulsus paradoxus (> 20-40 mmHg) consequent to inspiratory decline in systolic blood pressure;
Altered sensorium and diminished level of consciousness, e.g., agitation, confusion, lethargy and coma;
Presence of high-risk factors.

Pulsus paradoxus is the clinical correlate of cardiopulmonary interaction during severe asthma. It is defined as the exaggeration of the normal inspiratory drop in systolic blood pressure: normally < 5 mmHg, but > 10 mmHg in pulsus paradoxus.

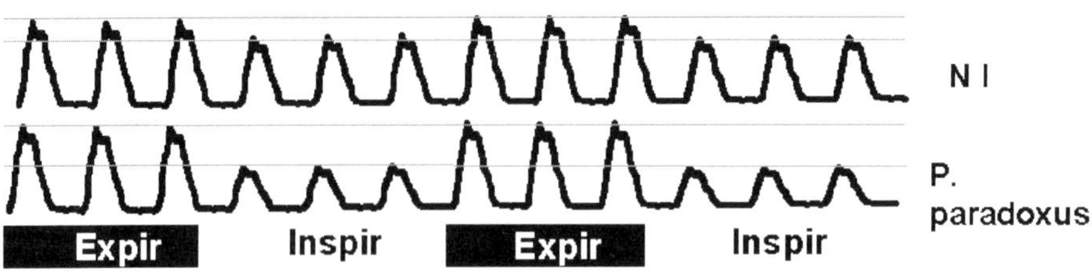

Figure 6. Pulsus paradoxus: Superior part of the graph is normal condition. Inferior part shows situation during asthma attack.

Tracheal intubation

Oral and nasal-tracheal intubation during inhaled anesthesia is a safe method that allows spontaneous breathing to be maintained. Halogenated anesthetics (e.g. isoflurane, enflurane, sevoflurane and halothane) can be used to obtain a suitable level of anesthesia to perform intubation with reduced risks (laryngospasm, vagal stimulation, vomiting). Hypotension and cardiac dysrhythmias can be observed in hypoxemic and hypovolemic patients during the maneuver.

An easy intubation can be performed using a short or middle duration neuromuscular blocking agent (succinylcholine, rocuronium, vecuronium), which allows a rapid return to spontaneous breathing in cases where it is impossible to intubate. When the patient is paralyzed and ventilation is impossible, the risk of severe hypoxia (neurological damage) and the patient dying is increased [21]. Bradicardia is frequent after succinylcholine in children and can be prevented by atropine. Neuromuscolar blocking agents which liberate histamine (e.g. pancuronium) are not indicated in asthmaticus status.

Manipulation of the airway can worsen the bronchospasm and provoke laryngospasm that can be prevented by topical application of local anesthetics at glottis level. Paralysis with curare drugs suppresses the tonic part of hyperinflation (accessory muscle paralysis) and may aggravate the obstructive syndrome leading to unventilatable thorax.

Bag-and-mask ventilation can be difficult or even impossible after muscle paralysis in acute asthma patient because of severe lung hyperinflation. Gastric distension connected with ineffective manual lung ventilation can increase the risk of aspiration of gastric content.

Immediately after intubation overzealous manual ventilation that can lead to emphysema and pneumothorax must be avoided. If manual ventilation is prolonged and dry gases are used endotracheal tube obstruction can appear due to consolidation of abundant secretion. Tube obstruction is more frequent at pediatric age using tube of reduced internal diameter. After intubation and uncontrolled mechanical ventilation (tidal volume) a rapid reduction of $paCO_2$ can result in hypotension and circulatory failure.

Mechanical ventilation strategies

The goals of mechanical ventilation include: [4, 15, 19]
reduction of dynamic lung hyperinflation, using a low tidal volume and high respiratory frequency, with an extended expiratory time;

permissive hypercapnia, as consequence of reduced tidal volume;
avoiding intrinsic positive end-expiratory pressure (auto PEEP) connected with bronchoconstriction and dynamic hyperinflation;
removal of secretion and restoration of airway patency.

At present no ventilation mode has been proved to be superior in status asthmaticus. Adults and children have been ventilated both in volume controlled and in pressure controlled ventilation. Volume controlled ventilation appears to be more effective because this mode guarantees stable tidal volume despite variations in airway resistance. A reduction of inspiratory pressure can be obtained by using low tidal volume and high respiratory rate, maintaining adequate minute volume necessary to obtain normal gas exchanges. Pressure controlled ventilation has been used for a long time to reduce barotrauma connected with peak inspiratory pressure in newborns and infants. This method has been under review in recent years primarily because volume (volutrauma) rather than pressure (barotrauma) has been shown to determine the damage of the lung. This method does not guarantee stable tidal volume as it is influenced by the resistance of the airways and can lead to hyper- or hypo-ventilation.

Ventilator settings

In an emergency it is essential to control hypoxia and lung dynamic hyperinflation.High inflating pressures are usually required to ventilate the lungs because of increased airway resistance. When the emergency phase is resolved, recruitment of lung areas and maintaining the patency of bronchioles and alveoli become indispensable to reducing lung barotrauma and improving lung pathology. When bronchospasm has been reduced and the mucous largely aspirated, the pressure needed to open the airways during the inspiratory time must be reduced. Bronchioles and alveoli recruited have to be maintained continuously open in order to reduce shear forces which hinder opening of alveoli. Mucous and secretion removal are facilitated when bronchial obstruction is resolved [21-24].

For many years, "permissive hypercapnia" [25-28] or "controlled hypoventilation" has been considered a suitable way of providing adequate oxygenation and ventilation while minimizing high airway pressures and barotrauma. It involves the administration of as high an inspired oxygen fraction (FiO_2) as is necessary to maintain normal PaO_2, acceptance of hypercapnia (maximum 90 mmHg, without rapid rise) and treatment of respiratory acidosis (pH > 7.20 minimum level accepted) with intravenous sodium bicarbonate. Ventilator rate is maintained at no more than 12/minute, with a slow inspiratory flow rate and a long expiratory time (1:3 or 1:4 inspiration-expiration [I-E] ratio) to facilitate complete emptying of lung during expiration. With low ventilatory rates and prolonged expiratory time, breathing can be very uncomfortable for the patient and syncronisation with the ventilator difficult to obtain. Deep sedation is necessary and neuromuscular blocking agents are often required.

Permissive hypercapnia can expose the patient to:
intracranial hemorrhage;
worsening of pre-existent intracranial hypertension in the presence of cerebral vasodilation and edema;
myocardial depression;
pulmonary vasoconstriction;
cardiac arrhythmia connected with hypokalemia.

It is important to monitoring of flow-volume loops to ascertain whether adequate time has been provided for expiration to avoid breath stacking, which will occur if the next breath is delivered before expiration is completed. It is also important to monitor exhaled tidal volume and auto-PEEP.

This ventilator strategy is not uniformly successful in critically ill asthma patients. Reasons for this include the difficulty of synchronizing ventilation, of eliminating secretion, of weaning the patient from the ventilator and the frequent appearance of complications such as pneumonia and atelectasis which are the result of the consolidation of secretions and inability to cough

A new ventilatory mode [21, 29] which involves ventilator settings being adapted in order to maintain $PaCO_2$ near to normal limits using a low tidal volume (8-10 ml/kg) and high respiratory rate (> 20-40% of expected for the patient's characteristics age) according to preset minute volume, and an I:E ratio 1:2 is gaining consensus . A PEEP level of
5 cm H_2O is applied at the beginning of ventilation if the cardiovascular system is stable and when hypovolemia is resolved. The PEEP level is progressively increased in accordance with the re-opening of bronchioles and when the bronchospasm is reduced. This technique reduces the risk of hypercapnia to a minimum, resolves the cerebral edema frequently present in these patients, and reduces the need for deep sedation and the use of neuromuscular blocking agents, with beneficial effect on mobilization and aspiration of broncho-alveolar mucous.

Epinephrine (intramuscular or intravenous), salbutamol (intravenous) and inhaled anesthetics are recommended when patients are difficult to ventilate [30].

When the emergency phase is resolved, it is necessary to focus on recruiting a large part of the lung areas to ventilation using a ventilatory model that opens the lung and keeps it open during the entire respiratory phase. This goal can be achieved by using an end inspiratory pause > 10% in order to favor gas redistribution in the alveoli at the end of inspiration and a PEEP level increased progressively over the inflation point of pressure volume curve (until 10 cm H_2O) according to hyperinflation and hemodynamics. This approach appears to be effective in resolving severe cases and reduces the duration of artificial ventilation with consequent barotrauma and volutrauma [31, 32].

Weaning from ventilator can begin when bronchoconstriction has been reduced and dynamic hyperinflation has decreased. The PEEP level must be reduced progressively to 5 cm H_2O, in order to avoid alveolar collapse. A short period of pressure support ventilation and volume support ventilation alternating with mechanical ventilation appears to be the best weaning method because it does not create fatigue, does not increase the effort to breath or consume oxygen. T-tube trial should be avoided in order to reduce the tendency to alveolar collapse and increase the possibility to create stiffing secretions when the patient is ventilated through endotracheal tube in air ambient. Extubation maintaining the catheter in continuous aspiration should be avoided in order to reduce the risk of a collapse of alveoli and bronchioles especially in very young children.

PEEP

Controversies remain around the role of PEEP and its use in status asthmaticus. PEEP has generally been considered to be contraindicated in asthmatics because of hyperinflation of the lung. In acute asthma, however, "hyperinflation" may not result from air-trapping so much as from the voluntary or reflex attempts of the patient to maximize the diameter of the small airways during expiration by maintaining a large functional residual capacity (FRC). Those who do not support its use point out that PEEP should not be applied during mechanical ventilation (0 - 3 cm H_2O maximum) in order to decrease the risk of auto-PEEP and any reduction of venous return and cardiac output. On the other hand, those who support its use stress the advantages to be derived from decreasing the effort to breath if the extrinsic PEEP is not higher than the intrinsic PEEP (< 10 cm H_2O), preventing the airway collapsing during expiration and maintaining bronchioles and alveoli open. They also point out that it facilitates the recruitment of non-ventilated areas and mucous elimination. [33-36].
Assisted and spontaneous ventilatory support are not indicated in the first phase of treatment, especially if the patient is comatose or deeply sedated.

Recently, non-invasive positive pressure ventilation (NIPPV) such as continuous positive airway pressure (CPAP) or bi-modal positive airway pressure (BiPAP) with a facemask have been used to support of status asthmaticus. NIPPV has been shown to "splint" the airways, allowing better exhalation and emptying to take place. These methods need more confirmation from clinical data. At present clinical data is insufficient to propose the use of these methods in routine clinical practice.

Complications from mechanical ventilation

The asthmatic patient is at great risk from barotrauma during mechanical ventilation because of the high pressure required and the uneven distribution of ventilation. Barotrauma is severe if the patient is not completely synchronized with the ventilator, in the presence of a sudden cough or if a very tight cuffed tube is used. Interstitial emphysema is frequent and cannot have clinical evidence. Monolateral or bilateral pneumothorax is the most feared complication. Sudden deterioration of either respiratory or cardiovascular function should suggest this complication. It can

be life-threatening and requires rapid chest drainage. Pneumo-mediastinum can appear either on its own or along with pneumothorax. If it is small it can escape detection, be barely visible on a chest X-ray and not lead to important cardiocirculatory alterations. When present, however, the prognosis is severe due to the reduction in cardiac output. Treatment is difficult and reappearance after apparent resolution is frequent. Pneumoperitoneum is a rare complication and is present when air migrates from the thorax to the abdominal cavity. It is generally treated by the application of a thoracic drainage that can be used to deal with the pneumothorax that usually accompanies it [22, 37].In addition, mechanical ventilation can be responsible for:
Hemodynamic variations. After intubation hypotension is linked to hyperinflation, hypovolemia and sedation [22]. The treatment consists of replacing intravascular volume. In addition mechanical ventilation can decrease venous return, increase tissue hypoxia, and lead to the development of cardiac arrhythmia and edema.

Mucous plugs. Incorrect or inadequate humidification/heating of ventilated gases may result in worsening of bronchospasm and may lead to the formation of dense secretions that cannot be easily drained. Such a situation may increase airway obstruction, create atelectasis, favor the appearance of pneumonia and worsen gas exchange.

Atelectasis. Micro, segmental or lobar atelectasis may be present at the beginning of treatment or can appear during treatment. It is related to non-idoneous ventilatory mode, use of unsuitable PEEP or with bronchial secretions that are difficult to remove, e.g. inefficacy of humidification and insufficient warming of ventilated gases, deep sedation and/or muscular paralysis and non-mobility of patient (prone positioning and postural drainage).

Nosocomial infection. Nosocomial infections such as pneumonia and sinusitis are frequently seen in artificially ventilated patients. These infections develop more frequently in cases of inadequate nursing or when the patient is deeply sedated.

Unventilatable thorax. *Paralysis with curare drugs suppresses the tonic part of hyperinflation (accessory muscle paralysis) and may cause the thorax to become unventilatable due to an aggravation of the obstructive syndrome. Transitory solutions to break this vicious circle (i.e. reducing hyperinflation) are the use of progressive high PEEP levels (> 10 cm H_2O) and alternative techniques such as halogenated gases. The use of FiO_2 1 for several minutes can help during hypoxia but increases atelectasis tendency from reabsorption. High oxygen concentration used for prolonged time can favor surfactant inactivation.*

Alternative ventilation techniques

Various techniques have been attempted when bronchoconstriction worsens and seriously affects ventilation and oxygenation.

Ketamine. A dissociative anesthetic with strong analgesic effect and direct bronchodilating action. Its use can be proposed for intubation (2mg/kg i.v.) as well as in continuous infusion (0.5 -2 mg/kg/hr). It induces bronchorrea and can provoke emergency reaction [38-40].

Halogenated gases. Inhalation anesthetics have been used in severe status asthmaticus for direct relaxant effects on airway smooth muscle. Isoflurane, sevoflurane or halothane added to a ventilator gas flow of 0.5-2 % can improve bronchoconstriction due to bronchodilatory effect of the gas itself, as well as its sedative effects. Use is generally recommended only for patients with severe status asthmaticus who are intubated and unresponsive to conventional therapy [41-43].

Magnesium. Magnesium produces bronchodilation by competing with calcium at its binding sites. Although no controlled studies have been conducted with children, one study of adults indicated no difference in admission rates and duration of ED stay between patients treated with magnesium and those treated with a placebo [44].

Heliox. Some studies have shown that a mixture of helium/oxygen (60 - 40 %) reduces PCO_2 significantly, improves pH, decreases the work of breathing by decreasing inspiratory and expiratory resistance, and improves clinical symptoms. Because of the physical properties of helium, adding it to inhaled oxygen streamlines an otherwise turbulent airflow. The only limitation to helium use is the amount of supplemental oxygen required by the patient to maintain satisfactory oxygen saturation. Its use is limited for patients who have higher oxygen requirements. The efficacy of heliox has not been confirmed in a randomized controlled trial. Benefits were reported in an observation study. At present it has a role in refractory asthma or in specific circumstances [45-48].

Nitric oxide. Nitric oxide has recently been tried successfully in a child with refractory asthma. The future role of this therapy remains to be determined [49].

Surfactant. The possible involvement of pulmonary surfactant in the pathophysiology of respiratory diseases with predominant disturbance in the conducting airways, such as asthma and bronchiolitis, has only recently been addressed and treated [11, 50-54]. Airway obstruction with increased airway resistance in asthma may partially be due to the poor functioning of pulmonary surfactant. Analysis of phospholipid molecular species from BAL and plasma in asthmatic patients suggested that changes in the phosphatidylcholine composition was due to infiltration of plasma lipoproteins, but not to phospholipid catabolism [55]. Thus, the most likely reason for surfactant inactivation was connected with proteins invading airways as they reached a 10-fold increase in concentration. There is extensive proof that proteins inhibit surfactant function [56,57]. In animal models it has been shown that the prophylactic treatment of sensitized animals with intratracheal instillation of surfactant reduces the deterioration in lung function [58], and reduces airway resistance [12].

ECMO (Extra corporeal membrane oxygenation via veno-arterial or veno-venous by-pass). The method has been proposed [59] in order to obtain lung rest. The high cost of therapy and not controlled positive results limit its use in clinical practice.

REFERENCES

Brenner BE. Bronchial asthma in adults: presentation to the emergency department. Part I: Pathogenesis, clinical manifestations, diagnostic evaluation, and differential diagnosis. Am J Emerg Med 1983; 1:50-70

Brenner BE. Bronchial asthma in adults: presentation to the emergency department. Part II: Sympathomimetics, respiratory failure, recommendations for initial treatment, indications for admission, and summary. Am J Emerg Med 1983b; 1:306-33

FitzGerald JM, Macklem P. Proceedings of a workshop on near fatal asthma. Can Respir J 1995; 2:113-26

Williams TJ. Risks factors for morbidity in mechanically ventilated patients with acute severe asthma. Am Rev Respir Dis 1992; 146(3):607-615

DeNicola LK. Treatment of critical status asthmaticus in children. Pediatric Crit Care 1994; 41(6):1293-1324

McFadden ER Jr, Warren EL. Observations on asthma mortality. Ann Intern Med 1997; 127(2): 142-7

Newcomb R. Respiratory failure from asthma. A marker for children with high morbidity and mortality. Am J Dis Child. 1988; 142(10):1041-1044

Spitzer WO. The use of b-agonist and the risk of death and near death from asthma. N Engl J Med 1992; 326(8):501-506

Tobin MJ. Asthma, airway biology, and nasal disorders in AJRCCM 2001. Am J Respir Crit Care Med 2002; 165: 598-618

Hohlfeld JM, Ahlf K, Enhorning G, et al. Dysfunction of pulmonary surfactant in asthmatics after segmental allergen challenge. Am J Respir Crit Care Med 1999; 159:1803-1809

Hohlfeld JM. The role of surfactant in asthma. Respir Res 2002; 3:4-12

Kurashima K, Fujimura M, Tsujiura M, Matsuda T. Effect of surfactant inhalation on allergic bronchocontriction in guinea pigs. Clin Exp Allergy 1997, 27:337-342

British Thoracic Society, et al. Guidelines for the management of asthma: a summary. BMJ. 1993 Mar 20; 306(6880):776-782

Huber P et le Groupe Francophone de Réanimation Pédiatrique. Les asthmes aigus graves en reanimation pediatrique. Arch Pediatr 1994; 1:346-349

Corbridge TC. The assessment and management of adults with status asthmaticus. Am J Respir Crit Care Med 1995; 151(5):1296-1316

Downes JJ. Arterial blood gas analysis and acid base desorders in infants and children with status asthmaticus. Pediatrics 1968; 42(2):238-249

Hoonig EG. Chronic obstructive pulmonary disease and asthma. In: Handbook of Mechanical ventilation support. Ed: Perel A and Stock MC. Williams & Wilkins, Baltimore. 1992; pp 227-245

Kurth CD. Obstructive airway diseases in infants and children. In: Neonatal and Pediatric Respiratory Care. Beck Koff Eitzman and Neu. 2nd Edition. Mosby, St Louis. 1993; 102-127

Jain S. Ventilation of patients with asthma and obstructive lung disease. Critical Care Clinics 1998; 14(4):685-705

Cox RG. Efficacy, results and complications of mechanical ventilation in children with status asthmaticus. Pediatr Pulmonol 1991; 11:120-126

Marraro G. Airway management. In Bissonnette B, Dalens BJ eds. "Principles and practice of pediatric anesthesia" McGraw-Hill Comp. New York 2001; pp778-814

Tuxen DV. The effects of ventilatory pattern on hyperinflation, airway pressures and circulation in mechanical ventilation of patients with severe air-flow obstruction. Am Rev Respir Dis 1987; 136:872-879

Tuxen DV. Use of measurement of pulmonary hyperinflation to control the level of mechanical ventilation in patients with acute severe asthma. Am Rev Respir Dis 1992; 146:1136-1142

Slutsky AS. Mechanical ventilation. Chest 1993; 104:1833-1859

Darioli R. Mechanical controlled hypoventilation in status asthmaticus. Am Rev Respir Dis 1984; 129(3):385-387

Hicking K, Joyce C. Permissive hypercapnia in ARDS and its effect on tissue oxygenation. Acta Anaesthesiol Scand 1995; 39:201-208

Feihl F, Perret C. Permissive hypercapnia: how permissive should we be? Am J Respir Crit Care Med 1994; 150:1722-1737

Tuxen D. Permissive hypercapnia ventilation. Am J Respir Crit Care Med 1994; 150:870-874

Marraro GA. Innovative practices of ventilatory support with pediatric patients. Pediatric Critical Care Medicine 2003; 4(1): 8-20

Appel D, Karpel JP, Sherman M. Epinephrine improves expiratory flow rates in patients with asthma who do not respond to inhaled metaproterenol sulfate. J Allergy Clin Immunol 1989 Jul; 84(1): 90-98

Dreyfuss D, Saumon G. Ventilator-induced lung injury:lesson from experimental studies [State of the Art]. Am J Respir Crit Care Med 1998; 157:294-330

Lachmann B. Open up the lung and keep the lung open. Intensive Care Med. 1992; 18(6):319-321

Qvist J. High-level PEEP in severe asthma. N Engl J Med 1982; 18:1347-1348

Shivaram U. Cardiopulmonary responses to continuous positive airway pressure in acute asthma. J Crit Care 1993; 8:87-92

Smith TC. Impact of PEEP on lung mechanics and work of breathing in severe airflow obstruction. J Appl Physiol 1988; 65:1488-1499

Ranieri VM. Physiologic effects of positive end-expiratory pressure in patients with chronic obstructive pulmonary disease during acute ventilatory failure and controlled mechanical ventilation. Am Rev Respir Dis 1993; 147:5-13

Bierman CW. Pneumomediastinum and pneumothorax complicating asthma in children. Am J Dis Child 1967; 114:43-47

Van der Touw T. Cardiorespiratory consequences of expiratory chest wall compression during mechanical ventilation and severe hyperinflation. Crit Care Med 1993; 21:1908-1914

Rock MJ. Use of ketamine in asthmatic children to treat respiratory failure refractory to conventional therapy. Crit Care Med 1986; 14(5):514-516

Strube PJ. Ketamine by continuous infusion in status asthmaticus. Anesthesia 1986; 41(10):1087

L'Hommedieu CS, Arens JJ. The use of ketamine for the emergency intubation of patients with status asthmaticus. Ann Emerg Med 1987; 16:568-71

Bierman MI. Prolonged isoflurane anesthesia in status asthmaticus. Crit Care Med 1986; 14:832-833

Saulnier FF. Respiratory and hemodynamic effects of halothane in status asthmaticus. Intensive Care Med 1990; 16:104-107

Bloch H, Silverman R, Mancherje N, et al. Intravenous magnesium sulfate as an adjunct in the treatment of acute asthma. Chest 1995; 107(6): 1576-81

Kass JE, Castriotta Rj. Heliox therapy in acute severe asthma. Chest 1995; 107:757-760

Downey P, Cox R.. Update on the management of status asthmaticus. Curr Opin Pediatr 1996; 8:226-233

Gluck EH, Onorato DJ, Castriotta R. Helium-oxygen mixtures in intubated patients with status asthmaticus and respiratory acidosis. Chest 1990; 98:693-698

Manthous CA, Hall JB, Caputo MA, et al. Heliox improves pulsus paradoxus and peak expiratory flow in non intubated patients with severe asthma. Am J Respir Crit Care Med 1995; 151:310-314

Rishani R, El-Khatib M, Mroueh S. Treatment of severe status asthmaticus with nitric oxide. Pediatr Pulmonol 1999; 28(6):451-453

Luchetti M, Casiraghi G, Valsecchi R, Marraro G. Porcine-derived surfactant treatment of severe bronchiolitis. Acta Anaesthesiol Scand 1998; 42(7):805-810

Luchetti M, Gallini C, Pigna A, et al. Multicenter, randomized, controlled study of porcine surfactant in severe respiratory syncytial virus-induced respiratory failure. Pediatric Critical Care Medicine 2002; 3(3): 261-268

Tibby SM, Hatherill M, Wright SM, et al. Exogenous surfactant supplementation in infants with respiratory syncytial virus bronchiolitis. Am J Resp Crit Care Med 2000; 162:1251-1256
Marraro GA. Surfactant use in severe bronchiolitis. Biol Neonate 2002; 81 (suppl 1):28
Kurashima K, Ogawa H, Ohka T, et al. A pilot study of surfactant inhalation in the treatment of asthmatic attack. Aerugi (Jpn J Allergol) 1991, 40:160-163
Heeley EL, Hohlfeld JM, Krug N, Postle AD. Phospholipid molecular species of bronchoalveolar lavage fluid after local allergen challenge in asthma. Am J Physiol (Lung Cell Mol Physiol) 2000, 278:L305-L311
Fuchimukai T, Fujiwara T, Takahashi A, Enhorning G. Artificial pulmonary surfactant inhibited by proteins. J Appl Physiol 1987, 62:429-437
Seeger W, Grube C, Günther A, Schmidt R. Surfactant inhibition by plasma proteins: differential sensitivity of various surfactant preparations. Eur Respir J 1993, 6:971-977
Liu M, Wang L, Li E, Enhorning G. Pulmonary surfactant given prophylactically alleviates an asthma attack in guinea-pigs. Clin Exp Allergy 1996, 26:270-275
King D. Extracorporeal membrane oxygenation as emergency treatment for life-threatening acute severe asthma. Postgraduate Med J 1986; 62:855-857

AFFILIATIONS

Giuseppe A. Marraro, MD works at the Anesthesia and Intensive Care Department, Pediatric Intensive Care Unit, Fatebenefratelli and Ophthalmiatric Hospital, Milan, Italy
E-mail: gmarraro@picu.it

GERRIT BORCHARD

UNDERSTANDING THE MECHANISMS OF LOCAL PULMONARY DRUG ABSORPTION AND METABOLISM: AN IN VITRO MODEL OF THE AIRWAY EPITHELIUM

1. INTRODUCTION

Following the image created by John Patton [1], the trachea and bronchi of the lung may be compared to the stem and branches of a tree. Their function, of course, is not limited to supporting and supplementing the tissues of gas-exchange (leafs and alveoli, respectively); the conducting airways are also "more than just a barrier", as recently stated by Folkerts and Nijkamp [2]. With every breath, the airway epithelium is exposed to the environmental air, and has to cope with airborne microorganisms and pollutant particles, which have to be filtered out before reaching the lower respiratory tract. Throughout evolution, defense mechanisms have been developed to face this challenge. These include the production and secretion of mucus, which entraps foreign airborne particles, and is expelled by the action of epithelial cilia (mucociliary escalator). A range of degrading enzymes and antibacterial agents are found in the airway surface liquid (ASL), complementing the effect of the mucociliary escalator. As a mucosal tissue, the airway epithelium is also part of the mucosal associated lymphoid tissue (MALT) [3].

The airway epithelium and its underlying tissues are the site of a number of severe diseases, such as asthma, chronic obstructive pulmonary diseases (COPD), infections and genetic dysfunctions such as cystic fibrosis (CF). It is not surprising that the majority of pharmaceuticals delivered to the lung are targeted at the upper airways for the therapeutic approach of these diseases. In addition, however, the pulmonary route of drug delivery has found considerable attention amongst pharmaceutical scientists as an alternative to parenteral application of macromolecular drugs, such as peptides and proteins. This route of delivery has not been fully optimized yet, so that only a mere 20% of the administered dose reaches the alveolar compartment of the deep lung [4]. A large percentage of the dose is therefore deposited at the level of the conducting airways, which may raise safety concerns towards pharmacologically active compounds as well as excipients used in the aerosol formulation.

In order to assess absorption, metabolism and probable toxicity of drugs and excipients at the epithelial surfaces, cell cultures have been developed that serve as in vitro test models. These models have to be characterized thoroughly, and results

L. Gradoń and J. Marijnissen (eds.), Optimization of Aerosol Drug Delivery, 295–307.

obtained in vitro must be verified in vivo as much as possible (in vitro/in vivo-correlation). Although naturally limited to a certain extent, these models offer the possibility to closely examine drug absorption and metabolism mechanisms at a cellular, or even sub-cellular level [5]. With regard to the airway epithelium of the lung, drug absorption and metabolism have been studied along with non-viral gene delivery using a small number of cell culture models. This chapter will provide a brief summary on studies done with one of the most important cell cultures, the human submucosal cell line Calu-3.

In vitro models of the bronchial epithelium

Data on the permeability of drugs at the airway epithelium from *in vivo* studies are scarce, because of the inaccessibility and the delicate structure of the lung. In order to predict absorption *in vivo*, cell cultures modeling absorptive epithelia were shown to be useful tools (table 1). Until recently, merely a few drug transport studies involving airway epithelial cell cultures appeared in literature [6]. These studies involved both established cell lines, such as human 16HBE14o- [6] and Calu-3 [7,8], and primary cell cultures derived from a number of species, such as rabbits [9], guinea pig [10], rat [11], hamster [12] and human [13]. While cell lines do not afford painstaking isolation procedures and suffer from occasionally low availability, they may not present features (receptor expression, gene products, enzyme production, etc.) of the epithelium *in vivo*.

Primary cell cultures of the airway epithelium were shown to better resemble the *in vivo* situation, featuring tight cell layers, apical cilia, and mucus production under certain culture conditions [9]. However, as primary cultures are prone to infection with bacteria or mycoplasms, specific pathogen-free animals should be the preferable source of material for isolation of primary cells.

Human Calu-3 cells as an in vitro model of the airway epithelium

Several cell culture models of the bronchial epithelium, such as explant outgrowth cultures [14], primary cell cultures [9,10, 11] and cell lines [13], have been described. Among the cell cultures representing the airway epithelium, Calu-3 cells appear to be one of the best-suited models for drug transport and metabolism studies [15]. Calu-3 is an adenocarcinoma cell line, derived from a 25-year old Caucasian and established by Fogh in 1975 [16]. When seeded onto microporous membranes, Calu-3 cells form confluent monolayers interconnected by tight intercellular junctions [17]. This was shown by staining of the ZO-1 protein, which is associated with functional epithelial tight junctions [18]. Calu-3 cells also express desmosomes and zonulae adherentes, as detected by immunostaining of desmoplakin and E-cadherin, respectively [19].

Measurement of transepithelial resistance (TEER) is routinely used to determine integrity and tightness of epithelial monolayers in vitro. When TEER was measured using a Voltohmmeter equipped with chopstick electrodes, maximum values between 400 Ωcm^2 [20] and 600Ω cm^2 [21] were reported for Calu-3 cells. Development of TEER is dependent on the differentiation status of the cell

monolayer, with maximum TEER values leveling off between day 10 and 14 after seeding onto filter membranes coated with collagen.
In addition to TEER, the measurement of paracellular permeability of specific marker substances supplies an overall view of the tightness of the cell monolayer. The paracellular permeability of hydrophilic markers, such as radiolabeled mannitol or fluorescently labeled dextrans, is expressed as the apparent permeability (P_{app}) of the marker substance, according to the following equation:
Eq. 1: $P_{app} = dQ/dt /(60 \times A \times C_0)$
With DQ/dt : amount marker transported with time; 60: conversion factor(min->sec); A: surface area of filter; C_0: initial mannitol concentration in donor.

Typical values for P_{app} of ^{14}C-mannitol in the range of 10^{-7} cm sec^{-1} [8,19] show that Calu-3 cell monolayers represent fairly tight barriers against solute permeability. This barrier function, paralleling the diseased state of the airway epithelium in vivo, may be challenged by a number of factors. In a recent study, the mechanism of airway epithelial barrier function and wound healing in asthma [22] was examined using Calu-3 cells. It was shown that the epithelial barrier function of tight junctions between Calu-3 cells is disrupted by pro-inflammatory Th2-type cytokines IL-4 and IL-13. The cytokines bound to the IL-4 receptor complex and IL-13 receptor, respectively, which lead to an increase in paracellular mannitol flux, decrease in TEER and inhibition of Calu-3 cell migration. On the other hand, cell migration and barrier function were enhanced by Th1-type cytokine IFN-□. Another report examined the effect of the major allergen fraction of house dust mite on the permeability of Calu-3 and 16HBE14o$^-$ cell monolayers [17]. These studies indicated that the Calu-3 cell line might be a useful model to examine the pathogenesis of inflammatory and allergic pulmonary diseases such as asthma.
Culture conditions have a considerable impact on the differentiation of airway epithelial cells and have therefore to be optimized for each cell model individually. Both submerged (i.e. culture medium present on both sides of the cell monolayer) and air interface (i.e. removal of apical culture medium one day after seeding) culture conditions have been described for Calu-3 cells [8]. In particular, ciliogenesis on the apical (mucosal) side of the Calu-3 cell layer is influenced by the incubation method applied. In an air interface cell culture, about 30% of the cells in a Calu-3 monolayer develop apically located cilia. Under submerged conditions, however, these structures appear to be shorter and thicker than under air interface conditions, rendering the latter culture conditions more favorable. Coordinated ciliary activity would allow the modeling of the mucociliary escalator as one of the major defense mechanisms of the conducting airways. Its presence in Calu-3 cells, however, has not been verified yet.

Calu-3 cells produce and secrete mucus-like glycoproteins
Intracellular mucus vesicles in Calu-3 cells were detected by specific staining. In a number of studies Calu-3 cells are reported to secrete mucins as well as antibacterial proteins. Berger et al. [23] examined six different lung carcinoma cell lines (Calu-3, A549, NCI H292, Calu-6, RPMI 2650, A-427) with respect to expression of

members of the *MUC* gene family as well as mucin biosynthesis and secretion. In human adult respiratory tissue, *MUC*1, *MUC*2, *MUC*4, *MUC*5, *MUC*5B, *MUC*7 and *MUC*8 are expressed on messenger RNA level [24]. It was found that Calu-3 cells express two different transcripts of *MUC*1, reflecting *MUC*1 allele variability, as also observed in primary bronchial and nasal cells [25]. *MUC*4 and *MUC*5B mRNA is expressed at low and *MUC*5 at high levels. *MUC*5 gene expression is found predominantly in goblet cells of normal respiratory tissue [26], and in metastasizing non-small-cell lung cancer [27].

In order to correlate *MUC* gene expression to actual mucin biosynthesis, secretions of the cell lines were analyzed. All of the six cell lines tested secreted proteoglycans, however, only three cell lines (A549, H292 and Calu-3) secreted sulfated mucins. Some of the glycoproteins secreted by Calu-3 cells were identified as MUC5 mucins by immunoprecipitation [23]. The Calu-3 and A549 cell lines were suggested as suitable tools to study regulation of mucin secretion and gene expression in lung diseases. However, as A549 cells do not form confluent and tight monolayers [17], and considering other features of Calu-3, this cell line might be the better choice for an in vitro model of the airway epithelium.

Mucus is part of the airway surface liquid (ASL) covering the airway epithelium. ASL contains antibacterial agents such as □-defensin-1 and -2 [28,29], cathelicidin LL-37 [30], lactoferrin and lysozyme [31]. Due to an increased ASL salt concentration, antibacterial activity is reduced in cystic fibrosis patients [32], promoting bacterial colonization of the airways by opportunistic pathogens [33]. Bacterial infection leads to chronic inflammation in cystic fibrosis lungs, which successively causes epithelial destruction and airway obstruction. Calu-3 cells have been employed to study bacterial binding to airway epithelial cells, and regulation of antibacterial activity of ASL. Zhang and co-workers [34] recently showed that ASL secreted from Calu-3 cells exerted salt-sensitive bactericidal activity against *Escherichia coli* and *Pseudomonas aeruginosa*. Antibacterial activity was enhanced by incubation with the glucocorticosteroids budesonide (most pronounced), triamcinolone, or beclomethasone, whereas presence of □$_2$-agonists reduced antibacterial activity. By Reverse Transcriptase/Polymerase Chain Reaction (RT-PCR) assay of the isolated total Calu-3 cell RNA, cDNA of antibacterial compounds □-defensin-1 and 2, cathelicidin LL-37, lactoferrin, and lysozyme was obtained.

Another study reported that *P. aeruginosa* bound to a very small extent (<1 *P. aeruginosa*/100 cells) to apical surfaces of confluent Calu-3 and primary bovine tracheal cell monolayers [35]. Binding of bacteria to cells at free edges of sub-confluent or injured monolayers was increased significantly, due to the accessibility of basolateral membranes. This was also verified in monolayers treated with EGTA, which caused opening of intercellular tight junctions and access to basolateral membranes. However, the actual binding site of *P. aeruginosa* on Calu-3 cells, which appears to be located between adjacent cells, remained undetermined. Another study showed that infection of Calu-3 cells with *P. aeruginosa* upregulated expression levels of matrilysin, a metalloproteinase involved in the innate host defense (activation of □-defensins) and re-epithelialization of mucosal tissues after injury [36]. Further studies involving Calu-3 cells may clarify the role of matrilysin in the balance between injury and repair in the CF lung.

Calu-3 cells show metabolic activity

Metabolic activity in the airways has a large impact on the activity and retention time of locally applied therapeutics. Lipophilic compounds, such as inhaled glucocorticosteroids (GCS) are prone to metabolization by isoforms of the pulmonary cytochrome P450 (P450 or CYP) enzyme system. Major CYP450 activity is found in Clara cells [37], Type II pneumocytes and alveolar macrophages [38]. A recent study by Lee et al. [39] in rats and rhesus monkeys demonstrated regiospecific and species-specific differences in the activities of pulmonary P450 isoforms. In addition, the inducibility of pulmonary P450s in the rat lung was clearly shown to be dependent on the enzyme isoform and its location.
The report showed that CYP1A1 and CYP2B dealkylase activities were highest in rat parenchyma, whereas CYP2E1 hydroxylase activity was highest in the airways. Rat CYP1A1 activity was induced by □-naphthoflavone in the airways and lung parenchyma, whereas CYP2B and CYP2E1 activities were not induced by oral application of phenobarbital. P450 reductase activities were similar in airways and parenchyma and were lower in trachea. In monkeys, no significant site-selective differences in CYP1A1 and CYP2B1 activities were found. In contrast, CYP2E1 activity was higher in the distal bronchioles and parenchyma than in the proximal airways, P450 reductase activities were similar in microsomes prepared from all subcompartments of monkey lung.
In a study on the activity of CYP450 isoforms in rabbits about 90% of total CYP450 activity was accounted to CYP2B1 and CYP4B1 isoforms [40]. In primary human bronchial epithelial cells basal expression of mRNAs coding for CYP2B7, CYP2F1 and CYP2E1 were detected by RT-PCR analysis [41]. CYP2E1protein content was low and in most cultures of generations 2-5 could not be detected by immunoblot analysis of native protein extracts. Nevertheless, in some cases immunoreactive CYP2E1 protein was present in monolayers obtained even after several passages after 18 weeks. Treatment with ethanol increased CYP2E1 activity by 5-fold, showing differences for individual donors.
In another study [42] the expression and localization of CYP2E1 in primary pulmonary carcinomas and surrounding normal bronchial tissue was examined. Seromucous glands and bronchial epithelial cells showed expression of CYP2E1. Whereas no association was observed between the expression of CYP2E1 in tumour tissue and normal bronchial tissue, there was a significant correlation between the expression of CYP2E1 in seromucous glands and bronchial epithelium of normal tissue.
The presence and inducibility of CYP enzymes in Calu-3 cells in vitro was recently investigated by Foster et al. [21]. The authors found evidence for the presence of CYP1A1 and CYP2B, but not for CYP3A4 isoenzymes. CYP1A1 activity in Calu-3 cells was comparable to the activity found in A549 cells, a human Type II pneumocyte cell line, and in human lung microsomes. CYP2B activity, however, was lower than CYP2B1 activity found in rabbit microsomes. Both isoenzymes

were not inducible by pretreatment with micromolar concentrations of 3-methylcholanthrene in combination with dexamethasone and β-naphthoflavone (CYP1A1), or phenobarbitol (CYP2B).

ATP-dependent transport of drug metabolites in Calu-3 cells

Research aimed at identifying membrane proteins mediating the ATP-dependent transport of drug conjugates with glutathione, glucuronate, or sulphate has led to the identification of the multidrug resistance-associated proteins (MRP). MRP's belong to and share genetic homologies with a family of ATP-dependent membrane transporters, which includes P-glycoprotein and cystic fibrosis transmembrane regulator (CFTR) protein [43]. Both are also expressed in Calu-3 cells. MRP2 is the apical MRP isoform, whereas MRP1 and MRP3 are both basolaterally expressed. The substrate spectrum of MRP1 is similar to that of MRP2, comprising amphiphilic anions, particularly conjugates of lipophilic compounds with glutathione, glucuronate, or sulfate, including cysteinyl leukotrienes, bilirubin glucuronosides, 17β-glucuronosyl estradiol, and sulfatolithocholyl-taurine as endogenous compounds [44,45].

Evidence for the presence of MRP1 in Calu-3 cells was shown by Hamilton et al. [46]. Reverse transcription-polymerase chain reaction revealed *MRP1* gene expression, and the gene product was identified by indirect immunofluorescence at the basolateral membranes of Calu-3 cells.

Indirect evidence for the presence of MRP1 at the basolateral membrane of Calu-3 cells was obtained by our group upon studying the mechanism of estradiol transport across Calu-3 cell monolayers [47]. Previous pharmacokinetic studies described the metabolism of 17β-estradiol [48,49]. The major metabolic pathway is the oxidation of estradiol to estrone and successive conjugation. The ratios of estrone to estradiol obtained after nasal administration are higher than those obtained after i.v. administration, indicating that estradiol undergoes metabolism to estrone in the nasal cavity. The examination of intracellular estradiol metabolism and MRP-mediated transport in Calu-3 cells showed evidence of oxidation of estradiol to estron in Calu-3 cells. This correlates with the concept of uptake of estradiol via the organic anion transporting peptide (oatp1) receptor and its intracellular biotransformation in Calu-3 cells. Other metabolites of estradiol were found to be present in the acceptor chamber after the transport experiments.

In summary, evidence exists that major metabolic pathways present in the lung are also expressed and (partially) inducible in Calu-3 cells.

Multidrug-resistance efflux system activity in Calu-3 cells

The presence of P-glycoprotein (Pgp) efflux pumps in mammalian lung tissue, which might play a key role in the active transport and retention of GCS in the airway epithelial tissue, has previously been demonstrated [50-52]. As members of the ATP-Binding Cassette (ABC) superfamily, P-glycoproteins are ATP-dependent drug efflux pumps initially identified as proteins that confer multidrug resistance (MDR) [53]. Based on the observation that Pgp is expressed in the apical membrane of mucosal cells in the intestine, in the brush border of the renal proximal tubules, in

the biliary membrane of hepatocytes and in capillary endothelial cells of brain and testes, a role for Pgp in the protection of the organism from xenobiotics has been proposed [54].
In a study examining the interaction of glucocorticosteroids with P-gp [46], however, an effect on the accumulation of a known P-gp substrate, rhodamine 123 (Rh123) into Calu-3 cells was not observed at BUD concentrations lower than 100 μM (about 2.4-fold increase compared to the untreated control). The most potent inhibitor of P-gp efflux resulting in increased Rh123 accumulation was progesterone (4.0-fold increase) at a concentration of 100 μM. Another GCS, flunisolide, showed an intermediate effect on Rh123 accumulation of merely 1.6-fold increase compared to the untreated control cell monolayers. Transport of flunisolide across Calu-3 cell monolayers was further examined by Florea et al. [20], revealing a polarised transport in the apical (luminal) to basolateral (serosal) direction. The results of this study showed that flunisolide is transported unmetabolised across Calu-3 cell monolayers. The polarised transport of flunisolide across Calu-3 cell monolayers was ATP dependent and susceptible to inhibition by specific Pgp inhibitors. In contrast to these studies, Hamilton et al. [55] showed by immunofluorescent staining that P-gp is predominantly expressed on the apical side of Calu-3 cells. This would suggest that the GCS flunisolide is transported by a Pgp-related transporter at the basolateral side of Calu-3 cell monolayer, which is modulated by P-gp inhibitors.

DRUG TRANSPORT STUDIES

Antibiotics

Observations from pharmacokinetic studies in humans showed that the fluoroquinolone antibiotic ciprofloxacin diffuses well into lung tissues. Drug levels comparable to serum concentrations in pulmonary fluid (sputum, bronchial secretions, saliva) were found even twelve hours after application [56]. The exact mechanism by which fluoroquinolones gain access to pulmonary tissues remained unknown. As one of the first drug transport studies in Calu-3 cells, Cavet et al. [57] therefore examined the transport of ciprofloxacin. They showed that the antibiotic is transported across Calu-3 cell monolayers primarily via the transcellular route. In contrast to the mucosal secretion pathway of ciprofloxacin in Caco-2 cells, the major part of transport of the antibiotic across Calu-3 cells appeared to be due to passive diffusion, and to a smaller extent to low-affinity mediated transport at both the apical and basolateral cell membranes. This mechanism was found to be in accordance with pharmacokinetic studies in humans. P_{app} values of ciprofloxacin across Calu-3 cell monolayers grown under submerged conditions were in the order of 7 x 10^{-6} cm sec^{-1} [57]. This single stduy reports the examination of transport of an antibiotic in the Calu-3 cell culture. Testing of other antibiotics used for infectious dieseases of the lung, e.g., tobramycin in the treatment of bystander infections in cystic fibrosis, appear to be appropriate.

Steroids

Inhaled corticosteroids (GCS) such as budesonide and fluticasone are therapeutics of first choice in the treatment of asthma, an inflammatory disease afflicting the airways. Favorable properties of these GCS with regard to lowering systemic side effects are the high hepatic first-pass extraction along with a combination of high receptor affinity and prolonged retention at the airway epithelium [58]. Increased retention time of the GCS budesonide (BUD) was found to be due to intracellular coupling to fatty acids to the hydroxyl group at the C_{21}-position [59,60]. These fatty esters of BUD form an intracellular depot airway epithelial cells, from which BUD is slowly released by the action of lipase. Our group found evidence for the presence of this metabolic pathway of BUD in Calu-3 cells [61]. Calu-3 cells incubated with 30 □M BUD solution for two hours retained the GCS at an average of 5.5 □g BUD/mg cell protein. In successive release experiments, BUD was secreted from the cells over a time period of ten hours, with a significantly higher amount being released to the mucosal side of the cell monolayer. BUD-oleate was identified as one major conjugate in Calu-3 cells by mass spectroscopy. In contrast to BUD, fluticasone propionate, which due to its structure cannot undergo coupling to fatty acids, was immediately released from Calu-3 cells. The apparent permeability of BUD across Calu-3 cell monolayers was determined to be in the range of 5×10^{-6} cm sec^{-1}. No significant difference was detected between P_{app} for apical-to-basolateral (ap-bl) and bl-ap directions, although BUD was shown to interact with an efflux pump, P-glycoprotein [46].

Amino acids, peptides and proteins

Merely a few studies on the transport of amino acids, peptides and proteins across Calu-3 cell monolayers exist in the present literature. A recent study by Mathias et al. [62] showed evidence for the active and energy-dependent transport of the amino acid leucine and the dipeptide glycyl-sarcosine (Gly-Sar). Leucine uptake into epithelial cells was shown to be both sodium-independent and, to a smaller extent, sodium-dependent in Caco-2 cells [63]. A similar mechanism may be present in Calu-3 cells, as well, however, this will need further functional assessment. Gly-Sar is a substrate for the proton-coupled peptide transporter (PepT1) that mediates the absorption of di- and tripeptides [64]. While PepT1 is found in the intestinal epithelium PepT2 is primarily expressed in tissues of the kidneys, lung, and in the central nervous system [65]. PepT1 and 2 show distinct differences in substrate affinity and substrate/H^+ stoichiometry [65]. As recent studies revealed that PepT2 transporters are present in alveolar type-II pneumocytes and within the respiratory epithelial cells in main bronchi [66], the nature of transport of Gly-Sar and of other substrates in Calu-3 cells should be further investigated in the future, especially with respect to establishing Calu-3 cells as a screening tool for pulmonary (peptide) drug absorption.

A single study examined the transport of insulin across Calu-3 cell monolayers [67]. In this study, a direction-dependent transport (basolateral-to apical) was noted for zinc-insulin, but not for a human monomeric insulin analog. The authors suggested this to be due to formation of insulin oligomers at the apical side of the Calu-3 cell monolayers, decreasing paracellular transport of these aggregates. The filtering effect of tight junctions located at the apical membrane of epithelial cells limiting the paracellular transport of macromolecules such as insulin, however, should be of equal effect in both directions. As the authors excluded active transport of insulin by insulin receptors, whose presence at both apical and basolateral membranes of Calu-3 cells were shown, they interpreted the directed transport of insulin as the result of the lower concentration of insulin, and thus a lower aggregation tendency in the paracellular space during basolateral-to apical transport experiments. Another explanation, also mentioned in the study, would be the presence of mucin molecules at the apical, but not at the basolateral, side of Calu-3 cells causing oligomerisation of insulin.

CONCLUSIONS

To understand absorption and metabolism of drugs applied locally to the lung, in vitro test systems like the Calu-3 cell line are valuable tools. By correlating data from in vitro test to in vivo studies, the predictive potential of the model is confirmed. Moreover, the combination of both approaches is bound to create synergistically new knowledge with regard to the improvement of aerosol delivery of pharmacologically active agents.

ACKNOWLEDGMENTS

María Luz Cassará, Clare Meaney and Bogdan I. Florea are acknowledged for fruitful discussions and experimental work that lay the scientific basis for this chapter.

REFERENCES

1. PATTON J.S. – Mechanism of macromolecule absorption by the lungs [Review]. – Adv. Drug Del. Rev., **19**, 3-36, 1996.
2. FOLKERTS G., NIJKAMP F.P. – Airway epithelium: more than just a barrier! [Review]. – Trends Pharmacol. Sci., **19**, 334-341, 1998.
3. MESTECKY J. – The common mucosal immune system and current strategies for induction of immune responses in external secretions. – J. Clin. Immunol., **148**, 265-276, 1987.
4. PATTON J.S., PLATZ R.M. – Routes of delivery: case studies. Pulmonary delivery of peptides and proteins for systemic action. – Adv. Drug Del. Rev., **8**, 179-196, 1992.
5. BORCHARDT R.T. - The application of cell culture systems in drug discovery and development. - J. Drug Target., **3**, 179-182, 1995.
6. FORBES B., LANSLEY A.B. - Transport characteristics of formoterol and salbutamol across a bronchial epithelial drug absorption model. - Eur. J. Pharm. Sci., **6**, S24, 1998.
7. CAVET M.E., WEST M., SIMMONS N.L. - Transepithelial transport of the fluoroquinolone ciprofloxacin by human airway epithelial Calu-3 cells. -Antimicrob. Agents Chemotherap., **41**,2693-2698,1997.

8. MEANEY C., FLOREA B.I., BORCHARD G., JUNGINGER H.E. - Characterization of a human submucosal gland cell line (Calu-3) as an in vitro model of the airway epithelium. - Proceed. Int'l. Symp. Cont. Rel. Bioact. Mater., **26**, 198-199, 1999.
9. MATHIAS N.R., KIM K.-J., LEE V.H.L. - Targeted drug delivery to the respiratory tract: Solute permeability of air-interface cultured rabbit tracheal epithelial cell monolayers. - J. Drug Target., **4**, 79-86, 1996.
10. ADLER K.B., CHENG P.W., KIM K.C. - Characterization of guinea pig tracheal epithelial cell maintained in biphasic organotypic culture: cellular composition and biochemical analysis of released glycoconjugates. - Am. Respir. Cell Mol. Biol., **2**, 145-154, 1990.
11. CLARK A.B., RANDELL S.H., NETTESHEIM P., GRAY T.E., BAGNELL B., OSTROWSKI L.E. - Regulation of ciliated cell differentiation in cultures of rat tracheal epithelial cells. - Am. J. Respir. Cell Mol. Biol., **12**, 329-338, 1995.
12. LEE T.C., WU R., BRODY A.R., BARRETT J.C., NETTESHEIM P. - Growth and differentiation of hamster tracheal epithelial cells in vitro. - Exp. Lung Res., **6**, 27-45, 1987.
13. GRUENERT D.C., FINKBEINER W.E., WIDDICOMBE J.H. - Culture and transformation of human airway epithelial cells. - Am. J. Physiol., **286**, L347-360, 1995.
14. WIESEL J.M., GAMIEL H., VLODAVSKY I., GAY I., BEN-BASSAT H. – Cell attachment, growth characteristics and surface morphology of human upper-respiratory tract epithelium cultured on extracellular matrix. – Eur. J. Clin. Invest., 13, 57, 1983.
15. FORBES B. – Human airway epithelial cell lines for in vitro drug transport and metabolism studies. – PSTT, **3**, 18-27, 2000.
16. FOGH J., FOGH J.M., ORFEO T. – One hundred and twenty-seven cultured human tumor cell lines producing tumors in nude mice. – J. Natl. Cancer Inst., **59**, 221-225, 1977.
17. WINTON H.L., WAN H., CANNELL, M.B., GRUENERT D.C., THOMPSON P.J., GARROD D.R., STEWART G.A., ROBINSON C. - Cell lines of pulmonary and non-pulmonary origin as tools to study the effects of house dust mite proteinases on the regulation of epithelial permeability. – Clin. Exp. Allergy, **28**, 1273-1285, 1998.
18. FURUSE M., ITOH M., HIRASE T., NAGAFUCHI A., YONEMURA S., TSUKITA S. – Direct association of occludin with ZO-1 and its possible involvement in the localization of occludin at tight junctions. – J. Cell Biol., **127**, 1617-1626, 1994.
19. WAN H., WINTON H.L., SOELLER C., STEWART G.A., THOMPSON P.J., GRUENERT D.C., CANNELL M.B., GARROD D.R., ROBINSON C. – Tight junction properties of the immortalized human bronchial epithelial cell lines Calu-3 and 16HBE14o-. – Eur. Resp. J., **15**, 1958-1068, 2000.
20. FLOREA B.I., VAN DER SANDT I.C.J., SCHRIER S.M., KOOIMAN K., DERYCKERE K., DE BOER A.G., JUNGINGER H.E., BORCHARD G. – Evidence of P-glycoprotein mediated apical to basolateral transport of flunisolide in human broncho-tracheal epithelial cells (Calu-3). – Brit. J. Pharmacol., **134**, 1555-1563, 2001.
21. FOSTER K.A., AVERY M.L., YAZDANIAN M., AUDUS K.L. - Characterization of the Calu-3 cell line as a tool to screen pulmonary drug delivery. – Int. J. Pharm., **208**, 1-11, 2000.
22. AHDIEH M., VANDENBOS T., YOUAKIM A. – Lung epithelial barrier function and wound healing are decreased by IL-4 and IL-13 and enhanced by IFN-□. – Am. J. Physiol. Cell Physiol., **281**, C2029-C2038, 2001.
23. BERGER J.T., VOYNOW J.A., PETERS K.W., ROSE M.C. – Respiratory carcinoma cell lines. *MUC* genes and glycoconjugates. – Am. J. Respir. Cell Mol. Biol., **20**, 500-510, 1999.
24. ROSE M.C., GENDLER S.J. – Airway mucin genes and gene products. In: Airway mucus: Basic mechanisms and clinical perspectives. D. Rogers, M. Lethem Eds., Birkhauser Publishing Limited, Basel, CH, 1997, pp. 41-66.
25. HOLLINGSWORTH M.A., BATRA S.K., QI W., YANKASKAS J.R. – *MUC*1 mucin mRNA in cultured human nasal and bronchial epithelial cells. – Am. J. Respir. Cell Mol. Biol., **6**, 516-520, 1992.
26. M.R., MADSEN C.S., JENNINGS A., KASPERBAUER J.L., GENDLER S.J. – Mucin mRNA expression in normal and vasomotor inferior turbinates. – Am. J. Rhinology, **11**, 293-302, 1997.
27. YU C., YANG P.C., SHEW J.H., HONG T.M., LEE Y.C., LEE L.N., LUH K.T., WU C.W. – Mucin mRNA expression in lung adenocarcinoma cell lines and tissues. – Oncology, **53**, 118-126, 1996.

28. MCCRAY P.B., BENTLEY L. – Human airway epithelia express a beta-defensin. – Am. J. Respir. Cell Mol. Biol., **16**, 343-349, 1997.
29. BALS R., WANG X., WU Z., FREEMAN T., BAFNA V., ZASLOFF M., WILSON J.M. – Human beta-defensin 2 is a salt-sensitive peptide antibiotic expressed in human lung. – J. Clin. Invest., **102**, 874-880, 1998.
30. BALS R., WEINER D.J., MEEGALLA R.L., WILSON J.M. – Transfer of a cathelicidin peptide antibiotic gene restores bacterial killing in a cystic fibrosis xenograft model. – J. Clin. Invest., **103**, 1113-1117, 1999.
31. COLE A.M., DEWAN P., GANZ T. – Innate antimicrobial activity of nasal secretions. Infect. Immun., **67**, 3267-3275, 1999.
32. SMITH J.J., TRAVIS S.M., GREENBERG E.P., WELSH M.J. - Cystic fibrosis airway epithelia fail to kill bacteria because of abnormal airway surface fluid. – Cell, **85**, 229-236, 1996.
33. RENDERS N., VERBRUGH H., VAN BELKUM A. - Dynamics of bacterial colonization in the respiratory tract of patients with cystic fibrosis. – Infect. Genet. Evol., **1**, 29-39, 2001.
34. ZHANG Y., REENSTRA W.W., CHIDEKEL A. – Antibacterial activity of apical surface fluid from the human airway cell line Calu-3. Pharmacologic alteration by corticosteroids and β_2-agonists. – Am. J. Respir. Cell Mol. Biol., **25**, 196-202, 2001.
35. LEE A., CHOW D., HAUS B., TSENG W., EVANS D., FLEISZIG S., CHANDY G., MACHEN T. – Airway epithelial tight junctions and binding and cytotoxicity of *Pseudomonas aeruginosa*. – Am. J. Physiol., **277**, L204-L207, 1999.
36. LÓPEZ-BOADO Y.S., WILSON C.L., PARKS W.C. – Regulation of matrilysin expression in airway epithelial cells by *Pseudomonas aeruginosa* flagellin. – J. Biol. Chem., **276**, 41417-41423, 2001.
37. JI C.M., CARDOSO W.V., GEBREMICHAEL A., PHILPOT R.M., BUCKPITT A.R., PLOPPER C.G., PINKERTON K.E. - Pulmonary cytochrome-P-450 monooxygenase system and Clara cell-differentiation in rats - Am. J. Physiol. Lung Cell. Mol. Physiol., **13**, L394-L402, 1995.
38. DEVEREUX T., DOMIN B., PHILPOT R. – Xenobiotic metabolism by isolated pulmonary cells. – Pharm. Ther., **41**, 243-256, 1989.
39. LEE C.H., WATT K.C., CHANG A.M., PLOPPER C.G., BUCKPITT A.R., PINKERTON K.E. - Site-selective differences in cytochrome P450 isoform activities - Comparison of expression in rat and rhesus monkey lung and induction in rats. - Drug Metab. Dispos., **26**, 396-400, 1998.
40. MA J., BHAT M., ROJANASAKUL Y. – Drug metabolism and enzyme kinetics in the lung. In: HICKEY A. (ed.) Inhalation Aerosols, Marcel Dekker, New York, 155-195, 1996.
41. RUNGE D.M., STOCK T.W., LEHMANN T., TAEGE C., BERNAUER U., STOLZ D.B., HOFMANN S., FOTH H. - Expression of cytochrome P450 2E1 in normal human bronchial epithelial cells and activation by ethanol in culture. – Arch. Toxicol., **75**, 335-345, 2001.
42. KIVISTO K.T., LINDER A., FRIEDEL G., BEAUNE P., BELLOC C., KROEMER H.K., FRITZ P. - Immunohistochemical localization of cytochrome-P450 2E1 in human pulmonary-carcinoma and normal bronchial tissue. – Virch. Arch., **426**, 243-247, 1995.
43. COLE S.P., BHARDWAJ G., GERLACH J.H., MACKIE J.E., GRANT C.E., ALMQUIST K.C., STEWART A.J., KURZ E.U., DUNCAN A.M., DEELEY R.G. - Overexpression of a transporter gene in a multidrug-resistant human lung cancer cell line. – Science, **258**, 1650-1654, 1992.
44. JEDLITSCHKY G., LEIER I., BUCHOLZ U., CENTER M., KEPPLER D. - ATP-dependent transport of glutathione S-conjugates by the multidrug resistance-associated protein. - Cancer Res., **54**, 4833-4836, 1994.
45. JEDLITSCHKY G., LEIER I., BUCHOLZ U., BARNOUIN K., KURZ G., KEPPLER D. - Transport of glutathione, glucuronate and sulfate conjugates byt the MRP gene-encoded conjugate export pump. - Cancer Res., **56**, 988-994, 1996.
46. HAMILTON K.O., YAZDANIAN M.A., AUDUS K.L. - Modulation of P-glycoprotein activity in Calu-3 cells using steroids and beta-ligands. – Int. J. Pharm., **228**, 171-179, 2001.
47. FLOREA B.I., CASSARA M.L., JUNGINGER H.E., BORCHARD G. - Drug transport and metabolism characteristics of the human airway epithelial cell line Calu-3. – J. Control. Rel., in press.
48. DEVISSAGUET J.P., BRION N., LHOTE O., DELOFFRE P. - Pulsed estrogen therapy: pharmacokinetics of intranasal 17-beta-estradiol (S21400) in postmenopausal women and

comparison with oral and transdermal formulations. - Eur. J. Drug Metab. Pharmacokinet., **24**, 265-271, 1999.
49. BAWARSHI-NASSER R.N., HUSSAIN A.A., CROOKS P.A. - Nasal absorption and metabolism of progesterone and 17□-estradiol in the rat. - Drug Metab. Dispos., **17**, 248-254, 1989.
50. OBERLI S.A., JONCOURT F., STADLER M., ALTERMATT H.J., BUSER K., RIS H.B., SCHMID U., CERNY T. – Parallel assessment of glutathione-based detoxifying enzymes, O6-alkylguanine-DNA alkyltransferase and P-glycoprotein as indicators of drug resistance in tumor and normal lung of patients with lung cancer. – Int. J. Cancer, **59**, 629-636, 1994.
51. ARVELO F., POUPON M.F., BICHAT F., GROSSIN F., BOURGEOIS Y., JACROT M., BASTIAN G., LE_CHEVALIER T. – Adding a reverser (verapamil) to combined chemotherapy overrides resistance in small cell lung cancer xenografts. – Eur. J. Cancer, **31A**, 1862-1868, 1995.
52. NOOTER K., STOTER G. – Molecular mechanisms of multidrug resistance in cancer chemotherapy. – Pathol. Res. Pract., **192**, 768-780, 1996.
53. GOTTESMANN M.M., PASTAN I. – Biochemistry of multidrug resistance mediated by the multidrug transporter. – Annu. Rev. Biochem., **62**, 385-427, 1993.
54. SCHINKEL A.H., SMIT J.J., VAN TELLINGEN O., BEIJNEN J.H., WAGENAAR E., VAN DEEMTER L., MOL C.A., VAN DER VALK M.A., ROBANUS M.E., TE R.H., ET A. – Disruption of the mouse mdr1a P-glycoprotein gene leads to a deficiency in the blood-brain barrier and to increased sensitivity to drugs. – Cell, **77**, 491-501, 1994.
55. HAMILTON K.O., BACKSTROM G., YAZDANIAN M.A., AUDUS K.L. - P-glycoprotein efflux pump expression and activity in Calu-3 cells. – J. Pharm. Sci., **90**, 647-658, 2001.
56. BERGAN T. - Extavascular penetration of ciprofloxacin. - Diagn. Microbiol. Infect. Dis., **13**, 103-114,1990.
57. CAVET M.E., WEST M., SIMMONS N.L. - Transepithelial transport of the fluoroquinolone ciprofloxacin by human airway epithelial Calu-3 cells. - Anitomicrob. Agents Chemotherap., **41**, 2693-2698, 1997.
58. DERENDORF H., HOCHHAUS B., MEIBOHM B., MOLLMANN H., BARTH J. - Pharmacokinetics and pharmacodynamics of inhaled corticosteroids. - J. Allergy Clin. Immunol., **101**, 440-446, 1998.
59. Tunek, K. Sjodin and G. Hallstrom, Reversible formation of fatty acid esters of budesonide, an antiasthma glucocorticoid, iin human lung and liver microsomes. Drug Metab. Dispos. 25 (1997) 1311-1317.
60. E. Wieslander, E.L. Deslander, L. Jarkelid, E. Hjertberg, A. Tunek and R. Brattsand, Pharmacologic importance of the reversible fatty acid conjugation of budesonide studies in a rat cell line in vitro. Am. J. Respir. Cell Mol. Biol. 19 (1998) 477-484.
61. BORCHARD G., LUZ CASSARA M., ROEMELE P.E.H., FLOREA B.I., JUNGINGER H.E. - Transport and local metabolism of budesonide in a human bronchial epithelial cell line (Calu-3). - J. Pharm. Sci., **91**, 1561-1567, 2002.
62. MATHIAS N.R., TIMOSZYK J., STETSKO P.I., MEGILL J.R., SMITH R.L., WALL D.A. - Permeability characteristics of Calu-3 human bronchial epithelial cells: in vitro-in vivo correlation to predict lung absorption in rats. - J. Drug Target., **10**, 31-40, 2002.
63. FRAGA S., SERRAO M.P., SOARES DA SILVA P. - L-type amino acid transporters in two intestinal epithelial cell lines function as exchangers with neutral amino acid. - J. Nutr., **132**, 733-738, 2002.
64. BOLGER M.B., HAWORTH I.S., YEUNG A.K., ANN D., VON GRAFENSTEIN H., HAMM-ALVAREZ S., OKAMOTO C.G., KIM K.J., BASU S.K., WU S., LEE V.H.L. - Structure, function, and molecular modeling approaches to the study of the intestinal dipeptide transporter PepT1. - J. Pharm. Sci., **87**, 1286-1291, 1998.
65. THEIS S., HARTRODT B., KOTTRA G., NEUBERT K., DANIEL H. - Defining minimal structural features in substrates of the H^+/peptide cotransporter PEPT2 using novel amino acid and dipeptide derivatives. - Mol. Pharmacol., **61**, 214-221, 2002.
66. GRONEBERG D.A., NICKOLAUS M., SPRINGER J., DORING F., DANIEL H., FISCHER A. - Localization of the peptide transporter PEPT2 in the lung: Implications for pulmonary oligopeptide uptake. - Am. J. Pathol., **158**, 707-714, 2001.
67. PEZRON I., MITRA R., PAL D., MITRA A.K. - Insulin aggregation and asymmetric transport across human bronchial epithelial cell monolayers (Calu-3). J. Pharm. Sci., **91**, 1135-1146, 2002.

AFFILIATIONS:
Division of Pharmaceutical Technology, Leiden/Amsterdam Center for Drug Research, PO Box 9502, 2300 RA Leiden, The Netherlands
Phone: ++31-71-527 4350, Fax: ++31-71-527 4565
E-mail: Borchard@lacdr.leidenuniv.nl

OVERVIEW

This book is a compilation of reports on the research done by participants of the *Workshop on the Optimization of Aerosol Drug Delivery*, which took place on June 13 to 15, 2002 at Jablonna Palace in Warsaw (Poland). The Workshop proved to be of rather intensive character, thanks to the location, the organising committee and the participants. Interdisciplinary discussions between the scientists, engineers and clinicians among the participants resulted in a creative atmosphere, although not always as harmonic a symbolised by the statue on the first page of this book.

An introduction on the *"Generation and administration of liquid inhalation medicine"* was given by Ola Nerbrink (Novo Nordisk, Denmark). From Aryuveda medicine about 400 years ago, inhalation therapy against local and systemic ailments was used throughout the millennia. Especially the inhalation of coarse droplet aerosols, based on the observation of seaside inhabitants being less prone to infectious diseases, was shown to be in use since Egyptian times, and culminated in the construction of vaporatoria in European cities in mid-1800's. Looking at the tremendous growth of interest in pulmonary delivery, partially initialised by the phase-out of CFC's as propellants (John Pritchard, GlaxoWellcome), it seems like it is history repeated for aerosol therapy. Aleksander Blum (Inhalation Systems, Seattle, USA), however, pointed out that about 50% of drugs that were supposed to be re-formulated as aerosol application fell off the development pipeline.

One of the major problems of aerosol delivery that we are still facing today was already discovered in 1872, when Waldenburg stated that merely 20% of an aerosol dose reaches the deep lung. One of the conclusions of Dr. Nerbrink's presentation was the fact that the principle of nebulisation was invented in 1850, and new devices to improve aerosol delivery are being developed since the 1930's.

During the course of the workshop it became clear that the three aspects of human physiology, aerosol characteristics and the role of the patient have to be addressed when optimisation of aerosol delivery is intended. Yung-Sun Cheng (Lovelace Respiratory Research Institute, Albuquerque, USA) and Eric Derom (University of Ghent, Belgium) pointed out that the "right" modelling of inhalation patterns is mandatory to predict clinical effects. Several studies have shown, however, that while prediction of deposition patterns for nebulizers correlate well with scintigraphy data, oropharangeal deposition after pulmonary application by pressurized metered dose inhalers (pMDI) and dry powder inhalers (DPI) is often underestimated. In addition, the correlation between local deposition and clinical effect is dependent on the pharmacokinetic properties (plasma half-life, absorption, metabolism) of the drug, and therefore not of linear character (Dr. Derom).

Another parameter, as pointed out by Rita Vanbever (University of Louvain, Belgium), is the loss of the drug delivered to the lung by enzymatic degradation and cell uptake. Drug absorption and metabolism might be modelled using cell culture models of the pulmonary epithelia (Gerrit Borchard, Leiden/Amsterdam Center for Drug Research, Leiden, The Netherlands). In a complementary approach, optimisation of pulmonary drug delivery and model drug absorption may even start

L. Gradoń and J. Marijnissen (eds.), Optimization of Aerosol Drug Delivery, 309–311.

at the "sub molecular level", through application of structure-based design, homology analysis, virtual combinatorial chemistry and QSAR (quantitative structure-activity relationship), as presented by Bogdan Lesyng from the University of Warsaw.

On the other hand, many parameters essential for optimised aerosol delivery, such as electrostatic, hydrophilic and hydrophobic forces leading to among others, agglomeration and growth of aerosol particles, as well as breathing patterns and diseased conditions have not been included in models for aerosol delivery (Frans de Jongh, Academic Medical Center, Amsterdam, The Netherlands).

Efficiency of aerosol delivery depends on the aerosol characteristics pre-determined by the delivery device used. Next to pressurized metered dose inhalers (pMDI) and dry powder inhalers (DPI), recent years have seen a "return of the nebulizers" (John Pritchard). Besides these delivery devices, several technologies have been developed to produce delivery systems for pulmonary application, such as mesoporous spherical particles (Tomasz Sosnowski, University of Warsaw), polymeric mucoadhesive particles for controlled drug release (Tomasz Ciach, University of Warsaw), large porous particles (Rita Vanbever), pulmospheres (Aleksander Blum), as well as particles produced by supercritical antisolvent technique (SAS; Leon Gradon, University of Warsaw) and by electro hydrodynamic atomisation (EHDA; Kees Geerse, Delft University of Technology, The Netherlands).

The importance of aerosol size distribution was regarded in a quite controversial way. Pieter Zanen (Academic Medical Center, Utrecht, The Netherlands) pointed out that monodisperse particles are mandatory for targeting of, e.g. corticosteroids to the airways to avoid side systemic effects. On the other hand, Joachim Heyder (GSF-Forschungszentrum für Umwelt und Gesundheit, München, Germany) mentioned that even aerosol particles of a narrow size range show a broad distribution pattern in the lung and that control of aerosol particle size would merely lead to a shift in this pattern of distribution. Hygroscopicity of excipients used in particle formulation does also play an important role, and may even lead to alveolar entrapment due to hygroscopic growth of particles (Joachim Heyder).

The most important "factor" for the improvement of aerosol delivery, of course, is the patient himself. Both Frans de Jongh and Giuseppe A. Marraro (Fratebenfratelli and Ophthalmiatic Hospital, Milan, Italy) pointed out that no model exists for aerosol delivery in individual collectives, such as children or even patients with a pulmonary disease. Differences in physiology, such as breathing patterns, tidal volume and tracheal inspiration velocity play a very important role in aerosol deposition. How the diseased state influences aerosol administration from the clinical point of view was impressively reported on by Dr. Marraro.

In summary, aerosol delivery must be oriented towards the patient and the disease, as indicated in the figure at the end of this overview. Tools for improvement of aerosol delivery are to be found in the modelling (including data from clinical experience), aerosol characteristics and the design of suitable devices. Naturally, the success of an aerosol delivery strategy will be dependent on the regulatory authorities and the patient and prescriber compliance.

One thing became very clear: just as we have to know the basics of a disease in order to be able to treat it, we have to know the basic principles that govern aerosol distribution in the patient. In order to optimise aerosol delivery especially in the diseased state, we also have to link characteristics of aerosols used in clinical trials to the therapeutic outcome. In order to achieve this goal, scientists, engineers and clinicians have to work closely together, in the spirit of the workshop at the Jablonna Palace.

Gerrit Borchard

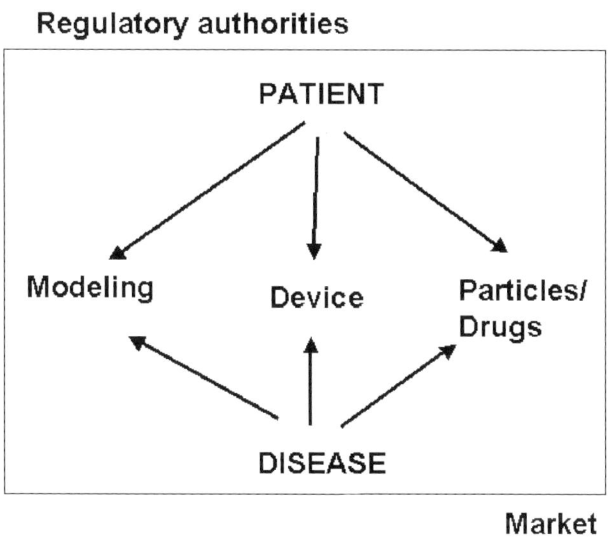

Figure 1. Aerosol delivery oriented towards patient and disease

KEYWORD INDEX

The page number refers to the beginning of the paper in which the keywords occur.

LIST OF AUTHORS